SCIENCE
AS SOCIAL EXISTENCE

Science as Social Existence

Heidegger and the Sociology
of Scientific Knowledge

Jeff Kochan

OpenBook
Publishers

https://www.openbookpublishers.com

Digital material and resources associated with this volume are available at https://www.openbookpublishers.com/product/670#resources

ISBN Paperback: 978-1-78374-410-7
ISBN Hardback: 978-1-78374-411-4
ISBN Digital (PDF): 978-1-78374-412-1
ISBN Digital ebook (epub): 978-1-78374-413-8
ISBN Digital ebook (mobi): 978-1-78374-414-5
DOI: 10.11647/OBP.0129

Cover image: Scanning electron micrograph of a cabbage white butterfly egg, very close up (colour-enhanced). Credit: David Gregory & Debbie Marshall, Wellcome Images, CC BY 4.0. Cover design: Anna Gatti.

All paper used by Open Book Publishers is SFI (Sustainable Forestry Initiative) and PEFC (Programme for the Endorsement of Forest Certification Schemes) Certified.

Printed in the United Kingdom, United States, and Australia
by Lightning Source for Open Book Publishers (Cambridge, UK)

Contents

Introduction

One sure-fire way to write an unsuccessful book is to try to make everyone happy. Because I had hoped to write a successful book, I started out by making a number of choices which I thought would make at least a few people unhappy. First, I chose to write a book promoting Martin Heidegger's existential conception of science. Second, I chose to write a book promoting the sociology of scientific knowledge (SSK). Third, I chose to argue that the accounts of science presented by SSK and Heidegger are, in fact, largely compatible, even mutually reinforcing. Hence, my choice of title: *Science as Social Existence*. In this book, I combine Heidegger's early view of science as a form of existence with SSK's view of science as a social activity. Through this combination, both accounts turn out to be more vital and interesting than they may have been when left to themselves. The book thus presents a tale of intellectual friendship between two perhaps unlikely companions. Of course, no friendship, no matter how promising, will please everyone. But this one happens to please me, and I hope that it will please you too.

SSK emerged in the 1970s, predominantly in the Science Studies Unit at the University of Edinburgh. The 'Edinburgh School' introduced what they called the 'strong programme' in SSK. This signalled a dramatic step beyond what was now, retrospectively, identified as the 'weak programme' in the sociology of science. The weak programme focussed mainly on institutional studies of the scientific community: how scientists were organised into groups; and the social relationships which existed between them. The actual products of scientific activity — theories and facts — and the means by which they are produced — techniques and methods — were excluded from sociological analysis. These were

   http://dx.doi.org/10.11647/OBP.0129.08

thought to form the hard centre of science, the rational core, which sociology was not meant to touch.

In the 1970s, SSK practitioners began to touch this core. This disturbed some people. In the view of critics, SSK was undermining the rationality of science by addressing its conceptual and methodological core in sociological terms. Effectively, this meant that scientific rationality was being treated, through and through, as a social phenomenon, a phenomenon necessarily dependent for its legitimacy on local social and historical circumstances. Critics of SSK urged that this was wrong-headed, and they educed diverse intellectual arguments to support their view. Perhaps more importantly, however, these critics *felt* it was wrong: their distaste was not just intellectual, it was also moral — it came from the gut. For SSK practitioners, none of this appears to have been surprising. They saw their critics as harbouring a quasi-religious desire to preserve the alleged 'sacredness' of scientific rationality against the secularising impulses of a self-consciously naturalistic and methodologically empiricist social science. As social scientists who set out to study science itself, SSK practitioners were determined to treat scientific rationality in wholly secular terms, as a completely natural phenomenon, produced by instinctively gregarious, historically embedded, and fundamentally biological creatures.

A proper disciplinary history of these events has yet to be written. My own suspicion is that SSK practitioners have tended to overplay the secularisation angle, no doubt because this bolsters their own self-presentation as hard-boiled scientific naturalists. Accusing your critics of theological tendencies is, at least in the current Euro-American academic context, a good way to score a few rhetorical points. In my view, however, questions about the sacred or secular nature of knowledge are, at base, questions about what it means to be a human being. To claim that scientific knowledge draws its authority from a source which transcends local social and historical circumstances is to make a substantive claim about human beings as the producers and carriers of that knowledge. Likewise for the contrasting claim, that the authority of scientific knowledge cannot be extricated from the social and historical circumstances in which that knowledge is produced and sustained. In the first case, some aspect of the human being — an aspect tied to knowledge — is thought to transcend its local circumstances. In the second case, such transcendence is deemed impossible.

For the critics, SSK's claim that there is nothing transcendent about scientific knowledge seems to make no sense. In their view, this amounts to a rejection of the objectivity of science. If the authority of knowledge is necessarily tied to local circumstances, then how does one explain the universal validity of, for example, simple rules of logic like those for deduction? From the critics' perspective, SSK practitioners appear to be rejecting the objectivity of logic and other unquestionably reliable techniques of knowledge production. Here, it may be useful to distinguish between descriptions and explanations of objectivity. If we consider our *experience* of objective knowledge production — for example, deducing from 'All humans are mortal' and 'Socrates is human' the conclusion 'Socrates is mortal' — then we seem to be faced with a procedure which cannot but be objective, regardless of local circumstances. The objective validity of deduction *feels* universal, as if it, necessarily, holds everywhere and at all times. In other words, it has normative force. This is a description of our experience — or, one may say, the phenomenology — of deductive inference. SSK does not dismiss this phenomenological description as false, but seeks to explain it without recourse to the notion that human knowers, when they engage in deductive reasoning, transcend their local circumstances. Hence, it is at the level of explanation, not description, that the dispute fundamentally operates. Whereas the critics seek to explain the normative force of deduction in terms of a transcendent feature of human cognition, SSK practitioners seek to explain it in wholly local and naturalistic terms. In the former case, our compulsive feeling that deduction must be objectively valid is the result of its transcendent nature. In the latter case, this feeling of compulsion, of logical necessity, is instead viewed as the result, in necessary part, of one's embeddedness in a particular social context, a context in which one learns and is afterward under recurring pressure to experience deduction, without deliberation, as an objectively valid technique of knowledge production. Normative force is thus social force rather than transcendental force.

Based on their radically different conceptions of what it means to be a human knower, these competing positions seem to lack sufficient common ground for their differences to be resolved through rational discussion. At least, the often acrimonious and mostly unproductive debates which have erupted with varying intensity over the last four decades would seem to suggest as much. I will have little more to say about this conflict in what follows. My own view is that, as more rigorously naturalistic models of human knowing continue to gain credibility across the disciplines, the original intellectual and moral motivations driving SSK will be largely vindicated. There is, however, another conflict, more central to my interests, which this first conflict helps to illuminate. This is a conflict between SSK practitioners and those in the slightly younger interdisciplinary field of science studies who argue that SSK did not go far enough in its rejection of past transcendental models of the scientific knower. Indeed, according to this line of criticism, the conception of the scientific knower promoted by SSK is still a transcendental conception. The only difference is that this knower is no longer viewed as an individual person, but has instead been replaced by society as a whole. On this reading, it is not, ultimately, the individual but society which develops and sustains knowledge of the natural world.

Central to this line of criticism is the claim that SSK trucks in a strong theoretical dichotomy between society, on the one hand, and nature, on the other. By allegedly taking this dichotomy for granted, SSK practitioners are said to gather all the activity relevant to knowledge production on the society side, leaving the nature side thoroughly inert or passive and, as a consequence, completely unnecessary for explanations of scientific knowledge production. But, so the science studies critics continue, it seems patently absurd to claim that nature plays no role in our knowledge of it. Such a claim amounts to a form of sociological idealism, where knowledge is explained solely in terms of the realm of ideas created and sustained by society, with the concrete reality of the natural world being left entirely out of the picture.

Interestingly, this criticism has much in common with the earlier criticism. In the earlier case, the worry was that SSK, by insisting that all knowledge must be explained in terms of local circumstances, fails to capture the universality of some well-established scientific knowledge claims. In other words, on this model, all that scientific knowledge ends up ultimately pointing to are the local social and historical situations which gave rise to and continue to sustain it. It does not, and cannot, point to the objective reality which exists independently of those situations. This too, then, is an accusation of a kind of idealism, where historical and sociological circumstances are placed front and centre, while the actual natural reality which science is purportedly meant to study is left to languish by the wayside. In the view of the first critics, the solution to this idealism is transcendence. Only by reference to an aspect of human cognition which transcends local circumstances can we explain how science succeeds in producing objective accounts of nature.

The more recent science studies critics employ a different strategy in response to SSK's alleged idealism. Like SSK, they too reject transcendence. From their perspective, to invoke transcendence is to offer an implausible solution to a pseudo-problem created by the dichotomous separation of society and nature. Rather than trying to resolve this supposed problem, they argue, we should simply reject the society-nature distinction which gave rise to it. No dichotomy, no problem. These critics propose that society and nature not be treated as fundamental resources in explanations of knowledge, but instead as topics which are themselves in need of explanation. As we will see later, their preferred alternative method is to explain society and nature in terms of the allegedly more fundamental concept of 'practice.' The idea is that stabilised phenomena like society and nature arise from the dynamic heterogeneity of ongoing practical activities which constitute the very fabric of existence. To remain stuck at the level of the society-nature distinction is to ignore practice as providing a more fundamental level on which to base explanations of scientific knowledge production.

My brief here is not to give a detailed account of, much less an extended critical commentary on, this alternative to SSK, although I will give it some further attention in Chapters Two and Three. For the time being, I would like to emphasise that this rejection of the society-nature distinction is intimately related to a more general critique of modernity which has been characteristic of this theoretical wing of science studies. In this context, the term 'modernity' is meant to pick out that aspect of our cultural condition which has given rise, above all, to ecological disasters. The connection between concrete ecological catastrophe and the abstract theoretical separation of society and nature seems to be that this abstract concept, in consequential part, enables human beings to view nature as a passive medium, devoid of intrinsic value and so freely available for manipulation in accordance with human imagination and intentions. By rejecting this distinction, these theorists hope to contribute to a reformulation of humanity's relationship with the rest of the natural world, a reformulation in which the threat of ecological catastrophe will be dramatically diminished.

As critics of modernity, these science studies theorists follow an intellectual path which had been cleared by scholars working earlier in the twentieth century, one of the most prominent of whom was Martin Heidegger. Yet, as we will see, an influential stream in practice-based accounts of science, while acknowledging a debt to Heidegger's earlier critique of modernity, also criticises Heidegger for not having gone far enough. In this respect, Heidegger is admonished for much the same reason that science studies scholars also admonish SSK. In both cases, an innovative step forward is acknowledged, but then immediately rebuked for nevertheless still falling firmly within the circle of an untenable modernist ideology.

One of my main objectives in this book is to demonstrate that these criticisms of SSK and Heidegger, despite their influence, are in fact largely mistaken. Indeed, both SSK and Heidegger have much more to offer a practice-based approach to science than has been allowed for by their critics. A key issue in this dispute is the methodological question of how best to address the conceptual problems generated by the modern theoretical separation of society and nature. This was, in fact, a question which, in a somewhat more abstract form, preoccupied Heidegger for much of his life. However, he responded to it in a dramatically different way than have many prominent science studies scholars. While the latter have counselled the rejection of the society-nature distinction, Heidegger instead advised its deconstruction. To this end, he spent much energy attempting to trace the history of this distinction back to its earliest conceptual manifestations. One principle guiding this methodology was Heidegger's conviction that human beings are fundamentally historical creatures. Hence, our present actions, including our conceptual acts, are inextricably bound together with the history of thinking and doing which informs the community to which we belong. For this reason, Heidegger was preoccupied with an intellectual excavation of the European intellectual tradition.

Science studies scholars who counsel the rejection of the society-nature distinction seem, in contrast, less convinced of the historical dependency of our thinking, believing instead that such traditional structures as the society-nature distinction may simply be sidelined in favour of radically new, historically unprecedented, intellectual tools. Once again, we see that an intractable theoretical dispute about knowledge may be rephrased as a fundamental disagreement about what it means to be a human being. The science studies scholars in question seem to believe that human beings can, at least in some aspect, liberate themselves from history. For Heidegger, in contrast, human existence is, before anything else, historical. From Heidegger's perspective, it follows that science, as a form of human existence, must also be a fundamentally historical phenomenon. As a result, Heidegger's largely philosophical account of science turns out to be highly compatible with the methods and goals of many historians of science. This compatibility with the history of science is yet another characteristic which Heidegger's conception of science shares with SSK.

One consequence of deconstructing the society-nature distinction is a recognition that it is but one special instance of a more general distinction between mind and body, or, in more theoretical terms, subject and object. It is towards this general distinction that both Heidegger, mainly in work preceding the Second World War, and more recent science studies scholars have directed most of their critical energy. In historical terms, the main lineage of the subject-object distinction emerges from the work of the seventeenth-century philosopher, René Descartes, as well as its subsequent formal elaboration in the eighteenth-century writings of Immanuel Kant. As we will see, Heidegger's deconstruction of this distinction involves a substantial critique of both Descartes and Kant. This deconstruction furthermore pushes Heidegger into a detailed engagement with the ancient Greek philosophers Plato and Aristotle. In Heidegger's view, the seventeenth-century subject-object distinction did not spring from nothing, but instead grew out of a specific set of intellectual possibilities introduced by ancient Greek thinkers. Heidegger's goal was to trace the roots of the distinction back through the history of philosophy, with the intention of disclosing new — potentially liberating — possibilities which were left latent in the work of earlier practitioners. His method is thus a deeply historical one,

one which acknowledges the inescapably historical nature of our forms of understanding, and one which also views history as a dynamic and heterogeneous means by which to overcome the potentially threatening limitations of the more orthodox, familiar, and often taken-for-granted threads of the European intellectual tradition.

SSK practitioners share Heidegger's desire for an alternative to the intellectual orthodoxy, an alternative which more accurately depicts the conditions of lived experience. Hence, they too adopt a critical stance towards the orthodox subject-object distinction, challenging, in particular, the individualism presupposed in its model of human subjectivity. As I will argue, however, SSK's challenge to individualistic models of the subject nevertheless leaves crucial aspects of the modern subject-object distinction intact. As a consequence, SSK practitioners have remained vulnerable to attacks from their allegedly more radical competitors in science studies, who exploit SSK practitioners' residual adherence to the subject-object distinction in promoting their own, quite different, accounts of scientific practice. I wish to demonstrate that SSK may be defended against these attacks through its combination with Heidegger's deconstruction of the subject-object distinction, as well as with his phenomenological analysis of the basic structures of human subjectivity. In turn, I wish to also demonstrate that Heidegger's theoretical position may be rendered more concrete, interesting, and useful through combination with empirical studies and theoretical insights already extant in the SSK literature. This will give grounds for my claim that SSK and Heidegger's early existential phenomenology present not just complementary but also mutually reinforcing models of the way scientists get things done.

Before moving into a summary of the chapters which follow, I should emphasise one last time that the goal of this book is a constructive combination of Heidegger's early existential conception of science with the sociology of scientific knowledge. In order to stay focussed on this goal, I have chosen, with some significant exceptions, to minimise critical engagement with the large secondary literature which has arisen in response to the works of both SSK practitioners and, more especially, Heidegger. This restriction has allowed me the freedom to develop my argument in a more straightforward and streamlined fashion, with the result being, I trust, of greater benefit to a majority of the book's readers.

Yet, I should also note that, particularly in the case of Heidegger, by sticking almost exclusively to primary texts, I have ended up with an interpretation which is sometimes at odds with the established scholarship. This is not what I had expected, but the outcome has, I must admit, been cause for some excitement. I hope that readers, in retracing my path through these texts, will also experience some of that same excitement.

Chapter One begins with a nod to the so-called 'science wars,' a heated intellectual dispute which took place in the 1990s. One battle in this multifaceted dispute was over the purported idealism of SSK practitioners. This charge of idealism was motivated by SSK's alleged philosophical scepticism about the existence of the external world. The assumption underlying this criticism was that science entails the existence of the external world, and so scepticism on that count amounts to an assault on the legitimacy of science. However, as I demonstrate, SSK practitioners have almost never denied the existence of the external world. On the contrary, they have often educed arguments against external-world scepticism, and they have usually insisted that a belief in the existence of the external world is central to SSK's method of social-scientific explanation. Nevertheless, I argue that SSK practitioners' attempts to deflect external-world scepticism are less successful than they could be, and hence that their method continues to be vulnerable to sceptical attack. The goal is not, however, to develop a more robust solution to the problem of the external world, but instead to question the very intelligibility of that problem. I suggest that external-world scepticism presupposes a specific model of human subjectivity, one in which the subject is separated from the world, a world external to it, and so it must then build a bridge to this external world in order to grasp it as an object of knowledge. In other words, external-world scepticism presupposes the fundamentality of the modern subject-object distinction. Although SSK practitioners have sought, in various ways, to shake off the more troublesome aspects of this distinction, I argue that they nevertheless have remained committed to it at a basic, tacit level. This commitment is evinced by their acceptance of external-world scepticism as a legitimate problem of knowledge. I attempt to help SSK out of this bind by combining it with Heidegger's phenomenology of the subject as 'being-in-the-world.' I suggest that by adopting Heidegger's alternative account of subjectivity, SSK practitioners will no longer be vulnerable to the threat of external-world scepticism, since they will no longer be wedded to the model of subjectivity which fuels that threat.

In Chapter Two, I address the question of 'realism' which emerges from the preceding discussion. Heidegger's diagnostic response to external-world scepticism is accompanied by an explicit rejection of both realism and idealism as legitimate theoretical positions. However, I argue that a 'minimal realism' may still be drawn from Heidegger's considerations. Heidegger affirms *that* things are, that they exist, independently of subjects, but rejects any attempt to determine *what* they are independently of subjects. This distinction between that-being and what-being gives grounds for minimal realism. It allows us to accept the core realist doctrine of independent existence (thatness), without also committing to the doctrine of independent essence (whatness). I then demonstrate that Heidegger's minimal realism is remarkably compatible with SSK's 'residual realism,' which affirms the independent existence of an external world, but rejects the claim that scientific truths are determined by that world. This compatibility can be further strengthened through the work already done in Chapter One: relieving SSK of its vestigial commitment to the orthodox model of subjectivity, and equipping it instead with Heidegger's alternative. With this combination in place, I go on to consider Joseph Rouse's criticisms of SSK and Heidegger. Rouse argues that both are committed to a theory-dominated account of science, and he instead promotes a practice-based account of science. I argue that Rouse has misunderstood Heidegger's account of science, not least because he overlooks Heidegger's distinction between that-being and what-being, existence and essence. Furthermore, although Rouse's criticisms of SSK do have some merit, I demonstrate that they are also marred by misinterpretation. Finally, Rouse's meritorious criticisms of SSK can also be deflected once SSK has been combined with Heidegger. Indeed, I conclude that this combination — along with the minimal realism accompanying it — offers a more coherent and serviceable basis for a practice-based account of science than does Rouse's alternative.

Chapter Three continues to develop the implications of minimal realism, largely through a discussion of the high-profile debate between the pioneering SSK practitioner, David Bloor, and the influential science studies scholar, Bruno Latour. At the centre of their dispute is the Kantian concept of the thing-in-itself, a thing to which we can attribute independent existence, but about whose independent qualities, or essence, we can know nothing. This concept is presupposed by minimal realism, and also by SSK. Latour attacks it as incoherent, and consequently rejects SSK as an unfit method for science studies. I begin by first reviewing Rae Langton's commentary on Kant's thing-in-itself. Langton argues that this concept follows from an acknowledgement of the finitude of human knowledge. To recognise the existence of things-in-themselves is to admit our inevitable ignorance in the face of nature. This recognition manifests itself in the humility we feel in our encounters with the natural world. I then turn to the Bloor-Latour debate. In Latour's view, Bloor's endorsement of the thing-in-itself fits hand in glove with his allegedly uncritical adoption of the Kantian subject-object distinction. Latour rejects this distinction, and the concept of the thing-in-itself along with it. Nature, on Latour's alternative account, does not outstrip our power to know it, but is itself a wholly constructed phenomenon, one constituted in a field of continuously circulating practices. As in the case of Rouse, Latour exploits weaknesses in SSK's treatment of the orthodox subject-object distinction. And, as in the case of Rouse, I argue that SSK, once combined with Heidegger, can successfully counter Latour's criticism. Indeed, Heidegger deconstructs the Kantian subject-object distinction, reformulating the thing-in-itself in a way commensurate with his own model of the subject. Crucially, the thing-in-itself correlates with the 'affectivity' of the subject. We know the thing exists because it affects us, because we experience that it is, even if we may fail to grasp what it is. Heidegger argues that this peculiar experience is marked by a feeling — an affective state — of anxiety. His reformulation of Kant preserves human finitude and humility, but rejects the Kantian notion of transcendence. It also preserves minimal realism. I conclude with a brief survey of clinical studies of anxiety which seem to provide empirical support for a belief in the thing-in-itself, as reformulated in the context of minimal realism.

Chapter Four begins a transition to themes more typical of the history of science. I start with a review of Heidegger's phenomenological history of logic, wherein logic is construed as the science of thinking. In Heidegger's view, this history is inextricably entwined with the

history of the modern subject-object distinction, in particular, and the
history of scientific subjectivity, more generally. He reads the history of
logic as growing out of earlier attempts to understand the fundamental
relation between thinking and things. This was viewed, above all, as an
intentional relation, a relation manifest in the subject's experience of its
being directed towards things. This relation then came to be construed
in the modern era as one between a propositionally structured mental
substance, on the one hand, and a property-bearing physical substance,
on the other. Heidegger locates the original impulse of logic in Plato's
claim that 'the good' guides thinking in its directedness towards
things. Aristotle then formalised this idea by modelling thinking on
the proposition, with the good now being denoted by the copula ('is'),
which combines subject and predicate in an intelligible sentence. This
move marks the beginning of logic as the formalising study of thinking.
Heidegger argues that Descartes later shifted the organising principle
of intelligibility from the 'is' to the subject position of the proposition,
above all, to the first-person singular subject, 'I.' Kant then submits the
Cartesian 'I' to a phenomenological critique, disclosing its content in
terms of rules of reason. These rules guide thinking in its directedness
towards things, ensuring that the relation is a 'good' one, productive
of intelligibility and understanding. According to Heidegger, this
history traces the way in which the informal and implicit rules guiding
thinking were first identified, and then formalised as a set of explicit
rules governing the structure of thought. He calls this formalisation
process 'thematisation.' Heidegger then offers his own contribution to
this history, arguing that the soil from which logic grows is thoroughly
historical, that the rules directing thinking are rooted in a shared
tradition, in the subject's inescapable 'being-with-others.' This move,
I argue, allows for a powerful point of contact between Heidegger's
phenomenology of logic and the sociology of logic. Indeed, SSK
practitioners also emphasise the rootedness of formal logic in the
informal rules of a shared tradition. Moreover, they have developed this
insight to a far greater extent than did Heidegger. Here, the combination
of SSK with Heidegger allows us to strengthen and expand on — to
more thoroughly thematise and articulate — Heidegger's somewhat
rudimentary considerations. At the same time, I argue that Heidegger
provides grounds for a non-propositional, naturalistic account of

intentionality which can help assuage the worry of SSK practitioners that intentionality, as a philosophical concept, conflicts with the naturalism of their own research methodology.

Chapter Five shifts focus from the history of formal science to the history of natural science, including medicine. In doing so, it builds on the argument from the previous chapter that science is a process of thematisation in which informal and indeterminate knowledge is thematised and articulated in a more formal and determinate way. This raises a concern, however, because it suggests that scientists only discover what they already know. Both SSK and Heidegger attribute a circularity to scientific reasoning. Yet, I argue, this circularity is not vicious. Indeed, it was already recognised by the second-century Greek physician, Galen of Pergamon, and became a topic of concentrated interest for physicians at the University of Padua during the Renaissance. These physicians argued that a determinate knowledge of the informal rules governing their medical practice could be articulated through an incremental process of working with things. The movement from informal to formal knowledge is thus an importantly empirical one. According to Heidegger, this process was carried over into the early-modern period, but not without radical transformation. He argues that, in this period, the rules guiding empirical thinking and doing were 'mathematicised,' that is, consolidated as a coherent set of basic principles, which Heidegger described as a 'basic blueprint' governing scientists' understanding of the thingness (whatness) of things. This process of mathematicisation grew from a 'reciprocal relation' between empirical work with things, on the one hand, and the metaphysical projection of the thingness of things, on the other. I thus argue that Heidegger offers an account of early-modern science which combines both mathematical and empirical elements, comparing his account to the respective metaphysical and empiricist accounts of the historians of science Alexandre Koyré and Peter Dear. For Heidegger, the emergence of early-modern science was neither an exclusively metaphysical nor an exclusively empirical event, but instead a radical transformation in the reciprocal relation between metaphysics and experience. I argue that this was, above all, a transformation in the role played by Aristotelian 'final causes' in early-modern natural philosophy. This challenges the historiographic commonplace that final causes were abolished from the new natural philosophy, a claim often supported by pointing to the alleged breakdown of the Aristotelian art-nature distinction. Extrapolating from Heidegger's work, I argue that there was no such breakdown, and that the art-nature distinction, as well as final causes, despite seventeenth-century rhetoric to the contrary, remained central to early-modern scientific practice. Indeed, both concepts figure as key resources in Heidegger's mathematical explanation for the emergence of early-modern science.

In Chapter Six, I undertake a discussion of the emergence of early-modern experimental philosophy, especially as exemplified in the work of Robert Boyle. I challenge SSK practitioner Steven Shapin's attempt to insulate Boyle from mathematical culture, arguing instead that Boyle was a mathematical philosopher in Heidegger's sense. First, however, I review Heidegger's claim that Newton's First Law is a formalisation of Galileo's mathematical conception of the thing as being 'left entirely to itself.' This conception provided the metaphysical blueprint for what I dub the Galilean First Thing, and I argue that, for Heidegger, the First Thing provided a condition of possibility for the early-modern experiment. This metaphysical blueprint emerged through its reciprocal relation with empirical experience. Drawing on recent work in the history of science, I develop this point through a discussion of late Renaissance and early-modern artisanal culture, with an emphasis on the uniform manufacture of pure metals. These metallurgical manipulations, I suggest, may have encouraged experimenters' metaphysical conception of the thing as a uniform and autonomous First Thing. On this basis, I propose that the fundamental aim of the early-modern experiment was to release things from environmental interference in order to let them be what they, essentially, are — that is, instances of the First Thing. This essential image thus operates as the final cause towards which physical things are naturally disposed, and towards which experimental manipulations seek to artfully direct them. I find support for these claims in Shapin and Simon Schaffer's classic SSK study of Boyle, focussing on Boyle's dispute with Francis Line. I demonstrate that Boyle's response to Line can be explained by attributing to Boyle a tacit commitment to the First Thing, as the blueprint or final cause guiding his experimental practice. I furthermore locate the difference between Boyle and Line in the fact that Boyle was committed to such a blueprint while Line was not, that Boyle experienced nature in terms of a uniform model while Line experienced it in a less unified, more heterogeneous way. This conclusion lends support to Heidegger's claim that the early-modern period saw experience as increasingly consolidated under a single 'world picture.' I conclude by comparing this claim with Bloor's observation that scientific knowledge is governed by 'social imagery,' that is, by images of society construed as a whole. On the one hand, Bloor's work suggests ways in which Heidegger's concepts of 'world picture' and 'basic blueprint' might be rephrased and further developed in a more sociological idiom. On the other hand, Heidegger's claim that these concepts apply only to the early-modern period and later suggests that Bloor's concept of 'social imagery' may prove useful only within a limited historical range.

Chapter Seven does double duty, first, as an unsystematic review of key themes from the preceding chapters, and, second, as a roughly sketched roadmap for future work. Here, I will discuss only the latter. Up to this point, my discussion of Heidegger will have been largely restricted to his work from the 1920s and 1930s. During this period, in my view, he is centrally concerned with the phenomenology of scientific subjectivity. Later, in the late 1940s and the 1950s, his attention shifts to more critical meditations on the dangers posed by scientific thinking to society in general. Indeed, he argued in the 1950s that modern science prepares the way for a comprehensive technologisation of society. I begin by reviewing Heidegger's friendship, from the mid-1930s until his death in 1976, with Carl von Weizsäcker, a noted physicist who had studied under Werner Heisenberg and Niels Bohr. Von Weizsäcker was convinced that Heidegger's analysis of subjectivity could help him to address conceptual problems resulting from the rejection, by the new physics, of the orthodox subject-object distinction. However, he also believed that Heidegger's own search for a solution was handicapped by Heidegger's superficial understanding of the new physics. Heidegger attributed the technologisation of society to what he called 'enframing,' a phenomenon which Heidegger felt limited the existential possibilities of the subject. Von Weizsäcker affirmed Heidegger's concept of enframing as an outgrowth of modern science, but insisted instead that it offered new, potentially liberating possibilities for humankind, especially in the form of systems theory, or cybernetics. While von Weizsäcker advocated for deeper engagement with cybernetics, Heidegger attempted to reconceptualise the thing in a way which radically departed from its conceptualisation by modern science. I argue that Heidegger's considerations may be usefully translated into the terms of an interactionist social theory, as commended by SSK pioneer, Barry Barnes. Enframing is thus viewed as a social phenomenon, constituted in the historically contingent interactions of naturally gregarious subjects. On von Weizsäcker's reading, in contrast, enframing is a system which organises autonomous subjects into a social whole. While the interactionist emphasises the subject over the system, the cyberneticist emphasises the system over the subject. I naturally opt for the former method, and conclude the chapter, and the book, by arguing for a strong compatibility between Heidegger's attention to the affectivity of the subject, on the one hand, and Barnes's interactionist attention to the internal emotional dynamics of 'status groups,' on the other. From this perspective, von Weizsäcker's commitment to enframing evinces his membership in a status group whose interpersonal dynamics enforce that commitment at an emotional level. A concentrated research focus on the emotional dynamics governing scientific status groups flows naturally from the arguments advanced throughout this book. The book thus sketches a road forward for those intrepid science studies scholars keen to produce innovative and exciting new work.

Chapter One

The Sociology of Scientific Knowledge, Phenomenology, and the Problem of the External World

1. Introduction

A leading contributor to the sociology of scientific knowledge (SSK), Harry Collins, invites us to consider the following parable.

> A scientist, a philosopher, a sociologist of scientific knowledge and a science warrior are aloft in a balloon. The balloon begins to deflate. The scientist says: 'A micro-meteorite might have punctured the envelope — do we have any sticky tape?' The philosopher says: 'My inductive propensities convince me that if the balloon deflates we will fall to earth — I must work out the rational basis for this belief.' The sociologist says: 'I wonder how they'll reach a consensus about the cause of our deaths.' The science warrior says: 'Told you so — there *is* an external reality!'[1]

No prize for guessing the odd person out here. The science warrior's non sequitur seems itself to be strangely disconnected from reality. For who among the other passengers challenged the existence of an

---

1   Harry M. Collins (1999), 'The Science Police,' *Social Studies of Science* 29(2), 287–94 (p. 287).

    http://dx.doi.org/10.11647/OBP.0129.01

external world? The answer is: no one. If, however, we instead ask who the science warrior *believes* to have challenged the existence of an external world, then we get a different answer. In this case, the culprit is the sociologist of scientific knowledge. And yet, the real peculiarity of the so-called 'science wars,' which erupted in the 1990s, is not so much that science warriors accused sociologists of denying the existence of an external world. We know, after all, that the first casualty in war is truth. The real peculiarity is just how many otherwise reasonable scholars imbibed this falsehood and hence felt compelled to also pick up the cudgel.

It has been common for philosophers, in particular, to think of SSK practitioners as radical sceptics who dismiss the very idea that nature has a role to play in the formation of scientific knowledge. The heat of the science wars only heightened their passion, and some of them became full-fledged warriors themselves. Philip Kitcher, for example, charged sociologists of science with a 'global skepticism,' because they 'inscribe on their hearts' the dogma that 'no system of belief is constrained by reason or reality.' Christopher Norris alleged that members of the 'Edinburgh school' in SSK 'routinely deny [...] the existence of a real-world (mind- and belief-independent) physical domain.' John Norton claimed that SSK endorses a 'complete scepticism' which rejects any role for evidence in scientific research.[2]

Strikingly, the natural scientists among the science warriors were more circumspect in their criticism. Indeed, the physicists Alan Sokal and Jean Bricmont, who distinguished themselves by their enthusiasm to serve repeatedly on the front line, only characterised SSK as 'ambiguous in its intent.' On the one hand, SSK practitioners appear to endorse a 'general' or 'radical' scepticism. On the other hand, they claim to be pursuing a genuinely scientific research programme.[3] Sokal and Bricmont argue that these two positions cannot be held together,

---

2   Philip Kitcher (1998), 'A Plea for Science Studies,' in *A House Built on Sand: Exposing the Myths about Science*, ed. by Noretta Koertge (Oxford: Oxford University Press), pp. 32–56 (pp. 46, 44); Christopher Norris (1997), *Against Relativism: Philosophy of Science, Deconstruction and Critical Theory* (Oxford: Blackwell), p. 314; John D. Norton (2000), 'How We Know about Electrons,' in *After Popper, Kuhn and Feyerabend*, ed. by Robert Nola and Howard Sankey (Dordecht: Kluwer), pp. 67–97 (p. 72).

3   Alan Sokal and Jean Bricmont (1998), *Fashionable Nonsense: Postmodern Intellectuals' Abuse of Science* (New York: Picador), pp. 92, 89.

because a general scepticism about the existence of an external world is unscientific: 'if one wants to contribute to science, be it natural or social, one must abandon radical doubts concerning the viability of logic or the possibility of knowing the world through observation and/or experiment.'[4] If SSK practitioners claim only that sociological principles must play a role in any causal explanation of scientific beliefs, regardless of whether we evaluate those beliefs as true or false, rational or irrational, then Sokal and Bricmont write that they would have 'no particular objection.'[5] However, if they furthermore insist that *only* social causes may enter into such an explanation, then Sokal and Bricmont say they would strenuously disagree.

Fortunately, SSK practitioners have never made anything more than the first claim, so the apparent ambiguity in their intent dissolves, and Sokal and Bricmont may thus rest content that SSK defends a theory of science to which they would, by their own admission, have no particular objection. Indeed, Barry Barnes, a co-founder of SSK's Edinburgh School, has more recently written that SSK, '[c]ontrary to what at one point was widely claimed by commentators and critics indifferent to what we had set down in print, [...] nowhere denies the existence of an external world.'[6]

Sokal and Bricmont draw a helpful distinction between 'specific scepticism' and 'radical scepticism.'[7] One may have, they say, legitimate doubts about a specific theory, but one should not use general sceptical arguments to support those specific doubts. For example, one may legitimately doubt a theory of evidence which explains evidential force by reference to a mind- and belief-independent world, but one should not try to support such doubt with a global scepticism about the very existence of that world. This distinction is helpful because it exposes the source of difference in the respective reactions to SSK of the scientists, Sokal and Bricmont, on the one hand, and the philosophers, on the other. SSK casts doubt not on the idea of evidence, as such, but instead on specific philosophical theories of evidence which insist that

---

4   Sokal and Bricmont (1998), *Fashionable Nonsense*, p. 189.

5   Sokal and Bricmont (1998), *Fashionable Nonsense*, p. 90.

6   Barry Barnes (2011), 'Relativism as a Completion of the Scientific Project,' in *The Problem of Relativism in the Sociology of (Scientific) Knowledge*, ed. by Richard Schantz and Markus Seidel (Frankfurt: Ontos Verlag), pp. 23–39 (p. 26 n. 3).

7   Sokal and Bricmont (1998), *Fashionable Nonsense*, p. 189.

evidential force must be explained in exclusively non-naturalistic and/ or non-social terms. Since Sokal and Bricmont, as natural scientists, have no vested interest in these particular philosophical theories, they can treat naturalistic and sociological explanations of evidence as unobjectionable. The philosopher warriors, in contrast, were largely trained and continue to work in a tradition deeply invested in individualistic and/or transcendental theories of evidence, and so their reaction to SSK has understandably been less relaxed. Furthermore, these philosophers have apparently had a hard time recognising the difference between their own specific theories of evidence and a general belief in the existence of an external world. Hence, they have tended to mistake a specific scepticism targeted at the former for a global scepticism also encompassing the latter.

Returning to Collins's parable, we see that philosophers are often in the business of working out the rational basis for the acceptance of belief. Sociologists, in contrast, seek to explain consensus concerning the acceptability of belief. These two approaches are closely related, and their proximity explains the friction between them. Both philosophers and sociologists investigate the *reasons* for accepting a belief.[8] For the sociologist, this entails describing the social negotiations through which reasons come to be agreed on. For the philosopher, in contrast, the focus is on the rational rules determining such agreement. Where the sociologist speaks of social negotiations, the philosopher speaks of rational rules. It is precisely on the question of how social negotiation and rational rules relate to one another that the two sides part company, for the sociologist insists that the validity of rules is a matter of social

---

8   This point has not always been appreciated. Jim Brown, for example, alleges that SSK practitioners refuse to admit reasons into their causal accounts of knowledge, writing that, for prominent SSK practitioner David Bloor, 'reasons simply aren't causes' (James Robert Brown (2001), *Who Rules in Science? An Opinionated Guide to the Science Wars* (Cambridge, MA: Harvard University Press), p. 151). However, Brown also admits that 'Bloor does not say this in so many words, but it is clearly implicit in all that he does' (p. 150). In fact, Bloor has explicitly affirmed the importance of reasons, and called for their sociological analysis (David Bloor (1984), 'The Sociology of Reasons: Or Why "Epistemic Factors" Are Really "Social Factors,"' in *Scientific Rationality: The Sociological Turn*, ed. by James Robert Brown (Dordecht: Reidel), pp. 295–324.). Barry Barnes has also written that 'there is no necessary incompatibility between reasons and causes as explanations' (Barry Barnes (1974), *Scientific Knowledge and Sociological Theory* (London: Routledge & Kegan Paul), p. 70). The real bone of contention between Brown and SSK is not whether reasons can be causes but whether reasons can be analysed in naturalistic and sociological terms.

negotiation, while the philosopher typically insists that it is not. In other words, the sociologist endorses, and the philosopher rejects, the view that rationality is a necessarily social phenomenon.

In the natural sciences, the reasons grounding a belief include the evidence educed in its favour. Empirical data, produced and selected using rational methods, may count as evidence in support of that belief. The job of the philosopher is to work out the rational basis for a scientific belief by demonstrating the rationality of the methods by which the evidence for it was educed. Only if those methods are deemed rational can one feel confident that the data successfully represents the world as it really is. Hence, from the philosopher's perspective, according to which the rational and the social must be strictly separated, the sociologist's attempt to model rational method in sociological terms is viewed as an attack on the ability of science to produce authoritative representations of the natural world. If scientific methods are stripped of their authority, then scientific beliefs will lose their purchase on the world. The result will be a global scepticism about the existence of an external world — that is, a world existing external to, or independently of, the system of beliefs and methods partly constitutive of the scientific enterprise. But SSK practitioners are not global sceptics. They do not reject science's authority to successfully represent an external world. They instead reinterpret that authority in sociological and naturalistic terms. For those philosophers whose confidence in science is heavily invested in a non-sociological and/or non-naturalistic conception of its methods and results, this reinterpretation is both objectionable and antiscientific. Hence, they mistake SSK practitioners' rejection of their specific philosophical conception of scientific authority for a more sweeping, global rejection of the authority of science, as such. Taking scientific method to be an instrument of theory, David Bloor, another co-founder of SSK's Edinburgh School, writes that '[i]t is not theories but theorists who generate the evidential force of experimental results.'[9] Bloor does not reject evidence; he rather advises its sociological reinterpretation.

---

9   David Bloor (2003), 'Skepticism and the Social Construction of Science and Technology: The Case of the Boundary Layer,' in *The Skeptics: Contemporary Essays*, ed. by Steven Luper (Aldershot: Ashgate), pp. 249–65 (p. 262). Together with David Edge, Bloor has also argued that something can count as evidence only within an agreed on theoretical framework. An account of the social processes through which such agreement is reached is thus a necessary, though not a sufficient, condition for

It is not clear that philosophers' worries about the allegedly antiscientific and objectionable nature of SSK also reflect the worries of scientists. Returning to our physicist warriors, Sokal and Bricmont, we find that they do not share the philosophers' need to rationally ground the belief in an external world. Indeed, Sokal and Bricmont even declare global scepticism 'irrefutable,' which implies that the philosophers are, from a scientific point of view, wasting their time in attempting such a refutation.[10] These physicists have no particular interest in justifying the authority of science by working out its rational basis, much less in ensuring that that rational basis is strictly protected from sociological study. They simply take it for granted that science rationally represents the world, and they get on with their research. Hence, there is, from their point of view, nothing particularly antiscientific, nor, as we saw above, otherwise objectionable, about SSK's move to introduce sociological categories into naturalistic explanations of scientific rationality.

As we will see in this chapter, SSK practitioners find themselves stuck somewhere between scientists and philosophers on these issues. As social *scientists*, they too are inclined to simply ignore the threat of global scepticism, taking for granted that their methods rationally represent the world, and so just getting on with their research. On the other hand, as social *epistemologists*, they also show signs of wanting to construct a global account of scientific knowledge which reveals its ineliminably social elements. The tension between these two goals has sometimes created confusion and conflict in SSK's ranks over the question of its relationship to scepticism.

I will not seek in this chapter to further defend SSK against the science warriors' erroneous accusations of global scepticism. I will instead take up the more interesting challenge of strengthening SSK's genuine but underdeveloped anti-sceptical orientation. First, I will outline the confusions and conflicts among SSK practitioners regarding scepticism; I will then identify the root cause of those confusions and conflicts; finally, I will suggest a resolution to these difficulties by drawing from the existential phenomenology of Martin Heidegger.

---

any adequate theory of evidence (David Bloor and David Edge (2000), 'Knowing Reality through Society,' *Social Studies of Science* 30(1), 158–60 (p. 159)).

10   Sokal and Bricmont (1998), *Fashionable Nonsense*, p. 189.

Although SSK practitioners have often represented their research as being committed to some form of scepticism, there is no consensus among them on what precisely underpins this commitment. Indeed, in some cases there is outright disagreement. This is most evident in their divergent attitudes towards the challenge presented by external-world scepticism. One camp defends an explicitly realist position regarding the existence of an external world, while the other camp shows no interest at all in defending such realism. I will argue that this disagreement is largely superficial. My argument turns on the idea, taken from Heidegger, that external-world scepticism is an epistemological problem which leaves unexamined a number of important metaphysical presuppositions. The most important of these presuppositions is that our experience of things is best interpreted in terms of a fundamental ontological distinction between a 'subject' and an 'object.' On this interpretation, the subject experiences itself as a discrete, cognising agent seeking access to the world experienced as an external object. The question of how such access may be achieved is often referred to as the 'problem of knowledge,' a core concern of orthodox epistemology. Crucially, the legitimacy of this problem presupposes the validity of the subject-object distinction. As we will see, a commitment to this distinction, and hence to the intelligibility of the question of access, is the engine driving external-world scepticism. In treating external-world scepticism as a legitimate threat, to which a response must be made, SSK practitioners of all stripes demonstrate their shared ontological commitment to the subject-object distinction. As a consequence, they are at perpetual risk of attack by the external-world sceptic. Their internal dispute over how to properly respond to the sceptic is a symptom of their residual adherence to an orthodox model of subjectivity, a model which asserts the fundamental separation of subject and object, mind and world.

After thus diagnosing the shared conceptual ailment of SSK practitioners, I will turn to the work of Martin Heidegger for a suitable treatment. In response to external-world scepticism, Heidegger launched a phenomenological inquiry into the basic ways in which a cognising subject experiences its relation to the world. He conceptualised this experience in existential terms as an experience of 'being-in-the-world.' On Heidegger's account, the most basic form of being-in-the-world is an experience of immersed involvement in a world of work.[11] The epistemological problem of how the subject gains access to an external world is neutralised once one recognises that subject and world were never separated in the first place. The chapter will conclude with the suggestion that, by adopting Heidegger's existential phenomenology, SSK practitioners can overcome the conflicts and confusions which have, until now, rendered their position vulnerable to sceptical attack.

2. Scepticism and SSK

Central figures in SSK have clearly emphasised the importance of scepticism for their work. Reflecting on the issue in his 1974 book, *Scientific Knowledge and Sociological Theory*, Barry Barnes writes that 'the epistemological message of the work [...] is sceptical.' Harry Collins has likewise applied 'philosophical scepticism' explicitly in his own research, and Steven Shapin has declared pointedly that 'SSK is [...] a form of scepticism.'[12] Yet, although Barnes, Collins, and Shapin have made striking use of sceptical techniques in their work, they have not offered any substantial reflections on scepticism as a method of sociological analysis. David Bloor has proven more forthcoming. His pioneering work in the methodology of SSK explicitly discusses and extensively builds on sceptical techniques. Given these credentials, it is noteworthy that Bloor offers a somewhat more guarded assessment

---

11   The word 'involvement' is a standard, if imperfect, translation for Heidegger's word *Bewandtnis*. In fact, one may argue that no single English word adequately translates *Bewandtnis*. Nevertheless, for present purposes, 'involvement' sufficiently captures the relevant meaning. Note, however, that *Bewandtnis* also carries a connotation of 'directedness,' which will prove central in later discussions. In Chapter Four, for example, I will translate *Bewandtnis* as 'assignedness.' In Chapter Five, I will introduce a highly specific, philosophically charged translation of *Bewandtnis* as 'end-directedness.'

12   Barnes (1974), *Scientific Knowledge*, p. 154; Harry M. Collins (1992), *Changing Order: Replication and Induction in Scientific Practice* (Chicago: University of Chicago Press), p. 3; Shapin (1995), 'Here and Everywhere: Sociology of Scientific Knowledge,' *Annual Review of Sociology* 21, 289–321 (p. 314). Benoît Godin and Yves Gingras reveal the striking parallels between Collins's position and the scepticism of Montaigne and Sextus Empiricus, a comparison which Collins describes as 'delicious' (Benoît Godin and Yves Gingras (2002), 'The Experimenter's Regress: From Skepticism to Argumentation,' *Studies in History and Philosophy of Science* 33(1), 133–48; Harry M. Collins (2002), 'The Experimenter's Regress as Philosophical Sociology,' *Studies in History and Philosophy of Science* 33(1), 153–60 (p. 153)).

of SSK's relation to scepticism than do Barnes, Collins, and Shapin. Rather than identifying SSK as a form of scepticism, Bloor draws a clear line between the two while at the same time stressing their productive interaction.

> Scepticism will always find the sociology of knowledge useful and vice versa. But there are profound differences between the two attitudes. Sceptics will try to use the explanation of a belief to establish its falsehood. [...] The conclusion will be a self-defeating nihilism or inconsistent special pleading. It is only an epistemological complacency, which allows us to feel that we can explain without destroying, that can provide a secure basis for the sociology of knowledge.[13]

Bloor rejects an identification of SSK with scepticism because SSK seeks to explain scientific knowledge whereas scepticism is, in his view, corrosive of all such explanatory attempts. According to Bloor, if SSK were itself a form of scepticism, then it would end up undermining its own explanatory project.

There appears, then, to be a significant disagreement between Barnes, Collins, and Shapin, on the one hand, and Bloor, on the other, over SSK's relation to scepticism. However, this apparent disagreement may be resolved by introducing a distinction between 'radical scepticism,' on the one hand, and 'mitigated scepticism,' on the other.[14] Radical scepticism is as Bloor describes it: a persistent acid of relentless doubt which dissolves any and all claims to knowledge. It endeavours to push us into a state of complete disbelief, leaving us without any signposts by which to take our bearings in the world. Mitigated scepticism, on the other hand, attempts to absorb the full impact of sceptical doubt without having to thereby relinquish all claims to knowledge. It relies on a distinction between knowledge in an absolutist and a relativist sense. Mitigated sceptics agree with radical sceptics that knowledge in the first sense is impossible, but they also argue that knowledge in the second sense is both possible and defensible. Hence, mitigated scepticism is not corrosive of belief in general; rather, it isolates and rejects the specific belief that knowledge, as such, must necessarily rest

---

13 David Bloor (1991 [1976]), *Knowledge and Social Imagery*, 2nd edn (Chicago: University of Chicago Press), p. 82.

14 I have taken the term 'mitigated scepticism' from Richard H. Popkin (1979), *The History of Scepticism from Erasmus to Spinoza* (Berkley: University of California Press).

on an absolute foundation, that is, a foundation which transcends any and every contingent social and historical circumstance.

Thus, when Bloor proposes that we exercise 'epistemological complacency' in the face of the sceptic's challenge, he is specifically concerned with radical scepticism. What Bloor proposes is not so much a direct defence against the sceptic as it is a strategy whereby the sceptic is simply ignored. He appears to hold that certain of our beliefs must be taken for granted, regardless of whether or not we can ground those beliefs in a way which satisfies the sceptic. Here Bloor seems to agree with Ludwig Wittgenstein, who observed that, when it comes to following the rules which guide thinking, just because a rule may lack a rational ground, this does not necessarily mean that we have no right to follow it.[15] In such cases, writes Wittgenstein, we follow the rule blindly. The philosopher Paul Boghossian describes this as a 'blind entitlement' to follow a rule or to assert a belief.[16] For example, as we shall see in the next section, Bloor claims that we are blindly entitled to assert a belief in the existence of the external world, and so scepticism regarding this belief should be met with a deliberate complacency.

When Barnes, Collins, and Shapin, on the other hand, urge that SSK be understood as a form of scepticism, they are specifically concerned with mitigated scepticism. According to them, SSK is sceptical because it rejects an understanding of knowledge in terms of absolute truth. This does not mean that knowledge becomes impossible, but only that it can never be rendered certain in an absolutist sense. For Barnes, Collins, and Shapin, SSK can be sceptical and yet still affirm the possibility of knowledge by accepting a more modest, or mitigated, conception of truth and validity.

The apparent disagreement between Barnes, Collins, and Shapin, on the one hand, and Bloor, on the other, thus turns out to be largely superficial. In their respective assessments of the relationship between SSK and scepticism, each side has a different brand of scepticism in

---

15   Ludwig Wittgenstein (1958), *Philosophical Investigations*, 3rd edn, trans. by G. E. M. Anscombe (Oxford: Basil Blackwell), §219.

16   Paul A. Boghossian (2006), *Fear of Knowledge: Against Relativism and Constructivism* (Oxford: Clarendon Press), p. 99. Bloor has deflected Boghossian's attempt to use blind entitlement against the sociology of knowledge (David Bloor (2007), 'Epistemic Grace: Antirelativism as Theology in Disguise,' *Common Knowledge* 13(2–3), 250–80 (pp. 259–61)).

mind. In fact, both sides endorse a mitigated scepticism which stands opposed to an attitude characteristic of those whom Bloor calls 'believers.' Believers, he writes, 'conflate the common currency of talk about the true and the good with specific theories of the real and ultimate nature of the True and the Good.'[17] In other words, believers reach beyond the realm of everyday experience in order to make absolutist claims about the nature of knowledge and reality. For this reason, they might also be described as fundamentalists, or dogmatists.[18] The benefit of scepticism for SSK has been its role in revealing the dogmatism at the heart of epistemic absolutism. SSK accepts the general sceptical claim that absolute knowledge is impossible, but rejects the radical sceptic's more thoroughgoing conclusion that knowledge, as such, is impossible. As Bloor remarks in the passage cited earlier, the radical sceptic's conclusion amounts to a self-defeating nihilism or an inconsistent special pleading. Indeed, it would seem that the radical sceptic helps herself to the very absolutism she is bent on destroying. It turns out, then, that to reject epistemic absolutism is also to reject a fundamental premise motivating the radical sceptic's own position. This is precisely what SSK does. The result is a mitigated sceptical position which endorses a non-absolutist theory of knowledge.

3. SSK and External-World Realism

SSK's rejection of radical scepticism is perhaps most evident in SSK practitioners' affirmation of the existence of an external world. In fact, they appear almost unified in asserting that a belief in the existence of an external world is a necessary condition for social life.[19] Shapin writes that such a presumption is 'common sense,' and 'a precondition for communication.' Barnes, Bloor, and John Henry claim that 'people everywhere' make reference to an external world, and that their mastery of 'the realist mode of speaking' serves them with 'marvellous

---

17   David Bloor (1998), 'A Civil Scepticism,' *Social Studies of Science* 28(4), 655–65 (p. 657).

18   Cf. Barry Barnes and David Bloor (1982), 'Relativism, Rationalism and the Sociology of Knowledge,' in *Rationality and Relativism*, ed. by Martin Hollis and Steve Lukes (Oxford: Blackwell), pp. 21–47 (p. 46).

19   As we shall see, Harry Collins is an exception warranting the qualified phrase '*almost* unified.'

efficiency.' They recommend that such realism be accepted as the standard for the sociology of knowledge. Bloor, for his part, asserts that 'we are all instinctive realists,' and that socialisation would be impossible in the absence of an external world. Barnes claims that 'we are obliged to *presuppose* an external world in order to act and interact.'[20] It seems clear, then, that SSK is strongly committed to the minimal realist doctrine that an external world exists independently of our interpretations and practices. This is made all the more evident in SSK practitioners' consistent efforts to defend themselves against charges of idealism. Indeed, Barnes, Bloor, and Henry even reserve the final lines of their book-length introduction to SSK for a repudiation of the claim that theirs is 'an idealist sociological account which denies the existence of an external world,' and they spend considerable time elsewhere in the book divorcing themselves from the 'methodological idealism' of the allegedly renegade SSK practitioner, Harry Collins. In addition, Bloor has offered his own lengthy defence of SSK against the charge of idealism.[21]

The *locus classicus* for SSK's position on realism is Barnes's 1992 paper, 'Realism, Relativism and Finitism.' Here, Barnes too is motivated by the need to secure SSK's realist credentials against charges of idealism. He begins by arguing that sociological relativists have been typically, but unjustifiably, lumped together with idealists because orthodox realists commonly exaggerate the minimum criteria which one must meet in order to be counted a legitimate realist. Not only do orthodox realists require that an external world exist independently of our interpretations and practices, they also claim that we can know specific features of that world independently of those interpretations and practices.[22] Barnes

---

20   Steven Shapin (1994), *A Social History of Truth: Civility and Science in Seventeenth-Century England* (Chicago: University of Chicago Press), pp. 29, 30; Barry Barnes, David Bloor and John Henry (1996), *Scientific Knowledge: A Sociological Analysis* (London: Athlone), pp. 88, 205 n. 3; David Bloor (1996), 'Idealism and the Sociology of Knowledge,' *Social Studies of Science* 26(4), 839–56 (p. 845); Barry Barnes (1992a), 'Relativism, Realism and Finitism,' in *Cognitive Realism and Social Science*, ed. by Diederick Raven, Lietke van Vucht and Jan de Wolf (New Brunswick, NJ: Transaction), pp. 131–47 (p. 139).

21   Barnes, Bloor and Henry (1996), *Scientific Knowledge*, pp. 202, 13–15, 75–76; Bloor (1996), 'Idealism and the Sociology of Knowledge,' passim.

22   Stathis Psillos argues that this second claim is 'a basic philosophical presupposition of scientific realism' (Stathis Psillos (1999), *Scientific Realism: How Science Tracks the Truth* (London: Routledge), p. xix). Ian Hacking disparages this claim with the

argues that this ambitious claim, quite apart from its plausibility, is simply unnecessary if all one wishes to do is affirm the existence, as such, of the external world. And this is all Barnes's relativistic realist wishes to do. The result is a minimal form of realism which recognises the independent existence of the external world while also declining to attribute any independent, or inherent, properties to that world.[23] Although it is less ambitious than the more robust position of many scientific realists, Barnes's modest, or, as he calls it, 'residual,' realism appears nonetheless sufficient to deflect the charge of idealism.

Problems arise, however, when Barnes attempts to justify this position. Under the heading 'Justifications for a Residual Realism,' he writes that '[t]here is nothing empty in the assertion that an external independent reality exists, underlying appearances. It is an assertion which does real work in a variety of contexts both in science and in philosophy.'[24] Note Barnes's endorsement, in this passage, of the ancient distinction between appearance and reality, a distinction which has played a central role in historical debates between idealists and realists. The idealist typically claims that appearances are the only things we can know exist, while the realist claims that we can also know that an external world exists and that it underlies appearances. Yet, note too that Barnes does not actually argue for the existence of the external world, but only for the utility of the assertion that such a world exists: asserting the existence of an external world has proven an effective strategy in diverse scientific and philosophical contexts. This agrees with Barnes, Bloor, and Henry's statement, cited above, that people everywhere use the realist mode of speaking with marvellous efficiency. Barnes makes this point forcefully with respect to explanations for changes in knowledge, arguing that 'primitive causal inputs from an external reality may operate on us so that we change our knowledge.'

---

deliberately unpleasant name 'inherent-structurism' (Ian Hacking (1999), *The Social Construction of What?* (Cambridge, MA: Harvard University Press), p. 83).

23   For this reason, Shapin would seem wrong to ascribe a 'robust realism' to SSK in general (Steven Shapin (1995), 'Here and Everywhere: Sociology of Scientific Knowledge,' *Annual Review of Sociology* 21, 289–321 (p. 315)). He furthermore appears to endorse the second, stronger, claim of the orthodox realist when he writes that 'the external world [...] has a determinate structure' (Steven Shapin (1994), *A Social History of Truth: Civility and Science in Seventeenth-Century England* (Chicago: University of Chicago Press), p. 4).

24   Barnes (1992a), 'Relativism, Realism, and Finitism,' p. 137.

The external world is the source of primitive, 'unverbalized' causes for 'dissatisfaction' with existing knowledge, and hence provides 'incentives' for changing that knowledge. Barnes favourably contrasts this position with idealism, which he argues cannot plausibly explain changes in knowledge. Specifically, he claims that idealists, because they eschew the concept of the external world, are unable to rationalise a 'sense of failure.'[25] The point seems to be that idealists have no way of explaining how one becomes dissatisfied with the state of one's knowledge and hence no way of explaining how one becomes motivated to change that knowledge. Leaving aside the question of whether or not Barnes has offered a fair description of the idealist's position, it should be clear that this is not an argument for the existence of the external world, but only for the efficaciousness of realist talk about the external world as compared with idealist strategies allegedly forbidding such talk. Hence, it is commensurate with Barnes's position that the distinctions between realism and idealism, and between reality and appearance, are distinctions made within the realm of discourse, and that, as such, they can tell us nothing about the discourse-independent existence of the external world. Barnes thus fails to provide a justification for external-world realism which accords with the realist's own minimal ontological commitments. In fact, he even concludes his discussion with an admission of this failure, thus leaving the issue unresolved.[26] As a result, Barnes leaves the door open for a sceptical construal of his position as a form of linguistic idealism.

Yet perhaps this need not trouble the SSK practitioner. Although Barnes has not successfully answered the sceptic's challenge to

---

25 Barnes (1992a), 'Relativism, Realism, and Finitism,' pp. 137–38.

26 Barnes (1992a), 'Relativism, Realism, and Finitism, p. 139. Note that, some years earlier, Barnes had written: 'I am not a realist, but an instrumentalist and a relativist' (Barry Barnes (1981), 'On the "Hows" and "Whys" of Cultural Change (Response to Woolgar),' *Social Studies of Science* 11(4), 481–98 (p. 493)). Yet, even back then, he enthusiastically endorsed 'a realist mode of speech' as 'a marvellous instrument' (Barnes (1981), 'On the "Hows" and "Whys,"' p. 493). In these earlier passages, Barnes seems to want to distance himself from the robust realism characteristic of scientific realists. Only later did he develop a more nuanced perspective, introducing the relativistic, or 'residual,' form of realism which is my primary interest here, and which I will discuss more thoroughly in Chapter Two. More on the topic of SSK and realism can be found in Jeff Kochan (2008), 'Realism, Reliabilism, and the "Strong Programme" in the Sociology of Scientific Knowledge,' *International Studies in the Philosophy of Science* 22(1), 21–38.

external-world realism, it may be that the failure lies not so much with his attempted justification as with the fact that he had even attempted to provide one. A more effective response to the sceptic may be found in Bloor's epistemological complacency. As cited above, Bloor holds that 'we are all instinctive realists,' Barnes that 'we are obliged to *presuppose* an external world in order to act and interact,' and Shapin that external-world realism is 'a precondition for communication.' If it were true that external-world realism is a matter of instinct or obligation, a necessary condition for social existence, then one might well wonder if radical scepticism about the external world is really worth the candle. Barnes, Bloor, and Henry make clear that their external-world realism is of a 'naive' sort, that it is, above all, a 'common-sense' kind of realism.[27] If this were indeed the case, then deliberate complacency with respect to the sceptic's challenge would surely be the most reasonable strategy. This is, however, far from the case.

External-world realism is neither as naive, nor as commonsensical, as it may at first seem. Not only does it take for granted the ancient distinction between appearance and reality, it also presupposes a particular model of the subject. Consider the sentence with which the philosopher Thomas Nagel begins his discussion of external-world scepticism: 'If you think about it, the inside of your mind is the only thing you can be sure of.'[28] As Nagel goes on to show, from this starting point the problem of justifying the existence of an external world naturally flows. For if the only thing that self-evidently exists is the inside of one's own mind, then it must follow, not only that there is likely to be an outside with respect to one's mind, but that the existence of this outside is not itself self-evident but in need of proof. The question of whether or not such a proof can be given forms the nucleus of external-world scepticism. These distinctions between mind and world, between an inside and an outside to one's own consciousness, between appearance and reality, between subject and object, together form a bundle of closely related and mutually supportive conceptual demarcations which are deeply rooted in the modern intellectual tradition. We thus seem to have at hand an explanation for the strange and apparently widespread

---

27 Barnes, Bloor and Henry (1996), *Scientific Knowledge*, pp. 88, 205 n. 3.
28 Thomas Nagel (1987), *What Does It All Mean? A Very Short Introduction to Philosophy* (New York: Oxford University Press), p. 8.

tendency, at least among academically trained scholars, to conceive of the physical world as a specifically *external* world. This tendency seems to be motivated, in significant part, by the prior, tacit interpretation of our own subjectivity as constituting an internal world, a world of the mind. The notion of an external world, a world out there, and the notion of an inside to our mind, a world in here, are thus as inextricably bound together as, say, the inside and outside of a glass bulb. On this model, the mind — as the seat of our experience — is like the interior of a sealed bulb, an autonomous substance existing in isolation from the bulb's exterior. The external-world sceptic accepts and exploits the glass-bulb model, challenging the credibility of modern epistemologies which claim that the interior of the bulb can access the exterior, that the mind, whether individual or collective, can penetrate the barrier separating it from the external world so as to achieve knowledge of that world. If this diagnostic model is correct, then it would seem that the struggle to meet the challenge of the external-world sceptic was lost even before it began. For if all one can be certain of is the 'inside' of one's own mind, and if the world is construed as being both external to and independent of that mind, then one will never succeed in proving beyond doubt that such a world exists. Indeed, Barnes has also endorsed what he calls 'external realism,' the position that our knowledge is of 'something *out there*,' but he also admits that this position 'cannot be justified.'[29]

Yet, following Bloor's strategy of epistemological complacency, if the intellectual conventions represented by the glass-bulb model were found to be wholly commonsensical, if not entirely naive, then the external-world realist may still claim a blind entitlement to these conventions even in the absence of rational justification. In other words, if we felt obliged to accept the glass-bulb model, if we felt ourselves under a powerful compulsion to adopt this model as a precondition for communication, if such acceptance were a matter of primitive instinct rather than of conscious deliberation, then we may well be justified in responding to the sceptic's challenge with nothing more than a complacent wave of the hand.

---

29 Barry Barnes (2004), 'On Social Constructivist Accounts of the Natural Sciences,' in *Knowledge and the World: Challenges beyond the Science Wars*, ed. by Martin Carrier, Johannes Roggenhofer, Günter Küppers and Philippe Blanchard (Berlin: Springer), pp. 105–36 (pp. 111, 119; emphasis added).

However, the glass-bulb model represents just one, albeit powerful, thread in the modern intellectual tradition. Well-established and increasingly influential alternatives to this model exist in the comparatively recent movements of American pragmatism and European phenomenology. What is more, these alternatives have already begun to earn a respected place within the broader field of science studies. As a consequence, science studies scholars can no longer take external-world realism for granted as a self-evidently valid position, nor can they reasonably respond to the sceptical challenge to this position with complacency. As a consequence, SSK is neither rationally justified in nor blindly entitled to maintain its commitment to external-world realism.

4. Phenomenology and the 'Natural Attitude'

In the remainder of this chapter, and, indeed, in all subsequent chapters, I will explore the benefits of combining SSK with the existential phenomenology of Martin Heidegger. My aim in this chapter is to demonstrate that Heidegger's early analysis of subjectivity can provide SSK with an effective response to the challenge posed by the external-world sceptic.

SSK is certainly no stranger to the methods of phenomenology. Indeed, several SSK practitioners have made significant use of the phenomenological concept of 'natural attitude,' that is, the idea that our conscious beliefs always presuppose a more fundamental, tacitly held attitude which must already be in place before we can even begin to make sense of our experiences, much less communicate those experiences to others. So, for example, within the context of scientific practice, Barnes, Bloor, and Henry describe the claim that an experiment proved a theory because the theory is true as 'a very natural attitude to adopt. [...] Indeed, it is *the* natural attitude.'[30] Yet, as they point out, the reasoning behind such an attitude is clearly invalid. The truth of the theory is explained by the success of the experiment, and the success of the experiment is explained by the truth of the theory. This kind of reasoning is most common with very well-established scientific theories,

---

30   Barnes, Bloor and Henry (1996), *Scientific Knowledge*, p. 30.

for example, electron theory. Barnes, Bloor, and Henry emphasise that it is wholly natural to explain the success of Robert Millikan's famous oil-drop experiment, which first measured the electron charge in the early 1910s, by reference to the truth of electron theory. Yet, Millikan's experiment is also accepted as an important confirmation of that theory. It turns out, then, that the natural attitude with respect to electron theory is not logically valid. However, this observation is not meant to discredit our belief in the truth of electron theory. On the contrary, it is consistent with this attitude that we are, under ordinary circumstances, blindly entitled to such a belief even if we cannot logically justify it. In other words, under such circumstances we have a right to be epistemically complacent about the truth of electron theory.

However, Barnes, Bloor, and Henry argue that the sociologist of knowledge is not working under ordinary circumstances, and hence she should not take the natural attitude for granted. As opposed to the physicist, who immerses herself in the practice of science, the sociologist's goal is to stand back from such practice in order to analyse how and why it works. Rather than adopting a natural complacency with respect to the truth of well-established theories, the SSK practitioner will instead thematise this complacency in an attempt to illuminate the important role it plays in the smooth operation of physical science. In the terminology of the phenomenologist, Barnes, Bloor, and Henry are recommending that the sociologist 'bracket' the scientist's natural attitude, that is, deliberately disengage from it, in order to more effectively analyse its structure. They suggest that it be turned from a resource into a topic for analysis.

SSK practitioners have also employed the phenomenological notion of natural attitude in the context of knowledge about the external world. Shapin, for example, declares that external-world realism is a direct consequence of the natural attitude.[31] However, rather than bracketing this attitude in order to illuminate its structure and the role it plays in social life, he simply takes it for granted, treating it as if it were a universal and inescapable fact of human experience. As a result, Shapin leaves unaddressed the sceptical threat to SSK's affirmation of external-world realism, as well as the ontological distinction between subject and object which gives rise to that threat. Collins likewise characterises

---

31 Shapin (1994), *A Social History of Truth*, pp. 28–31.

the natural attitude as an attitude 'taken to the external world in the normal way of things.' However, he rejects this attitude, using instead a 'philosophical scepticism' designed to initiate the methodological 'derailment of the mind from the tracks of common sense.'[32] In other words, unlike Shapin, Collins adopts a form of external-world scepticism. Yet, as a consequence, he nevertheless joins Shapin in tacitly reaffirming the bundle of distinctions which are represented and reinforced by the glass-bulb model. This has led to some confusion on the part of both Collins and his critics. Most importantly, Collins fails to distinguish sufficiently between external-world realism and realism as such. Thus, in recommending the suspension of belief in the external world, he in fact ends up going much further, arguing that 'all description-type language should be treated at the outset as though it did not describe anything real.' What Collins means, of course, is that language should not be taken to describe anything outside the social world. Indeed, he also writes that '[i]t is in the social world that the social scientist […] should find reality persuasively located.'[33] He calls this the natural attitude of the social scientist, and contrasts it with the natural attitude of the physical scientist, wherein the existence of a reality external to the social world is affirmed. Collins thus applies the term 'reality' in two quite distinct ways without always signalling this difference to his readers. In the context of the social sciences, the term refers to the interior of the intersubjective, social world. In the context of the natural sciences, the term refers beyond the social world to an external natural world. These applications of the term are consistent with idealism and realism respectively, and in both cases the glass-bulb model of subjectivity is taken for granted.

We are now in a position to shed further light on the long-standing dispute between Collins, on the one hand, and Barnes, Bloor, and Henry, on the other. As mentioned in the previous section, Barnes, Bloor, and Henry take Collins to task for eschewing external-world realism and espousing a form of idealism instead. Yet, as we have now seen, both sides are equally wedded to the glass-bulb model. This model is taken for granted by both, and it silently informs their respective arguments. It thus figures as a central background assumption, a key element in the natural attitude governing their respective positions. By failing

---

32   Harry M. Collins (1982), 'Special Relativism — The Natural Attitude,' *Social Studies of Science* 12(1), 139–43 (p. 140); Collins (1992), *Changing Order*, p. 1.

33   Collins (1992), *Changing Order*, p. 174; Collins (1982), 'Special Relativism,' p. 141.

to recognise that they hold this attitude in common, each side has misunderstood the nature and depth of its disagreement with the other. This is evident in the fact that Barnes's justification for external-world realism is largely consistent with the natural attitude Collins endorses on behalf of the social scientist. Collins argues that, for the social scientist, the term 'reality' takes its meaning from the social world. Similarly, Barnes justifies the social scientist's use of realist talk on the basis of the manifest utility of such talk in various scientific and philosophical contexts. There is, it seems, no substantial difference between these two positions. On the other hand, there does appear to be an important disagreement between the two sides with respect to the question of how seriously one should take external-world scepticism. Collins does take it seriously, and is thus willing to give up the idea of a world existing independently of our interpretations and practices. Barnes, Bloor, and Henry, in contrast, choose not to treat external-world scepticism as a serious threat, and Bloor advises that it be met with complacency. I have suggested that complacency does not provide an effective response to the external-world sceptic. Indeed, as Collins's own work shows, a clearly articulated commitment to external-world realism appears incidental to the production of successful SSK research. Nevertheless, Barnes, Bloor, and Henry strongly insist on rejecting the sociological idealism implied by Collins's method. In their desire to distance themselves from the taint of idealism, Barnes, Bloor, and Henry reaffirm external-world realism even in the absence of appropriate justification, relying instead on an ultimately unconvincing strategy of epistemological complacency.

Both sides of this dispute see no alternative between external-world realism, on the one hand, and sociological idealism, on the other. The narrowness of their vision is conditioned by their tacit adherence to a contingent bundle of conceptual distinctions represented by the glass-bulb model. This model is itself a central element in the natural attitude characteristic of these SSK practitioners, and, as such, the cause of some of their more persistent conceptual difficulties. I suggest that these difficulties may be solved by bracketing the glass-bulb model, by declining to take it for granted, and thus disengaging from it in order to better understand its role in modern theoretical practice. With this goal in mind, I now turn to the early phenomenological work of Martin Heidegger.

5. The Phenomenology of Subjectivity in Heidegger's *Being and Time*

So far, this chapter has largely focussed on a specific problem of knowledge, namely, the problem of how one can know that the external world exists. The concern has thus been an epistemological one. Yet, as we have also seen, in asking this epistemological question, certain ontological assumptions are also implicated. In particular, in asking 'How is knowledge of an external world possible?' the existence of a knower is being tacitly asserted. Furthermore, as long as the focus of enquiry lies solely on the epistemological question, ontological questions about the nature or 'being' of this knower — about the fundamental subjectivity of the subject — remain unasked. Under such circumstances, the enquiry both relies on and persistently reaffirms a prior, tacit understanding of the ontological structure of the subject. I have introduced the glass-bulb model in order to make this structure more explicit.

In his 1927 book, *Being and Time*, Martin Heidegger holds the orthodox model of the subject up to scrutiny, and seeks to explain it in terms of a more fundamental phenomenological model. The chief obstacle for such an alternative model is the self-evident character of the received view. Heidegger argues that, as long as the orthodox model is taken for granted, the fundamental 'phenomenal content' [*phänomenale Bestand*] of the subject — our basic experience of our own subjectivity — remains hidden. In this section, we will review some key aspects of Heidegger's account of this phenomenal content, and in the next section we will consider the ways in which he brings these to bear on the challenge posed by external-world scepticism.[34]

Heidegger attempts to loosen up intuitions about the self-evidence of the orthodox model of the subject by tracing its early-modern

---

34 Martin Heidegger (1962a [1927]), *Being and Time*, trans. by John Macquarrie and Edward Robinson (Oxford: Blackwell), p. 72 [46]. (Following scholarly convention, page numbers in square brackets refer to the original 1927 German edition of *Being and Time*.) The German word *Bestand* is a nominalisation of the verb *bestehen*, which can mean 'to exist,' 'to persist,' or 'to consist in.' It lacks the connotation of 'being contained within something' characteristic of the English word 'content.' In Heidegger's view, the subject is not a receptacle containing cognitive content; it is a self-aware, cognitively structured form of existence.

instantiation back to early sources in ancient philosophy and late Renaissance Christian theology.[35] In the former case, ancient Greek philosophers, most notably Aristotle, defined the subject as *zōon logon echon*, later interpreted to mean 'animal rationale,' that is, a living thing capable of reason. The first difficulty Heidegger notes is that the subject is here construed as a thing, a substance of some kind. The second is that this substance-subject is then endowed with a power of reason the nature of which is left no less obscure than the ontological structure of the compound entity taken as a whole. In the case of Renaissance theology, the ancient Greek definition becomes entangled with the Old Testament doctrine that human beings were created in the image of God, and the later Christian doctrine that human beings possess exceptional powers enabling them to transcend the physical realm. Here, Heidegger quotes two sixteenth-century claims: Johannes Calvin's claim that, in virtue of such faculties as reason, the human may 'ascend beyond [earthly life], even unto God and eternal felicity,' and Huldrych Zwingli's assertion that the human being is 'born somewhat closer to God, is something more *after his stamp.*'[36] Heidegger argues that these historical influences provide the departure point for early-modern interpretations of subjectivity. Although the modern notion of transcendence seems to have now lost its theological connotations, the assumption that humans may somehow reach beyond their finite incarnation as earthly things remains an enduring, if often implicit, theme up to the present day.

Heidegger thus locates in the prevailing attitude towards knowledge a self-evident description of the subject as a created thing, or creature. This creature possesses a superior power of reason which distinguishes it from other created things and allows it to transcend the finite conditions of its material existence. Heidegger argues that the dominant focus has been on the structure of this creature's essential relation to reason, as well as with the transcendent nature of this relation, while

---

35 Heidegger (1962a), *Being and Time*, pp. 74–75 [48–49].

36 Quoted in Heidegger (1962a), *Being and Time*, p. 74–75 [49]. The sources are Johannes Calvin's *Institutio* I, XV, Section 8, first printed in 1536, and Huldrych Zwingli's *Von der Klarheit des Wortes Gottes* (*Deutsche Schriften* I, 56), first printed in 1522. Heidegger's biblical reference is to Genesis 1:26. The English translations appear in the corresponding endnotes, Heidegger (1962a), *Being and Time*, p. 490, notes vii and ix.

the creature's existential status as a thing has been taken for granted.[37] His phenomenological move is to bracket this existential status and so submit the phenomenal content of the subject to systematic investigation.

Underpinning Heidegger's analysis is a distinction between existence and essence. He argues that the essence of the subject lies in its existence, that existence takes priority over essence.[38] He furthermore reserves the term 'existence' specifically for subjects, and introduces the term 'presence-at-hand' (*Vorhandenheit*) to designate the existence of everything else. This latter distinction is grounded, in significant part, in two naturally occurring grammatical distinctions: first, subjects are referred to as 'who,' while everything else is referred to as 'what'; second, only in addressing these subjects does one use a personal pronoun ('I am,' 'you are,' 'we are,' etc.).[39] The subject is thus a *person*, while all other entities are *things*. Heidegger uses the commonplace German term 'Dasein' as a general label for the person-subject. He emphasises that Dasein is not a thing, substance, or object; it is an accumulation of end-directed, or intentional, actions: 'The person is no Thinglike and substantial Being. [...] Essentially the person exists only in the performance of intentional acts, and is therefore essentially *not* an object.'[40] In undertaking a phenomenological analysis of the existential structure of the subject, Heidegger aims to unsettle the historically entrenched tendency to conceptualise it by analogy to things, with the unevenness of the analogy being smoothed over by the introduction of an incorporeal faculty of reason.

Heidegger begins his analysis of the subject by exposing one of its fundamental existential structures, namely, 'being-in-the-world.' This term refers to a unitary phenomenon which can be analysed in terms of three constitutive elements, the most important of which are, for

---

37   Heidegger (1962a), *Being and Time*, p. 75 [49].
38   Heidegger (1962a), *Being and Time*, pp. 67 [42], 68 [43]. With this, Heidegger inverts the relation between existence and essence introduced by medieval Christian metaphysicians on the basis of the Biblical doctrine of creation. For them, God adds existence to those things whose essence God has determined in advance.
39   Heidegger (1962a), *Being and Time*, pp. 67 [42], 71 [45], 68 [42].
40   Heidegger (1962a), *Being and Time*, p. 73 [47–48]. Note also Heidegger's qualified remark that a person becomes present-at-hand only following her death (Heidegger (1962a), *Being and Time*, p. 281 [238]).

the present discussion, 'being-in' and 'world.'[41] 'Being-in' describes a fundamental relation between subject and world. Heidegger urges that this relationship should not be misunderstood as a case of one thing's being *in* another, like, for example, water in a glass. To do so would be to conceive of both subject and world in corporeal terms, as things which are present-at-hand. Yet, such a misunderstanding is natural, writes Heidegger, especially in such cases where the being-in relation is conceived in terms of the subject's knowledge of the world. Here, the subject is construed as an autonomous, isolated substance, the world as an external object, and knowledge thus as a relation between two things, a subject-thing and an object-thing. As a consequence, the fundamental relation between subject and world is obscured, because subject and world are not things. It must be emphasised, however, that Heidegger does not dismiss the orthodox subject-object distinction as a false account of the subject's relation to the world; rather, he points out that this account, whatever its merits, is a derivative picture resulting from an insufficiently critical analysis of subjectivity. It is, in other words, a 'founded mode' of being-in, that is, a mode of being-in which subsists only through its dependence on a more fundamental level of being-in-the-world. By obscuring the phenomenal basis of the subject's relation to the world, the substance ontology underpinning the orthodox subject-object schema recapitulates the very model of knowledge which Heidegger aims to bracket and then submit to rigorous phenomenological analysis.[42]

Heidegger writes that being-in-the-world may be experienced in a variety of different ways, for example, as 'having to do with something, producing something, attending to something and looking after it, making use of something, giving something up and letting it go, undertaking, accomplishing, evincing, interrogating, considering, discussing, determining...' All of these are experiences of being-in, and they all have, as their basis, 'concern,' a term Heidegger uses to denote the subject's active involvement with entities in the world, whether those entities are persons or things. In contrast to such involvement, Heidegger also considers ways in which the subject's being-in can manifest a

---

41   Heidegger (1962a), *Being and Time*, p. 78–79 [53]. The third element is Dasein's 'average everydayness.'

42   Heidegger (1962a), *Being and Time*, pp. 87 [60], 86 [59].

deficiency of concern: '[l]eaving undone, neglecting, renouncing, taking a rest.'[43] He argues that such deficient modes of concern are constitutive of the kind of knowledge which emerges, step-wise, through the passive observation of things. First, the subject 'holds back' from its active involvement with entities, and, as a result, is able to encounter them solely in the way that they look. Only through a deficiency of involvement can the subject just look at something, and nothing more. Second, pure looking then becomes 'thematising'; an entity is addressed and considered, thereby becoming an object of perception. Third, perceiving is a form of interpretation, and hence becomes a 'making determinate.' Fourth, what has been perceived and made determinate may now be expressed in propositional terms, that is, it may become the fixed subject matter for a knowledge claim.[44] Heidegger stresses that this step-wise process is a continuous one in which the subject's experience of being-in-the-world goes through successive modifications, from a basic concernful involvement with entities to a derivative 'at arm's length' observation and interpretation of entities as the determinate subject matter of propositional statements. Hence, the process should not be misunderstood as one whereby a subject produces internal representations which are then somehow brought into agreement with externally present entities. Such a misunderstanding ignores the phenomenal content exposed in the existential analytic of the subject, and reasserts the orthodox subject-object distinction as ontologically foundational. Indeed, even in those cases where the subject does no more than represent or think about entities, that is, even when it fails to physically engage with them, it is still in the world with those entities, it still has being-in as its basic structure. As Heidegger remarks, 'the perceiving of what is known is not a process of returning with one's booty to the "cabinet" of consciousness after one has gone out and grasped it.'[45] There is no 'returning' because there was never a 'going out' in the first place.

---

43 Heidegger (1962a), *Being and Time*, p. 83 [56–57].
44 Heidegger (1962a), *Being and Time*, pp. 88–89 [61–62]. I read step two in conjunction with Heidegger's later statement that 'Thematizing Objectifies' (Heidegger (1962a), *Being and Time*, p. 414 [363]).
45 Heidegger (1962a), *Being and Time*, p. 89 [62].

The modifications leading from immersed involvement to propositionally structured thinking are expressed phenomenologically in what Heidegger calls a 'change-over' in the subject's understanding of the world. Because the change-over from involvement to propositional thinking is specifically a change in the subject's mode of understanding, it follows that this change presupposes the prior existence of understanding in general. Moreover, Heidegger argues that the new mode of understanding ushered in by such a change-over has the potential to develop itself autonomously, and thereby to take over as the dominant attitude governing the subject's existence.[46] Thus, for example, as immersed involvement gives way to propositional thinking, the understanding implicit in such thinking, of the world as an external object and the subject as a discrete set of internally organised representations, may come to dominate the subject's way of understanding both itself and its relation to the world. As a consequence, the subject may mistake this new mode of understanding for a foundational one, thereby accepting that all investigations into the structure of subjectivity will finally be intelligible only against the backdrop of the prevailing subject-object distinction. In this case, the subject's basic state of being-in-the-world, its involved immersion in that world, falls into obscurity behind the suddenly pressing problem of what epistemic properties a substance-subject must possess in order to gain access to, and hence knowledge of, a world from which it would be otherwise separated. The irony, of course, is that this critical problem takes for granted a model of the subject which itself derives from a more fundamental mode of subjectivity. Propositional knowledge of the world cannot figure into a causal explanation of our immersed involvement in the world, because propositional knowledge depends for its very possibility on the fact of such involvement.[47]

---

46 Heidegger (1962a), *Being and Time*, pp. 200 [158], 161 [123], 90 [62].

47 Note that the historians of science Lorraine Daston and Peter Galison deliberately adopt the derivative model of subjectivity in their 2010 book, *Objectivity*: 'Because the word "subjectivity" is currently used to refer to conscious experience and its forms across cultures and epochs ("Renaissance subjectivity," "modern subjectivity"), we should make clear that we use the term historically: it refers to a specific kind of self that can first be widely conceptualized and, perhaps, realized within the framework of the Kantian and post-Kantian opposition between the objective and the subjective' (Lorraine Daston and Peter Galison (2010), *Objectivity* (New York: Zone Books), p. 199). Given this qualification, there would seem to be no *prima facie*

6. Heidegger's Response to External-World Scepticism

So far, I have argued that the dispute between external-world realists and sociological idealists unfolds against the backdrop of their shared acceptance of a bundle of distinctions represented by the glass-bulb model. These conceptual distinctions lie at the root of an apparently intractable philosophical problem, namely, the problem of 'epistemic access.' This problem is variously couched in terms of how the mind achieves access to the world, how an epistemic agent breaks through appearances and grasps onto reality, and, perhaps most familiarly, how a subject gains epistemic access to an object. All of these variants take for granted a disjuncture between an inside and an outside, and thus address the question of how this disjuncture might be overcome and knowledge thereby achieved. The external-world sceptic may therefore be interpreted as challenging the claim that a subject can gain access to a world from which it is separated and which exists independently of that subject. As we saw in the last section, Heidegger provides grounds for arguing that the glass-bulb model, implicitly deployed by external-world realists, idealists, and sceptics alike, takes for granted a specific model of the subject, a model which fails to capture the phenomenal content of the subject's basic experience of its own subjectivity. In other words, the model incorporates an unanalysed presupposition that propositional thinking is a basic existential state of the subject (Dasein). Heidegger responds by arguing that such thinking is a founded mode of the subject's being-in-the-world, that it is the result of a *post hoc* change-over from the subject's phenomenologically more original existential state of immersed involvement.

When Heidegger turns specifically to the challenge posed by external-world scepticism, he applies this same analysis. His argument is brief: '[t]he question of whether there is a world at all and whether its Being can be proved, makes no sense if it is raised by Dasein as

---

conflict between their analysis and the one offered by Heidegger. Hence, I now withdraw my previous criticism of their analysis (see Jeff Kochan (2015b), 'Putting a Spin on Circulating Reference, or How to Rediscover the Scientific Subject,' *Studies in History and Philosophy of Science* 49, 103–07 (p. 105 n. 3)).

Being-in-the-world; and who else would raise it?'[48] Heidegger observes that the question of epistemic access, of whether or not one can know that the external world exists, can only make sense if one has already accepted the conceptual distinctions at play in what I call the glass-bulb model. His aim here is not to challenge the truth or falsity of assertions made about the existence of the external world; it is, rather, to point out that such judgements can only be made about assertions which have already been recognised as intelligible. Heidegger argues that the realm of intelligibility in which the concept of the external world makes sense is a derivative one resulting from a change-over in the way the subject understands itself and its world. That this mode of understanding may appear self-evident, that it may have become the prevailing attitude governing our modern self-understanding, is a consequence of our having mistaken the glass-bulb model for a fundamental representation of our basic existential state. Heidegger does not so much refute the external-world sceptic as point out the derivative, superficial nature of her purportedly fundamental challenge.

Just as he had earlier argued that propositional thinking is a founded mode of the subject's being-in, Heidegger now argues that such thinking is also 'a *founded* mode of access to the Real,' and, furthermore, that it is only through this derivative mode of understanding that an analysis of reality becomes possible. The idea seems to be that only once our understanding has changed over to a propositionally structured thinking does it become possible for us to interpret the world as 'Reality,' which for Heidegger also means 'substantiality.' Two steps are involved in this process. First, with the change-over in its mode of understanding, the subject begins to encounter the real in a new way, that is, in terms of 'beholding' (*das anschauende Erkennen*, 'visual cognition').[49] Second, as this beholding, this pure looking which holds back from involvement, comes to dominate the subject's way of relating to entities in the world, it begins to take over as the subject's prevailing mode of understanding that world. It is under these circumstances, argues Heidegger, that we begin to interpret the world, as a whole, in terms of substantiality, as reality. We see here the emergence, once again, of those derivative phenomena represented in the glass-bulb model. The subject's holding

---

48   Heidegger (1962a), *Being and Time*, pp. 246–47 [202].
49   Heidegger (1962a), *Being and Time*, pp. 245–46 [201–02].

back from immersed involvement in the world, so as to then step back and visually examine its surroundings, leads to a perceived separation between subject and world. Mistaking this separation for a fundamental structure in its basic relation to the world, the subject then faces the vexing question of how it may overcome this separation, of how we may, in general, achieve access to a world which we now understand to lie beyond our reach. Heidegger therefore takes the problem of reality, precisely because it is the problem of whether or not an external world exists, to rise out of our failure to recognise being-in-the-world as a central aspect of our fundamental existential state. He writes:

> The 'problem of Reality' in the sense of the question whether an external world is present-at-hand and whether such a world can be proved, turns out to be an impossible one, not because its consequences lead to inextricable impasses, but because the very entity which serves as its theme, is one which, as it were, repudiates any such formulation of the question.[50]

The subject's failure to understand its fundamental relation to the world as one of being-in means that it also fails to understand the basic structure of the world itself. As it is led astray by the conceptual distinctions represented by the glass-bulb model, the subject begins to see the world as a thing standing outside of itself, and this world-thing subsequently gets buried in an epistemological problematic which first puts the world's existence into question and then demands that its existence be proved. Heidegger reckons that this epistemological problematic, the 'problem of Reality,' lies at the heart of the protracted dispute between realists and idealists, and he criticises both sides for having mistaken their derivative understanding of world for an ontologically foundational one. Both sides fall short, Heidegger claims, because neither has brought sufficiently to light the basic phenomenal content of the subject. Both have, in other words, given too much attention to epistemology and not enough to phenomenological ontology.

Heidegger provides clear grounds for distinguishing his own position from both realism and idealism, as he understands them.[51] With respect to realism, he fully agrees with the realist's claim that there is a world in

---

50   Heidegger (1962a), *Being and Time*, p. 250 [206].
51   Heidegger (1962a), *Being and Time*, pp. 250–52 [206–08].

which things exist, in the sense of being present-at-hand. However, he argues that realism goes too far when it interprets the world itself as a present-at-hand thing, that is, in terms of a reality existing independently of the subject. On the basis of this misinterpretation, Heidegger furthermore argues, the realist makes the additional problematic claim that a proof of the existence of the world is both necessary and possible. Heidegger does not follow the realist down this road because for him the world is not a present-at-hand thing separate from the subject. As regards idealism, Heidegger is in full agreement with the idealist's claim that reality cannot be understood on the model of the present-at-hand thing, noting that, with this insight, idealism has an advantage in principle over realism. Where idealism goes astray, Heidegger claims, is when it makes the psychologistic supposition that reality must reside 'in the consciousness' of a subject. Heidegger observes that, so long as this claim leaves unexamined the phenomenal content of consciousness itself, it will fail to advance a properly articulated concept of reality.

Placed against the backdrop of the orthodox subject-object distinction, Heidegger's analysis of the errors of realism and idealism would seem to be as follows. The realist errs in construing the world as an object distinct from a subject, and then also in employing the term 'reality' to denote the 'objecthood' of a subjectless world. In contrast, the idealist errs in ignoring the phenomenal content of the subject, satisfying herself with the largely privative claim that the subject is not an object. She then assimilates reality to this ill-defined subjectivity, and, in the worst case, declares it a manifestation of the interior structures of a worldless subject. According to Heidegger, then, realism and idealism both come up short because neither has recognised being-in-the-world as one of the subject's fundamental ontological structures.

In this chapter, one of my chief aims has been to elucidate Heidegger's response to the external-world sceptic. However, it also seems appropriate, in the present context, to very briefly highlight another crucial aspect of Heidegger's proposed alternative to the bundle of concepts employed by the external-world realist and idealist alike, and so also exploited by the sceptic. A more extended discussion will follow in Chapter Two. The crucial aspect in question is Heidegger's distinction between reality and the real. Recall Heidegger's claim that propositional thinking is a founded mode of access to the real, and that it is only through this derivative mode of understanding that the reality of the real may be grasped as an object of analysis. This suggests that the real may be encountered in a more fundamental way, one which does not entail an accompanying concept of reality. Heidegger's idea seems to be that, when the real is interpreted in terms of reality, it is encountered as an object, with reality signifying its objecthood. The reality of the real is thus the objecthood of the object. Yet, as we have seen, for Heidegger entities are encountered as objects only following a change-over in the subject's being-in-the-world from immersed involvement to the detached thematisation and determination characteristic of propositional thinking. Heidegger argues that the being of entities, *what* they are, depends on the way in which they are understood by the subject, but that the existence of those entities, the brute fact *that* they are, is not dependent on the subject's understanding. He furthermore asserts the more specific proposition that, while reality depends on the subject's understanding of being, the real does not.[52] In this way, Heidegger prepares the conceptual ground on which to assert that the real exists independently of the subject's understanding. Indeed, for Heidegger, the term 'the real' appears to signify independently existing entities. In Chapter Two, I will suggest that this feature of Heidegger's analysis provides the basis for a minimal form of realism which both escapes Heidegger's critique of external-world realism, as explicated above, and proves amenable to SSK's own minimal realist commitments. In the meantime, let us consider how Heidegger's response to the external-world sceptic might bear on the responses made to the sceptic by the SSK practitioners surveyed earlier in this chapter.

7. A Heideggerian Critique of SSK's Response to External-World Scepticism

The principal response of SSK to the external-world sceptic is to eschew the requirement that a belief in the existence of the external world must be absolutely justified. SSK practitioners accept the sceptical argument that absolute knowledge is impossible, but reject the more radical conclusion that knowledge, as such, is impossible. Instead, they endorse a mitigated form of scepticism which allows room for a non-absolutist, or relativistic, conception of knowledge.

---

52   Heidegger (1962a), *Being and Time*, p. 255 [212].

It should be clear that, from the standpoint of Heidegger's own response to the external-world sceptic, the distinction SSK practitioners draw between absolute and relative knowledge is somewhat beside the point. Both absolutist and relativist approaches remain firmly rooted in an epistemological problematic which takes for granted the intelligibility of the sceptical challenge; they differ only in the strategies they deploy when addressing that challenge. Heidegger argues that the intelligibility of external-world scepticism entails the prior acceptance of a set of conceptual distinctions which I have represented with the glass-bulb model. As argued earlier, SSK practitioners take this model for granted even while they reject an absolutist notion of knowledge. As a consequence, they accept as foundational what is, in fact, a derivative conceptualisation of the subject's relation to the world, one which does not sufficiently recognise that one of the subject's basic existential states is being-in-the-world. Only following a change-over in understanding, in which the subject retreats from its original immersed involvement in the world, does it begin to view its access to the things around it as an epistemological problem, that is, a problem of whether or not, as well as how, one may come to know such things in their reality.

The dispute between Barnes, Bloor, and Henry, who assert the existence of the external world, and Collins, who rejects the existence of such a world, can be re-examined in this light. Because they each either affirm or deny the possibility of knowledge of the external world, both sides reveal their shared acceptance of the intelligibility of such a possibility, and thus their tacit reliance on the glass-bulb model. The ensuing debate, though it has produced insights of genuine epistemological interest, remains ontologically adrift insofar as both sides have failed to expose and clarify the basic phenomenological experience of the subject as such. Collins's idealism may have an advantage over the realism of Barnes, Bloor, and Henry, since it acknowledges that there is no sense in speaking of the world as a thing existing independently of the subject's understanding. However, because Collins leaves the ontological structure of this understanding unexamined, he is unable to articulate a sufficiently clear account of the relation between subject and world. As a consequence, he has chosen instead to develop a method in which the world, as well as the things in it, are simply left out of the picture. In contrast, Barnes, Bloor, and

Henry preserve the important insight that things exist independently of the subject's way of understanding them, but they then abrogate this insight by interpreting the world itself as an object, a present-at-hand thing, which exists independently of the subject. They must then face the intractable problem of how to justify the claim that the subject can, in fact, know that this world, as well as the things within it, actually exists.

I have already argued that the justifications they have offered are insufficient. They assert that a belief in the existence of the external world is a presupposition which must necessarily precede any action taken within that world. Yet such arguments only serve to underwrite a realist mode of discourse rather than to establish the existence of the external world. Moreover, as Heidegger points out, when one asserts the need for such a presupposition, one tacitly affirms the derivative notion of the subject as worldless. After all, if one had already recognised being-in-the-world as belonging the subject's basic existential state, then one would not feel obliged to presuppose the existence of the external world.[53] The same goes for Bloor's strategy of epistemological complacency. This strategy takes for granted the epistemological problematic and responds to it by recommending complacency. Such a strategy makes sense only if one has already agreed with the external-world sceptic that subject and world are separated, and that the subject can only achieve knowledge of, or gain epistemic access to, this world by overcoming that separation. Bloor expresses his belief that such knowledge is possible, but responds with complacency to the sceptic's demand for a justification of that belief. Heidegger argues that such a strategy, because it fails to render transparent the subject's basic ontological structure, treats the subject as a wordless thing which must first assure itself, somehow, of a world. As such, the strategy is itself an expression of a founded mode of understanding, a mode in which the glass-bulb model is taken for granted, and hence one in which a derivative mode of understanding is mistaken for one in which the fundamental ontological relation between subject and world is originally revealed.[54]

---

53 Heidegger (1962a), *Being and Time*, p. 249 [205–06].
54 Heidegger (1962a), *Being and Time*, p. 250 [206].

8. Conclusion

This chapter has been concerned with the threat posed to SSK by external-world scepticism. Although SSK practitioners have made effective use of sceptical techniques in their analyses of scientific knowledge, their methods are best seen, not as sceptical, but as advocating a mitigated response to the radical sceptical claim that knowledge of the external world is impossible. With the exception of Collins, SSK practitioners have attempted to advance a minimal realist position which asserts the existence of an external world without also feeling obliged to meet the sceptic's demand that such an assertion be absolutely justified. I have argued that they have not been successful. The key obstacle preventing SSK practitioners from developing a defensible realist position is their preoccupation with epistemological, at the expense of ontological, issues. Indeed, despite the long dispute between the realist and idealist wings of SSK, both sides have failed to adequately address the way in which their taken-for-granted ontological commitments inform the content of their epistemological arguments. I have used Heidegger's phenomenological analysis of the subjectivity of the subject to expose the nature of those commitments.

A phenomenological analysis of the subject's basic state of being-in-the-world reveals that external-world scepticism makes no sense as a fundamental challenge to the subject-world relation. External-world scepticism depends for its dialectical force on a derivative understanding of that relation, a conceptualisation of it in terms of a distinction between subject and object. The weakness at the heart of SSK's responses to the external-world sceptic is its tacit adherence to the metaphysical image of the subject which underpins the orthodox subject-object schema.[55] It is only within the epistemological problematic generated by this schema that external-world scepticism becomes intelligible and so comes to threaten the realist ambitions of SSK. If SSK practitioners wish to avoid the debilitating challenge posed to their work by the external-world sceptic, then I recommend that they divest themselves of their residual commitment to orthodox ontology and adopt the position advanced

---

55   In Chapter Three, I will give detailed attention to the way Bloor attempts to transform, without wholly rejecting, the Kantian version of the orthodox subject-object distinction.

by Heidegger. Yet, this recommendation comes with a worry. If SSK were to adopt a Heideggerian ontology, which is openly critical of both realism and idealism, would it not lose the grounds for its realism? In this chapter, I have already suggested that Heidegger, despite his abnegation of realism, nevertheless provides the basis for a minimal realist doctrine which both escapes his own criticism and is compatible with the main tenets of SSK's realist commitments. It lies with Chapter Two to make good this claim.

Chapter Two

A Minimal Realism for Science Studies

1. Introduction

One of the most ridiculed concepts in Heidegger's work is his 'question of being.' An unlikely collection of critics, ranging from the philosopher Simon Blackburn to the science studies scholar Bruno Latour have exercised considerable rhetorical flair in roundly repudiating the significance of this question. Blackburn pokes fun at those who 'flutter around the flame of Being.' Latour lampoons Heidegger's 'epigones [who] do not expect to find Being except along the Black Forest Holzwege,' and he burlesques their alleged claim that '[w]e are keeping the little flame of Being safe from everything, and you, who have all the rest, have nothing.'[1] For the incorrigibly counter-suggestive, like me, such enthusiastic denunciations from on high encourage the thought that Heidegger's question of being may warrant close attention after all. Indeed, as I hope to demonstrate in this chapter, Heidegger's question yields resources for a minimal realism compatible with the social constructivism of science studies.

---

1   Simon Blackburn (2004), 'Lights! Camera! Being!' *New Republic* (February 23); Bruno Latour (1993), *We Have Never Been Modern* (Harvard: Harvard University Press), pp. 65, 66. I discuss Latour's criticism in Jeff Kochan (2010b), 'Latour's Heidegger,' *Social Studies of Science* 40(4), 579–98 (pp. 587–88).

   http://dx.doi.org/10.11647/OBP.0129.02

Heidegger's question of being can be more fully described as the question of the *meaning* of being. The words 'meaning' and 'being' may raise expectations that the question will lead us into deep and mysterious philosophical waters, but, in fact, we may profitably address it on the most superficial and mundane of levels. The word 'being' translates the German infinitive *sein*, which can be more strictly rendered as 'to be.' Hence the German sentence 'Alle wollen glücklich sein' means 'Everybody wants to be happy.' Note that, unlike the other terms in this sentence, the verb 'to be' does not refer to anything. Its role is rather to bind together and give an overall meaning to the sentence. The word 'being' should, therefore, not be mistaken for the name of an entity, or thing. As Heidegger writes, '[t]he Being of entities "is" not itself an entity.'[2] If 'being' names anything at all, then it names the way in which things gather together and so acquire meaning. According to Heidegger, this is an ontological event with a temporal structure. The question of the meaning of being thus motivates an enquiry into the way meaning takes place as this temporal event. Heidegger's question is not 'What meaning does "being" have?' Meaning is not a thing being possesses. The question is rather 'How does "being" mean?' Meaning is an event, something being does. Grammatically, the phrase 'the meaning of being' is similar in structure to the phrase 'the thrill of a lifetime.' The thrill is not the property of a lifetime, because a lifetime is not a thing with properties. A lifetime is a historical-existential space wherein thrills can happen. Likewise, being is a historical-existential space wherein meaning can happen.

Heidegger observes that, because the word 'being' and its cognates play such a ubiquitous role in our language, we tend simply to take them for granted, without giving them a second thought. Yet, he argues, useful insights may be won by turning 'being' from a taken-for-granted resource into a topic for investigation. One such insight will be especially crucial for this chapter: the polysemy of 'being.' The word 'being' carries connotations of both existence and essence. By attaching the name of a thing to the verb 'to be,' one may then say of the thing *that* it is, or *what* it is, or both. As will be discussed later, Heidegger marks

---

2   Martin Heidegger (1962a [1927]), *Being and Time*, trans. by John Macquarrie and Edward Robinson (Oxford: Blackwell), p. 26 [6]. (Following scholarly convention, page numbers in square brackets refer to the original 1927 German edition of *Being and Time*.)

this as a distinction between the 'that-being' and the 'what-being' of a thing, a move which furthermore distinguishes between the thing's existence and its essence. We have already encountered this lattermost distinction in Chapter One.

The distinction between existence and essence, between the that-being and the what-being of a thing, is an ancient one, and it also features prominently in several of Heidegger's works. Nevertheless, the distinction has been largely overlooked by those writers concerned with explicating Heidegger's views on science and realism. One such writer, Joseph Rouse, stands out as being both a highly regarded expositor of Heidegger's philosophy of science, on the one hand, and a key contributor to theoretical debates in science studies, on the other. The latter half of this chapter will give focussed attention to his work on both counts. Two other writers, while not having bridged between Heidegger and science studies to the same degree as Rouse, also bear mentioning: Trish Glazebrook and Dimitri Ginev.[3] Like Rouse, neither Glazebrook nor Ginev have recognised the important role played by Heidegger's distinction between existence and essence. Glazebrook has come the closest, correctly observing that Heidegger was vexed by the problem of how a worldly thing may be acknowledged to exist independently of the subject when its intelligibility nevertheless depends on that subject. As we will see, Heidegger uses the distinction between existence and essence to solve this problem, recognising the independent existence of a thing while maintaining the necessary dependence of its essence, construed broadly to include its core meaning or basic intelligibility, on the subjectivity of the subject. Glazebrook, in contrast, argues that Heidegger solves this problem by differentiating between a thing and its being, what is conventionally called the 'ontological difference.'[4] But this cuts the knot in the wrong place. The distinction between independent existence and dependent essence is a

---

3   There is an extensive literature more generally addressing the topic of Heidegger on realism and science. However, a discussion of it would carry us too far beyond the narrow scope, and specific goals, of the present chapter. Curious readers may consult the Appendix at the end of this chapter (p. 106).

4   Trish Glazebrook (2012b), 'Why Read Heidegger on Science?,' in *Heidegger on Science*, ed. by Trish Glazebrook (Albany: SUNY Press), pp. 13–26 (p. 20); Trish Glazebrook (2001a), 'Heidegger and Scientific Realism,' *Continental Philosophy Review* 34(4), 361-401 (p. 368). The ontological difference is the difference, mentioned earlier, between being and entities.

distinction *in* the being of a thing, not between a thing and its being: 'it is precisely the two of them that make up the structure of being.'[5] As will be explained below, this distinction stems from the difference between a thing's existing but meaning nothing, and its existing and meaning something. By picking up the wrong distinction, Glazebrook is forced to grapple with a range of deep paradoxes, and it is not clear to me that she succeeds in resolving them. For example, on the one hand, she argues that 'for Heidegger, it is an incoherent demand to make of realists that they hold the independence thesis.' On the other, she also argues that Heidegger was a 'robust scientific realist.'[6] As I will argue, the independence thesis is the basic doctrine of realism, including scientific realism, and Heidegger was a realist just because he accepted this doctrine. He was, however, not a scientific realist; he was what I call a 'minimal realist.' Glazebrook's account of Heidegger's realism is an undoubtedly complex and difficult one. I commend my own account as a simpler, more modest, and more useful alternative.

This chapter begins with an explication of Heidegger's early existential conception of science. Heidegger introduced this conception as an alternative to the dominant logical conception, which views science as a conceptual system. He thus draws attention to the concrete, existential structures of scientific practice which are necessary for more abstract, theoretical reflection. Theory needs method, and method is concretely enacted in the world. Although Glazebrook is aware of this aspect of Heidegger's account of science, she nevertheless repeatedly attributes to him the view that science is a 'conceptual scheme.'[7] She shares this tendency with both Rouse and Ginev, but whereas Glazebrook attributes the position without criticism, the other two treat it as evidence for Heidegger's failure to have fully embraced scientific practice, and hence to have entirely freed himself from the orthodox trappings of a theory-dominant view of science.[8] Ginev credits Rouse with this criticism,

---

5   Martin Heidegger (1982a [1975]), *Basic Problems of Phenomenology*, trans. by Albert Hofstadter (Bloomington: Indiana University Press), p. 78.

6   Glazebrook (2001a), 'Heidegger and Scientific Realism,' pp. 386, 361.

7   Glazebrook (2012b), 'Why Read Heidegger on Science?,' pp. 20, 21; Glazebrook (2001a) 'Heidegger and Scientific Realism,' pp. 377, 381, 382, 389.

8   Dimitri Ginev (2005), 'Against the Politics of Postmodern Philosophy of Science,' *International Studies in the Philosophy of Science* 19(2), 191–208 (p. 199); Dimitri Ginev (2011), *The Tenets of Cognitive Existentialism* (Athens OH: Ohio University Press) pp.

which is based on the claim that the 'mathematical projection of nature,' which Heidegger located at the heart of modern science, is an inherently theoretical phenomenon. I will argue, to the contrary, that Heidegger introduced the mathematical projection as an existential phenomenon which serves as a condition of possibility for both theory and practice in the sciences.[9] Once again, Glazebrook comes the closest to my own view when she writes that Heidegger's strategy 'is not to establish a secure bridge between praxical involvement and theoretical analysis, but rather to trace both back to being-in-the-world.' This strikes me as largely correct.[10] Although scientific theory is necessarily enabled by practice, Heidegger resisted the urge to explain it reductively in terms exclusively of practice. This challenges the widespread view in science studies that theory can be unproblematically reduced to practice. By simply collapsing one side of the theory-practice divide into the other, the basic motivations which originally gave rise to and justified that divide are left hopelessly obscure. I will not, in this chapter, make any satisfying attempt to clear up this obscurity. Such an attempt will come later in Chapter Five, when we consider the relation between mathematical and empirical modes of scientific existence. Meanwhile, in this chapter, by at least drawing attention to this obscurity, I hope to begin undermining the unreflective business-as-usual attitude of many contemporary practice theorists towards the theory-practice divide.

Under the flag of a 'practical hermeneutics of science,' Rouse has been most enthusiastic about clearing the deck of theory and raising in its place an account of science based solely on the notion of practice. In the process, he has not only criticised the early Heidegger for allegedly retreating back into a theory-dominant account of science, but proponents of the sociology of scientific knowledge (SSK) as well. In the former case, Rouse's misconstrual of Heidegger's concept

---

5, 103; Joseph Rouse (1987a), *Knowledge and Power: Toward a Political Philosophy of Science* (Ithaca: Cornell University Press), p. 103.

9   I also makes this argument, with slightly more detail, in Jeff Kochan (2015a), 'Scientific Practice and Modes of Epistemic Existence,' in *Debating Cognitive Existentialism*, ed. by Dimitri Ginev (Leiden: Brill Rodopi), pp. 95–106.

10   Glazebrook (2001a), 'Heidegger and Scientific Realism,' p. 386. Unfortunately, Glazebrook seems to immediately lose grip on this insight when, on the same page, she concludes that 'the difference between theory and practice [is] the difference between two kinds of practice.' She credits Rouse, in part, for having influenced her on this point.

of mathematical projection, as referring to a theoretical rather than an existential phenomenon, leads him to exaggerate the extent to which Heidegger asserted the independence of theory from practice. Transfixed as he is by the theory-practice divide, Rouse fails to realise that a refusal to collapse theory into practice does not necessarily evince a counter-desire to collapse practice into theory. In the latter case, Rouse also exaggerates the commitment of SSK to a theory-dominant account of science, but he makes a good point that SSK remains debilitated by a vestigial commitment to a problematic theory of knowledge. I examined this vestigial commitment in Chapter One, suggesting there that SSK could overcome this epistemological problematic by adopting key aspects of Heidegger's existential phenomenology. In this chapter, I apply a similar strategy in order to defend SSK against Rouse's criticisms. On this basis, I conclude that Rouse's attempt to undermine SSK is not successful.

The gist of my argument is that Rouse's practice-based account of science poses no threat to the minimal realism which I draw out of Heidegger's work and recommend for science studies. Indeed, Rouse's attempts to close the door on realism appear unsuccessful even in the case of his own practical hermeneutics of science. As I will demonstrate, despite his theoretical attempts to keep the realist's basic independence thesis at bay, a close look at the way Rouse concretely articulates that theory reveals his own informal and unreflective commitment to that very thesis. The incoherent relationship between Rouse's theory and practice springs from his failure to recognize Heidegger's distinction between existence and essence. I suggest, then, that the practical hermeneutics commended by Rouse is best replaced with an existential phenomenology of science, because the latter is better able to accommodate the basic realist doctrine of independent existence.

Minimal realism is thus not a repudiation of practice and a flight back into theory. It is instead a recognition that theory and practice are phenomenologically distinct ways of actualising the range of possibilities opened up by the form of existence Heidegger dubbed the 'mathematical projection of nature.' Ginev criticises Rouse for not paying adequate attention to the existential basis, and especially the existential specificity, of science, and he deplores Rouse's consequent tendency to uncritically assimilate science into the broader cultural sphere in which it is embedded. According to Ginev, one can abandon a theory-dominant account of science while still viewing science as a unique and specifiable form of existence, one distinct from other forms of cultural existence.[11] As far as the minimal realist is concerned, this suggests that science may get at the real, at independently existing things, in ways characteristically distinct from the other forms of existence also enabled by our history and culture. This is an important point, which ultimately leads to political questions about the relationship between science and the broader social sphere. I will briefly comment on this in the concluding section of this chapter, and again in Chapter Seven of this book. For the time being, let us focus on the issue of realism and science, beginning with a discussion of Heidegger's existential conception of science, then exploring the significance of this conception for SSK, and finally defending the resultant account of minimal realism from the challenge posed to it by Rouse's practical hermeneutics of scientific practice.

2. Heidegger's Existential Conception of Science

In his 1927 book, *Being and Time*, Heidegger distinguishes between a 'logical' and an 'existential' conception of science.[12] The logical account understands science according to the representation of nature it produces, and the validity of this representation is itself defined as having been established on the basis of a coherent body of interconnected true propositions. On this account, then, science is taken to be a conceptual scheme. On the existential account, in contrast, science is understood to be a mode of existence, a way of being-in-the-world, which brings to light things for theoretical understanding. It is important to emphasise that these two conceptions of science are not opposed to one another. Heidegger commits himself to an existential conception of science, but he does not, in doing so, reject the logical conception as wrong or absurd. On the contrary, the existential account is meant to explain how the

---

11   Ginev (2005), 'Against the Politics of Postmodern Philosophy of Science,' p. 103.
12   Heidegger (1962a), *Being and Time*, p. 408 [357].

logical account is possible; it seeks to reveal the existential conditions necessary for the emergence of the theoretical attitude presupposed in the logical account. By undertaking a phenomenological investigation into the mode of existence which makes science possible, Heidegger shifts attention from science construed as a body of concepts and formal logical rules to science construed as an ongoing, goal-oriented human activity. In other words, he focuses on what scientists do, and also on the way they must experience and understand their relation to the world in order to do what they do.

By emphasising the actions of scientists, Heidegger would seem to adopt an approach similar to the one generally prevailing in science studies. Bloor, Barnes, and Henry, for example, write that '[f]or the scientist the world is the object of study; for the sociologist it is the scientist-studying-the-world that is the object.'[13] Yet Heidegger also stresses that he is primarily concerned neither with the historical development of science, nor with the particular goal-directed activities of scientists working in specific contexts. Hence, Heidegger's existential conception of science cannot be straightforwardly assimilated to the view of science favoured by SSK practitioners. Indeed, one crucial difference is that Heidegger, unlike Barnes, Bloor, and Henry, does not conceptualise the scientist-studying-the-world as an object. The reasons for this have already been covered in Chapter One. Rejecting a view dating back as far as Aristotle, Heidegger argues that the subject (Dasein), including the scientific subject, should not be conceptualised in fundamental terms as a thing, substance, or object. The subject is, rather, *existence*. By undertaking a phenomenological analysis of the basic existential structures of the subjectivity of the subject, Heidegger attempts to counteract the traditional metaphysical tendency to construe persons as special instances of a more general ontological category of 'thing.' In reserving the term 'existence' exclusively for the subject, and in order to guard against the subject's being conflated with a thing, Heidegger refers to the existence of things as 'presence-at-hand' (*Vorhandenheit*). His existential conception of science thus focuses on the activities of scientists, rather than on bodies of scientific knowledge, because such activities provide a necessary basis for scientific subjectivity, for the

---

13   Barry Barnes, David Bloor and John Henry (1996), *Scientific Knowledge: A Sociological Analysis* (London: Athlone), p. 30.

particular mode of existence within which scientific knowledge is produced and sustained.

Heidegger calls the ground state of the subject's elemental being-in-the-world 'circumspective concern.' This is what, in Chapter One, was referred to as immersed involvement. Determining the conditions of possibility for the theoretical attitude involves analysing the existential conditions under which theoretical thinking emerges from a basic everyday state of immersed involvement in a world. As discussed in Chapter One, Heidegger analyses this emergence in terms of four steps, which lead from immersed involvement to propositionally structured thinking. First, one holds back from immersed involvement so as to merely look at things, and no more. Second, pure looking becomes a thematising in which things are encountered as objects of perception. Third, perception interprets things so as to determine their properties. Fourth, determinate objects become the subject matter for propositional knowledge claims. Heidegger calls this transformation in the way things in the world are experienced a 'change-over' in the subject's mode of understanding. The phenomenological analysis of this change-over plays a central role in Heidegger's existential conception of science, and thus merits detailed examination.

Heidegger presents the change-over as a transition from an experience of things as 'ready-to-hand' to an experience of them as 'present-at-hand.' Things that are, in this context, present-at-hand are those encountered once one holds back from involvement with a thing and begins to interpret it as an object with determinable properties. In this context, then, a present-at-hand thing is called an 'object.' In contrast, things which are ready-to-hand are called 'equipment,' that is, things encountered in a basic existential state of immersed involvement in a world. On this account, the world in which one is always already immersed and involved is a world of equipment, what Heidegger also calls a 'work-world.'[14] He observes that, when we are absorbed in this work-world, our attention is not focussed on the equipment we use; rather, it is focussed primarily on the *work*.[15] For example, when one signs one's name, one does not focus on the pen in one's hand or the paper on which one signs, but on the act of signing. Similarly, when

---

14 Heidegger (1962a), *Being and Time*, p. 101 [71].
15 Heidegger (1962a), *Being and Time*, p. 99 [69].

a cyclist rides along a busy street, her attention is focussed not on her bicycle but on the task of cycling. Both pen and bicycle are, in these cases, experienced as things ready-to-hand, as equipment the significance of which lies in the task towards which it is, at that moment, being put. Pieces of equipment are put to use in an activity, but the activity is not about them. They are not the theme of the activity, much less its object. They are not, in other words, the topic of the activity, but a resource enabling that activity.

In clarifying how the change-over in understanding gets going, Heidegger must explain how a basic state of everyday immersed involvement could come to be disturbed or interrupted. He must, in other words, give some account of how one comes to leave behind this basic state, how one comes to hold back from involvement and begins to instead experience a thing as the theme of one's activity, and so as its object. This problem can be usefully contrasted with the problem of the external world, discussed in Chapter One. There the difficulty was to explain how a subject may gain access to a world from which it remains fundamentally separated. The solution demands an account of how the subject breaks free from the finite limits of its own internal state by building an epistemic bridge over to the external world which, in turn, exists in fundamental ontological separation from that subject. Traditionally, it is said to do this through the exercise of a transcendent reason. In the case of Heidegger's existential phenomenology, in contrast, the question of how the subject gains access to the external world never arises, because subject and world were never separated in the first place. In the former case, the problem is to explain how a basic deficiency in the subject's relation to the world may be overcome through the transcendent power of reason. In the latter case, the problem is to explain how one's basic existential absorption in a work-world may suddenly become deficient, how one might abruptly withdraw from the everyday work-world in which one is normally immersed. Only through this sudden deficiency in one's workaday relation to things can the change-over get going.

Heidegger addresses this problem by considering what happens when a 'breakdown' occurs in the smooth functioning of the subject's workaday world, when the circumspective concern characteristic of one's most immediate involvement with equipment becomes disturbed or interrupted. As a result of such an equipmental breakdown, '[t]he presence-at-hand of entities is thrust to the fore.'[16] One situation in which a breakdown may occur is when, in the course of everyday activity, something vital is suddenly found missing. In an obvious sense, the missing thing is not ready-to-hand; indeed, it is not 'to hand' at all. However, insofar as the readiness-to-hand of some other thing may depend on the missing thing, this other thing now loses its familiar readiness-to-hand and begins to obtrude as something present-at-hand. For example, my office door is ready-to-hand when there is a key to open it. If, however, I have forgotten my key, the door suddenly loses much of its equipmental significance, or meaning. If it is Sunday morning, when the administration is normally absent, then the readiness-to-hand of the door recedes still further. I encounter the door as ever more useless, a mere obstacle confounding the smooth running of the workaday context in which I normally find myself. If, furthermore, I have a flight leaving that morning, and my flight tickets and passport are locked in my office, then the door may lose entirely its significance as something ready-to-hand. I now encounter it in its brute existence, as a useless thing which just stands there confounding my travel plans. As Heidegger writes:

> The more urgently [Je dringlicher] we need what is missing, and the more authentically it is encountered in its un-readiness-to-hand, all the more obtrusive [um so aufdringlicher] does that which is ready-to-hand become — so much so, indeed, that it seems to lose its character of readiness-to-hand. It reveals itself as something present-at-hand and no more, which cannot be budged without the thing that is missing. The helpless way in which we stand before it is a deficient mode of concern, and as such it uncovers the Being-just-present-at-hand-and-no-more of something ready-to-hand.[17]

The sudden breakdown in the equipmental context of a work-world, which follows from the discovery that something vital to our operations in that world has gone missing, jars our attention from absorption in the task at hand, suddenly bringing forward the presence-at-hand of what is normally experienced as ready-to-hand.

Heidegger's phenomenological analysis of breakdowns in circumspective concern demonstrates how a deficiency can suddenly

---

16   Heidegger (1962a), *Being and Time*, p. 107 [76].
17   Heidegger (1962a), *Being and Time*, p. 103 [73].

appear in the subject's basic relation to a work-world, and hence also how its understanding of things as ready-to-hand may begin to change over to an understanding of things as present-at-hand within the world. In this first step, I encounter things as 'being-just-present-at-hand-and-no-more.' In some cases, the change-over may go no further than this: the overworked department head may suddenly appear in the hallway with a master key, allowing me to carry on with my travel plans. Yet, in other cases, one's understanding, rather than reverting back to a basic existential state of circumspective concern, may change over to a new mode of understanding through a process which Heidegger calls 'thematising.' This process is central to his existential conception of science.

We have already encountered Heidegger's concept of thematising in summarising the four-stage change-over from immersed involvement to the theoretical attitude. After immersed involvement switches to a state of pure looking (stage one), a thing becomes thematised as an object of perception (stage two). Indeed, Heidegger emphasises that: 'Thematizing Objectifies.'[18] The change-over marks a shift from understanding a thing in the world as ready-to-hand to understanding it as present-at-hand, as an object. This is a shift in the existential structure of understanding, a structure Heidegger calls 'projection.'[19] He argues that only on the basis of a projection do we encounter a thing as meaningful: 'The primary projection of the understanding of Being "gives" the meaning.'[20] The projection may thus be understood as providing the background of intelligibility against which things come to be perceived as objects with determinate properties (stage three), and thence as the subject matter for propositional knowledge claims (stage four). It should be noted, however, that the projection, as the basic structure of understanding, is present even when no change-over occurs; it resides just as much in the undisturbed practical understanding characteristic of immersed involvement in a work-world. Whether one understands a thing as ready-to-hand or as present-at-hand within-the-world, such an understanding will always have the structure of a projection.[21]

---

18   Heidegger (1962a), *Being and Time*, p. 414 [363].
19   Heidegger (1962a), *Being and Time*, pp. 185 [145].
20   Heidegger (1962a), *Being and Time*, p. 372 [324–25].
21   Heidegger (1962a), *Being and Time*, p. 371 [324].

Heidegger argues that scientific understanding is structured on the existential level by a particular kind of projection. The existentially decisive feature of science is, in his view, neither empirical observation nor mathematical modelling, but rather 'the way in which Nature herself is mathematically projected.'[22] This mathematical projection determines the range of possible ways in which nature may be intelligibly experienced, and so understood, both at the practical and the theoretical level. It furthermore serves as the existential source from which the logical conception of science draws its own currency. In the context of scientific activity, the change-over from immersed involvement to a theoretical conception of nature is a shift in the existential structure of the mathematical projection. According to Heidegger, within the scope of intelligibility opened up by this projection, pure looking shifts specifically towards a perception of things as objects which can be quantitatively determined in terms of such general categories as motion, force, location, and time. Only on the basis of this kind of projection can the scientist discover something like a 'fact' which may then be set up as part of an experimental investigation.[23] Heidegger suggests that this mathematical grounding of factual science was possible only because researchers recognised that there are, in principle, no bare facts. Science projects the factuality of things in terms of categories amenable to quantitative analysis. Furthermore, it does this in such a way that the measurability of those things is disclosed as an *a priori* feature of their being. Heidegger thus argues that the existential conditions of possibility for the empirico-mathematical sciences are manifest in the projection of nature as being essentially measurable in a fixed, quantitative sense. After these conditions have been fulfilled, the horizon within which the subject is able to intelligibly encounter things limits the possible ways in which things may be discovered within the world. Heidegger writes that the aim of thematising is 'to free the entities we encounter within-the-world, and to free them in such a way that they can "throw themselves against" a pure discovering — that is, in such a way that they can become "Objects."'[24]

---

22   Heidegger (1962a), *Being and Time*, pp. 413–14 [362].
23   Heidegger (1962a), *Being and Time*, p. 414 [362].
24   Heidegger (1962a), *Being and Time*, p. 414 [363].

Once there has been a disturbance in our immersed involvement with things as ready-to-hand, it becomes possible for that non-deliberative involvement to change over into thematising, and hence for us to begin experiencing things as objects rather than as equipment. Heidegger emphasises that the change-over in our understanding, from non-deliberative use of a thing in the course of everyday activity, at the one extreme, to thematising and then making propositional assertions about that thing, at the other, is marked by a number of intermediate steps. As thematising begins to objectify a thing, that thing acquires a more determinate meaning; it comes to be experienced as an object whose properties are an increasingly well-defined and stable subject matter for assertions, and thus better fitted to the propositional structure of theoretical and logical modes of understanding. Heidegger also calls this a process of articulation: 'thematizing modifies and Articulates the understanding of Being.'[25] By articulating the meaning of a thing in propositional terms, thematising may also affect the way we understand, and hence practically engage with, that thing. Heidegger furthermore warns that the intermediate stages of the thematising process cannot be understood in terms of the theoretical statements which emerge only at the conclusion of the process without dramatically perverting the meaning of those stages. Both these intermediate stages, as well as the theoretical assertions they finally constitute, have their 'source' in a circumspective, or practical, form of interpretation.[26] For this reason, Heidegger argues that logic is rooted in existence.[27] The more general conclusion to be drawn from this is that the logical conception of science, which views science as a coherent body of true propositions, has its own original source in a specific existential mode of understanding structured by the mathematical projection of nature.

The final implication of Heidegger's existential conception of science is that to construe science in purely theoretical terms, as a body of logically organised true propositions, as a conceptual scheme, is to ultimately misunderstand its significance as a human enterprise. He emphasises that scientific concepts cannot be understood independently

---

25   Heidegger (1962a), *Being and Time*, p. 415 [364].
26   Heidegger (1962a), *Being and Time*, p. 201 [158].
27   Heidegger (1962a), *Being and Time*, p. 203 [160]. This claim will be given more detailed attention in Chapter Four.

of scientific method, and that 'theoretical research is not without a *praxis* of its own.'[28] Unsurprisingly, then, his account of science gives a central and consequential place to scientists' practical manipulation of equipment:

> Reading off the measurements which result from an experiment often requires a complicated 'technical' set-up for the experimental design. Observation with a microscope is dependent upon the production of 'preparations.' Archaeological excavation, which precedes any Interpretation of the 'findings,' demands manipulations of the grossest kind. But even in the 'most abstract' way of working out problems and establishing what has been obtained, one manipulates equipment for writing, for example. However 'uninteresting' and 'obvious' such components of scientific research may be, they are by no means a matter of indifference ontologically.[29]

Indeed, equipmental manipulations play an integral role in the thematising process which gives rise to theoretical knowledge. It is not just linguistic practices but also concrete material practices which serve to more precisely articulate the meaning of the things taken up as a subject matter for science. To say that the instruments and material practices of a science are, in part, constitutive of its theoretical and logical content is to make a strong ontological claim about the interdependence of theory and practice. Yet interdependence is not identity. While theory cannot be understood independently of the linguistic and material practices which constitute it, it is clear that, for Heidegger, theory and practice remain different modes of scientific understanding. He views them as distinct but related existential modalities within which the intelligibility of things becomes possible.

Heidegger thus appears to suggest that the emergence of a new theoretical form of understanding, especially as exemplified historically in the development of mathematical physics, marks the emergence of a new ontological condition, a new form to human existence wherein the subject understands itself as a mental substance and the things surrounding it as objects with quantifiably determinable properties. In this process, the subject's own subjectivity likewise becomes increasingly modelled in accordance with the proposition: scientific

---

28 Heidegger (1962a), *Being and Time*, p. 409 [358].
29 Heidegger (1962a), *Being and Time*, p. 409 [358].

knowledge is propositionally structured knowledge. Heidegger's existential conception of science thus challenges the priority of the logical conception of science as propositionally structured theory, and seeks to reverse that priority by emphasising the concrete existential conditions on which theoretical knowledge ultimately depends. But this is not a reduction of theory to practice; it is the recognition that, although practical and theoretical behaviour are ontologically distinct, and although the latter emerges in a change-over from the former, each represents a distinct mode of the same existential projection of nature. Theory is distinct from, but not ontologically independent of, practice. Where the line between the two should be drawn, however, is a question Heidegger does not, and perhaps could not, answer. Indeed, he openly lamented that 'it is by no means patent where the ontological boundary between "theoretical" and "non-theoretical" behaviour really runs!'[30] That there is indeed a boundary running between them is, however, something he did not doubt.

3. Getting at the Real

In Chapter One, we briefly considered the way in which Heidegger differentiates his existential analytic of the subject from both realism and idealism. With the above discussion of Heidegger's existential conception of science now also behind us, it will be worthwhile to return to his comments on realism and idealism and considering them in greater depth.

Heidegger superficially agrees with the realist that things within-the-world are present-at-hand, in the sense that they exist. However, he criticises the realist for conceptualising the presence-at-hand of things in strictly epistemological terms, as the 'objecthood' of independently existing objects of knowledge. This conceptualisation takes it for granted that propositional thinking is itself a fundamental mode of the subject's existence, that the subject's relation to a thing within-the-world is fundamentally that of a substance-subject examining an independent object. Hence, Heidegger describes realism as the belief that 'the way to grasp the Real is by that kind of knowing which is characterized

---

30   Heidegger (1962a), *Being and Time*, p. 409 [358].

by beholding [das anschauende Erkennen].'[31] Yet, as we saw in the previous section, such knowing (or *cognising*) emerges from a change-over in the subject's understanding of things from things ready-to-hand to things present-at-hand, and so, for this reason, Heidegger goes on to question whether our 'primary access' to the real, that is, to existing things, can be suitably captured by the traditional epistemological conception of knowledge as rooted in the observational powers of a substance-subject positioned vis-à-vis an object.[32] Indeed, he argues instead that perceptual examination presupposes thematisation. The central failing of realism, according to Heidegger, is that it asserts the independent reality of objects while simultaneously projecting that reality as part of an objectifying thematisation which depends for its possibility on the subject's existence. In this regard, Heidegger views idealism as the more successful position, since it affirms the ontological dependency of objects on our understanding of them. In other words, idealism rejects the realist claim that our knowledge of objects within-the-world provides evidence for the independent existence of the real as such. However, Heidegger dismisses the conclusion which the idealist then draws from this insight: that the real must therefore exist only in consciousness, that it must be constituted solely by the subject.[33] On Heidegger's account, neither realism nor idealism offers a defensible position because they both remain entangled in the epistemological problematic, and hence they both fail to recognise the ontological basis for that problematic in the subject's own existence. As he puts it, realism and idealism 'can exist only on the basis of a neglect: they presuppose a concept of "subject" and "object" without clarifying these basic concepts with respect to the basic composition of Dasein itself.'[34]

Although Heidegger offers his existential analytic of the subject as an alternative to both realism and idealism, in this section I will argue that Heidegger's position is nevertheless compatible with a 'minimal' form of realism. I contrast this minimal realism with the 'robust' realism typically espoused by scientific realists. The difference between these

---

31 Heidegger (1962a), *Being and Time*, p. 246 [202].
32 Heidegger (1962a), *Being and Time*, p. 246 [202].
33 Heidegger (1962a), *Being and Time*, p. 251 [207].
34 Martin Heidegger (1985), *History of the Concept of Time*, trans. by Theodore Kisiel (Bloomington: Indiana University Press), pp. 222–23.

two doctrines becomes clear once we recognise that robust realism is comprised of two distinct theses. The first thesis declares that things can exist independently of our own existence, that they are not the products of our interpretations, theories, or practices. The second thesis makes the more complex assertion that the determinate properties of these things, including their relational or structural properties, can also exist independently of our own existence. Robust realism affirms both the first and the second thesis, while minimal realism affirms only the first thesis. I will call this first thesis the 'basic independence thesis.' As we will see in the next section, this twofold account of realism has striking similarities with SSK practitioner Barry Barnes's account of 'double-barrelled' realism.

That Heidegger's existential conception of science is compatible with the basic independence thesis can be seen in his careful discrimination between reality, on the one hand, and the real, on the other. He writes, for example, that 'Being (not entities) is dependent on the understanding of Dasein; that is to say, Reality (not the Real) is dependent on care,' with care being a fundamental existential structure in the subjectivity of the subject. He furthermore emphasises the dependency of reality on the subject when he says of reality that 'ontologically it has a definite connection in its foundations with Dasein, the world, and readiness-to-hand.' Finally, he argues that, when entities are conceived as a 'context of Things (*res*),' by which he means a context of 'substances,' the being of those entities acquires the meaning of 'Reality,' or 'substantiality.'[35] Again, the idea is that things can be held distinct from the way in which they are experienced and conceptualised by the subject, including their conceptualisation as property-bearing substances. In short, one may interpret Heidegger as arguing that reality depends on the subject's understanding. In the absence of such understanding, there can be no reality. However, the real, in contrast, is independent of the subject's understanding, and hence may exist in the absence of such understanding. Note, furthermore, that Heidegger is careful to distinguish the independent existence of the real from the *assertion* of its independent existence.

---

35   Heidegger (1962a), *Being and Time*, pp. 255 [212], 245 [201].

> Of course only as long as Dasein *is* (that is, only as long as an understanding
> of Being is [...] possible), 'is there' [*gibt es*] Being. When Dasein does not
> exist, 'independence' 'is' not either, nor 'is' the 'in-itself.' In such a case
> this sort of thing can be neither understood nor not understood. In such
> a case even entities within-the-world can neither be discovered nor lie
> hidden. *In such a case* it cannot be said that entities are, nor can it be said
> that they are not. But *now*, as long as there is an understanding of Being
> and therefore an understanding of presence-at-hand, it can indeed be
> said that *in this case* entities will still continue to be.[36]

The point here is that, in the absence of the subject, there would be
nobody around to assert the independent existence of the real. The
real could thus not be understood to exist independently of the subject
because, in such a case, understanding itself would be absent. However,
in the context of the current discussion, where there is understanding, it
becomes possible to assert that the real does indeed exist independently
of our understanding, and furthermore that it will continue to so exist
even once we, and hence our understanding, are gone. The *assertion* that
the real exists independently of the subject, and even the *fact* that it so
exists, entails the existence of the subject, but the independent existence
of the real does not.

This issue may be further illuminated by introducing a distinction
between a thing's existence and its intelligibility. There is, according
to Heidegger, an internal relation between the subject's being-in-the-
world, on the one hand, and the intelligibility of the real, on the other.
Only things disclosed within the context of the subject-world relation
may be encountered as intelligible. In other words, intelligibility can
be a feature only of things *within-the-world*; a thing without-the-world
cannot be understood by the subject, and, for this reason, it cannot be
intelligible.

In *Being and Time*, Heidegger explores two phenomenologically
distinct ways in which we may encounter the real as intelligible: either
in terms of 'readiness-to-hand' (*Zuhandenheit*, or 'equipmentality'),
or in terms of 'presence-at-hand' (*Vorhandenheit*). In the former case,
equipment is that *with which* we are involved in our workaday dealings
in the world. In the course of our ongoing immersed involvement
with equipmental things, we understand the world as a work-world,

---

36   Heidegger (1962a), *Being and Time*, p. 255 [212].

a totality of interrelated equipment available for our use. In the latter case, substances, or objects, are those things *about which* we concern ourselves when we take up a spectator's position in the world. Through our encounters with things as objects of observation, we may thus develop an understanding of the world as an object-world, a totality of substance with thematically determinable properties. According to Heidegger, the world encountered in this second way is what realists refer to as 'reality.' Realists thus violate the basic independence thesis insofar as they identify the real, as such, with the way it is encountered 'in reality,' that is, in the world constituted by a particular mode of the subject's projective understanding.

From this it should be clear that to distinguish the existence of the real from its intelligibility is to assert that the real may exist without the world, which is just to say, without the subject. This assertion forms the basis for the minimal realism which I suggest is present in Heidegger's early work. The assertion is most powerfully evinced in *Being and Time*, in the different ways in which Heidegger uses the term 'present-at-hand,' and in *The Basic Problems of Phenomenology*, with the distinction he draws between the existence and essence of a thing. Let us now consider the evidence from these two texts.

That Heidegger uses the term 'present-at-hand' in different ways has often been overlooked by commentators, and this has led to significant confusion over his intentions in *Being and Time*. One influential example of such confusion, present in the work of Joseph Rouse, will be addressed later in this chapter. In the meantime, it should be noted that Heidegger himself did not articulate these different uses as explicitly as he might have, and so responsibility for the subsequent confusion must lie, in some considerable part, at his own feet. Possibly the best study seeking to clarify these tricky exegetical matters comes from Joseph Fell, who detects at least four distinct senses for the term 'present-at-hand' in Heidegger's early work. Only two of these need worry us here.[37] In the first case, a thing is encountered as present-at-hand following a local

---

37   Joseph P. Fell (1989), 'The Familiar and the Strange: On the Limits of Praxis in the Early Heidegger,' in *Heidegger and Praxis*, ed. by Thomas J. Nenon (*The Southern Journal of Philosophy* 28, Spindel Conference Supplement), pp. 23–41. Of the two other senses, the first is an 'improper' sense in which all entities, including the subject, are referred to as present-at-hand things. Obviously, this is not a use to which Heidegger puts the word. The second is a sense in which all referentiality

breakdown in the subject's workaday world. In the second case, the term denotes those things which have been thematised as objects. In the first case, one encounters the real as something which exists but which cannot be understood, something which lacks intelligibility. In the second case, one encounters the real as something which both exists and is intelligible. We have already met both of these modes of being present-at-hand in the previous section. According to Heidegger's phenomenological description of equipmental breakdowns, when I stand desperate and discombobulated in front of my locked office door, I experience the door deficiently as a thing 'just-present-at-hand-and-no-more.' In contrast, when I encounter a thing thematically, as an object of perception, and hence also as the potential subject matter for a propositional assertion, I experience it as possessing 'a definite character in its being-present-at-hand-in-such-and-such-a-manner.'[38] These two ways in which the real can be present-at-hand — as either present-at-hand-in-such-and-such-a-manner or present-at-hand-and-no-more — correspond, respectively, to things present-at-hand within-the-world and things present-at-hand without-the-world. The second is phenomenologically available only when there is a deficiency in our understanding, a breakdown in the subject-world relation. Under these circumstances, we can still say *that* the thing exists, but we cannot say *what* it is. When such a deficiency occurs, the real can still be experienced as something which exists, but it lacks any determinate character by which we could make sense of it.

This crucial distinction is a central outcome of Heidegger's enquiry into the 'question of being,' that is, into the meaning of the infinitive verb 'to be' and its cognates. In *Being and Time*, Heidegger lists a number of 'prejudices and presuppositions which are constantly reimplanting and fostering the belief that an inquiry into Being is unnecessary.'[39] One such prejudice is the assumption that the meaning of 'being' is self-evident. He notes that '[w]henever one cognizes anything or makes an assertion, whenever one comports oneself towards things, even towards oneself, some use is made of "Being"; and this expression

---

      fails and the world as a whole becomes unintelligible. Attention will be given to this underappreciated use of 'present-at-hand' in the latter part of Chapter Three.

38 Heidegger (1962a), *Being and Time*, p. 200 [158].

39 Heidegger (1962a), *Being and Time*, p. 22 [2–3].

is held to be intelligible "without further ado," just as everyone understands "The sky *is* blue," "I *am* merry," and the like.'[40] Heidegger resists the impulse to treat the meaning of 'being' as self-evident. One important observation he makes is that the word 'being' is polysemic. In *The Basic Problems of Phenomenology*, a set of lectures he delivered in 1927, the same year in which *Being and Time* was published, Heidegger identifies at least two basic meanings for the term 'being' — existence and essence — and he argues that both belong to the being of a thing. Furthermore, the existence of a thing answers to the question of *whether* it is, and its essence answers to the question of *what* it is.[41] Two years later, in a 1929–1930 lecture course, Heidegger would restate the whether and what of a thing in terms of a thing's 'that-being' and 'what-being.'[42] In most cases, these two meanings will combine in a single proposition: 'The sky is blue' tells us both that there is a sky, that it exists, and that it has the property, the whatness or *quidditas*, of being blue.

The distinction between existence and essence is an ancient one. Indeed, Heidegger points out that its roots can be traced back to the biblical notion of a divine Creator: 'The ancient distinction runs thus: Since every entity that is actual comes from God, the understanding of the being of entities must ultimately be traced back to God.'[43] This ancient doctrine was transformed by medieval Christian metaphysicians into the idea that entities exist only as the creatures of God, that is, as produced. Hence, essence — as pure potentiality in the 'mind' of God — takes priority over existence — as God's idea made actual. Echoing Plato, the philosopher Charles Kahn has offered an apt description of this doctrine: 'existence now tends to be thought of as the final push into actual being provided by the Demiurge, as He sends things forth from His pre-cosmic workshop of logical possibilities.'[44]

As we saw in Chapter One, Heidegger reversed this ontological order in his existential analysis of the subject. For him, '[t]*he essence*

---

40   Heidegger (1962a), *Being and Time*, p. 23 [4].

41   Heidegger (1982a), *Basic Problems of Phenomenology*, pp. 78, 88.

42   Martin Heidegger (1995a [1983]), *The Fundamental Concepts of Metaphysics: World, Finitude, Solitude*, trans. by William McNeill and Nicholas Walker (Bloomington: Indiana University Press), p. 331.

43   Heidegger (1982a), *Basic Problems of Phenomenology*, p. 81; translation modified.

44   Charles H. Kahn (1966), 'The Greek Verb "To Be" and the Concept of Being,' *Foundations of Language* 2, 245–65 (p. 264).

of Dasein lies in its existence.'[45] We can now see that Heidegger made a similar move with respect to present-at-hand things, that is, things which exist but are not subjects, insofar as he took their existence to also be phenomenologically prior to their essence. However, such things are unlike the subject in that their essence does not lie in their own existence; it lies rather in the world as constituted by the existence of the subject. Hence, the essence of such things depends on our existence, but their existence does not. It is important to note that Heidegger does not so much reject the ancient productionist metaphysics as challenge its Christian interpretation. As the source of essence and meaning, human beings now assume the role of creator. Yet, notwithstanding such fictional things as Don Quixote or Daffy Duck, neither of which is present-at-hand, human beings do not produce things *ex nihilo*. Indeed, drawing from a pre-Christian productionist metaphysics, the early Heidegger argues that when something is produced it is always produced from something else; the notion of production thus always presupposes the prior existence of some material: 'If we bring to mind productive comportment in the scope of its full structure we see that it always makes use of what we call *material*, for instance, material for building a house. On its part this material is in the end not in turn produced but *is already there*.'[46] From this observation, Heidegger draws the more general conclusion that, when considering any productive activity, 'matter' necessarily arises as a basic phenomenological concept.[47] The subject discovers this indeterminate material in use as equipment ready-to-hand, or the subject may step back and observe it as an object present-at-hand. In either case, the material is experienced as something within-the-world, as 'intraworldly.' Yet Heidegger also makes it clear that this material, which he also calls 'nature,' does not depend on the subject for its existence:

> [I]ntraworldliness does not belong to nature's being. Rather, in commerce with this entity, nature in the broadest sense, we understand that this entity *is* as something extant, as an entity that we come up against, to which we are delivered over, which on its own part already is. It is, even if we do not uncover it, without our encountering it within our world.

---

45 Heidegger (1962a), *Being and Time*, p. 67 [42].
46 Heidegger (1982a), *Basic Problems of Phenomenology*, p. 115.
47 Heidegger (1982a), *Basic Problems of Phenomenology*, p. 116.

> Being within the world *devolves upon* this entity, nature, solely when it is
> *uncovered* as an entity.[48]

So as to leave no doubt that this indeterminate thing, nature, can
exist independently of subjectivity and world, Heidegger repeats the
point several more times: '[n]ature can also be when no Dasein exists';
'[n]ature can also be without there being a world, without a Dasein
existing.'[49] My argument in this section has been that Heidegger's
concepts of pure extantness or existence, of presence-at-hand without-
the-world, of indeterminate matter, and of independent nature can all
be read as various attempts to get at 'the real,' that is, at that which
exists independently of our theoretical and practical activity. Together,
they provide a richly articulated argument in defence of the basic
independence thesis, and hence for the position I call 'minimal realism.'

In the next section, we will return to a discussion of the sociology
of scientific knowledge. Specifically, I aim to show that SSK's oft-
overlooked endorsement of realism is importantly similar to the
minimal realism I have now drawn out of Heidegger's early texts. By
exploiting these similarities, it becomes possible to free SSK's realism
from the difficulties arising from its lingering adherence to the ontology
implicit in the orthodox subject-object distinction. In the next but one
section, I will then demonstrate the virtues of minimal realism in critical
comparison with an influential, alternative interpretation of early
Heidegger's philosophy of science, that of Joseph Rouse.

4. A Phenomenological Reformulation
of SSK's Residual Realism

As discussed in Chapter One, SSK practitioners are often criticised
by their opponents for allegedly subscribing to sociological idealism.
The underlying premise driving such criticism seems to be that social
constructivism is incompatible with realism. There is a puzzle here,
however, as leading SSK practitioners have consistently insisted on
their credentials both as social constructivists and as realists. For them,
the two positions are not mutually exclusive. Nevertheless, there can be

---

48   Heidegger (1982a), *Basic Problems of Phenomenology*, p. 169; translation modified.
49   Heidegger (1982a), *Basic Problems of Phenomenology*, p. 170, 175.

no doubt that social constructivists do reject something that scientific realists hold dear. So, what is it? Barry Barnes provides an answer. He suggests that realism, as it is usually defined in the literature, is actually a 'double-barrelled' realism, that is, a combination of two distinct claims: (1) that an external reality exists independently of our beliefs and theories; and (2) that the truth of specific beliefs and theories is determined by that reality.[50] Barnes claims that most realists in the philosophy of science take both of these claims for granted, but that the first alone provides sufficient grounds for claiming the credentials of a realist. It is this first claim which many SSK practitioners accept, making them adherents to a position which Barnes dubs 'residual realism.' He argues that residual realism is compatible with social constructivism.

Barnes's distinction immediately recalls the distinction I drew in the previous section between two theses: the independent existence of nature; and the independent existence of the determinate properties or structures of nature. I called the first of these the 'basic independence thesis' and defined minimal realism as a position which affirms this first thesis while rejecting the second. Barnes's first claim appears almost identical with the basic independence thesis, and thus his residual realism would seem very close to my minimal realism. Indeed, at one point in his discussion, Barnes even describes residual realism as 'minimal realism.'[51] Yet, there is an important difference between the two. Barnes's residual realism asserts the independence of an 'external reality.' Minimal realism, in contrast, asserts the independence of an indeterminate and undifferentiated nature. As we have seen, there is an important conceptual difference between the notion of an indeterminate nature and the notion of an external reality: the latter implies a theoretical commitment not found in the former. This is a commitment to what, in Chapter One, I called the glass-bulb model. Because he takes the glass-bulb model for granted, Barnes's residual realism has more in common with standard forms of realism than does minimal realism.

Heidegger argues that the orthodox realist asserts the independence of reality while unwittingly projecting that reality as part of an objectifying

---

50   Barnes (1992a), 'Relativism, Realism and Finitism,' in *Cognitive Realism and Social Science*, ed. by Diederick Raven, Lietke van Vucht and Jan de Wolf (New Brunswick, NJ: Transaction), pp. 131–47 (p. 132).

51   Barnes (1992a), 'Relativism, Realism and Finitism,' p. 133.

thematisation which itself depends on the subject's being-in-the-world. 'Reality' is thus a concept whose meaning derives from our picture of the world as a totality of objects standing in ontological separation from a cognising subject. On this basis, the traditional realist must now explain how the ontological divide between subject and object may be crossed, how one may transit from the inside to the outside of the glass bulb, how, in short, knowledge of the external world is possible. The prevailing tendency of orthodox realists is to root knowledge of the external world in the observational powers of the subject. Underlying this tendency is the assumption that the subject is itself a special sort of object, a substance with an added perceptual power which gives it access to the objects populating a world beyond itself. Heidegger, of course, rejects the primacy of this epistemological model, analysing it in existential-phenomenological terms as depending on a more basic subject-world relation in light of which the epistemological problematic no longer carries force. In this way, as we saw in Chapter One, Heidegger is able to short-circuit sceptical doubts about the existence of an external world by deconstructing the premises uncritically adopted by the sceptic and traditional realist alike. As we also saw in Chapter One, many SSK practitioners, Barnes included, join traditional realists in uncritically adopting those premises, and so they are perpetually vulnerable to sceptical attack.

My argument in this section is that SSK need only defend the basic independence thesis in order to achieve its goals. This thesis is what remains once one has stripped Barnes's first claim — that an external reality exists independently of our beliefs and theories — of the additional theoretical premises to which it also needlessly commits itself. Because SSK practitioners have failed to sufficiently recognise the contingency of those premises, thus tacitly accepting the fundamentality of the glass-bulb model, they end up defending an unnecessarily robust position which renders their approach ineluctably vulnerable to sceptical attack. It must be emphasised, however, that, in recommending that SSK practitioners trade their theoretically-loaded residual realism for a more phenomenologically modest minimal realism, I am not suggesting that the orthodox subject-object distinction should simply be abandoned as a useless bit of conceptual confusion. Indeed, one can easily agree that the style of thought which takes this distinction as its foundation has produced valuable results. The point is that this style of thought, despite its success in specific areas, has proved incapable of defending itself against sceptical doubt. Indeed, it may well be that the inescapable possibility of such doubt is an inherent feature of that very style. If the orthodox distinction were accepted as the conceptual bedrock for our way of understanding ourselves and the world in which we live, then we would need to simply accept sceptical doubt as an inevitable feature of our very existence. But neither this distinction nor the ontological presuppositions underlying it form the conceptual bedrock of our understanding, and hence they need not ground our conviction that nature exists independently of our theories, interpretations, and practices. If we wish only to defend the indubitability of the basic independence thesis, then there is no reason why we should also saddle ourselves with the more onerous, and probably fruitless, task of defending the indubitability of an allegedly fundamental distinction between subject and object.

Yet this is just what David Bloor has attempted to do. He argues that the received subject-object distinction, once freed from individualism, is a foundational concept, and he does this on the basis of a theory of reference. In his view, '[r]eference is an intentional state demanding intentional, conceptual and propositional content, that is, things which require an explanation of their normativity and objectivity.'[52] Bloor naturally favours a sociological explanation of the normativity and objectivity of such content. For him, reference is a collective achievement made possible by social interaction. This sociological theory of the normativity and objectivity of conceptual content is a central pillar of SSK. Bloor furthermore claims that reference is an intentional state demanding intentional, conceptual, and propositional content. For the purposes of the present analysis, whether that content is best explained in collectivist or individualist terms is beside the point. In the remainder of this section, the meaning of the term 'subject' should thus be treated as neutral between the terms 'group' and 'individual.'

The principal problem with Bloor's theory of reference is that he takes it to apply, not just to objects within the world, but to the world as such. He writes of 'genuine reference to an external reality.'[53] In addition,

---

52   David Bloor (2001), 'What Is a Social Construct?,' *Facta Philosophica* 3, 141–56 (p. 148).

53   Bloor (2001), 'What Is a Social Construct?,' p. 149.

he seems to think that knowledge of an external reality is a necessary condition for reference to particular objects. Yet, from the perspective of existential phenomenology, Bloor has got things backwards. In fact, knowledge of particular objects is a necessary condition for referring to the world, as such, as an external reality. Only once we have experienced things within-the-world as objects, and hence as distinct from a subject, can we then conceptualise the world itself as a reality which stands externally to a subject; only under these conditions does reference to an external reality become possible. Furthermore, reference to an object depends on a specific existential mode of being-in-the-world. We can refer to objects only because we are already *in* the world; the subject-world relation is ontologically prior to an encounter with things within-the-world as objects. Thus Bloor is right that acceptance of the subject-object distinction is entailed by the claim that an external reality exists, but he is wrong that this claim must be endorsed out of necessity. That he believes we cannot help but accept this claim is implied in his assertion that 'we are all instinctive realists.'[54] But the belief in an external reality is not hardwired into our brains; it is the result of a change-over in the subject's mode of understanding, a change from immersed involvement with things to a thematising projection of things, and then of the world itself, as objects of knowledge standing in separation from an autonomous subject. Belief in an external reality thus presupposes the diagnostic model of the glass bulb.

The key point here is that the subject-object distinction is a modification of the more fundamental subject-world relation. Moreover, this second relation is internal; there can be no world which exists independently of the subject. The implication is that the subject-object relation is thus also an internal relation; there can be no object which exists independently of the subject. However, this is not to say that nothing at all exists independently of the subject. As Heidegger writes, '[n]ature can also be without there being a world, without a Dasein existing.'[55] Hence, nature should not be confused with the world, including the world projected thematically as an external reality. Yet this is what Bloor does. He writes that 'nature, in our ordinary way of thinking, is the object of knowledge, the thing that is known, while science is the knowledge we have of it, our

---

54   David Bloor (1996), 'Idealism and the Sociology of Knowledge,' *Social Studies of Science* 26(4), 839–56 (p. 845).

55   Heidegger (1982a), *Basic Problems of Phenomenology*, p. 175.

theories about it and our description of it.'[56] Bloor appeals to our 'ordinary' way of thinking in order to maintain a strict distinction between nature and its scientific description, between an object of cognition and its conceptualisation by a cognising subject. Yet it is not clear why we should accept Bloor's implicit assumption that ordinary thinking demands acceptance of the subject-object distinction, that is, the theory-laden view that nature itself is an object of observation. It seems more 'ordinary' to say that nature is just that which exists independently of our descriptions and theories. This is precisely what is claimed in the basic independence thesis of minimal realism, and it has the advantage of avoiding the sorts of sceptical problems inevitably attracted by a foundational commitment to the subject-object distinction. As long as Bloor insists on calling nature an object, on conceiving of it in terms of one side of the subject-object schema, as long as he takes the glass-bulb model for granted, he cannot comfortably maintain the independence thesis that is basic to any genuine realist position. However, as soon as he gives up conceptualising nature as an object, he can no longer include nature, as such, under the umbrella of his theory of reference, because under that theory reference is always reference to an object, to a thing present-at-hand within-the-world.

There are some signs that Bloor has recognised this lattermost problem. In remarkable coincidence with Heidegger's comments on production, Bloor observes that the idea of construction has connotations of 'building and making': 'What is built, must be built from something: construction needs materials. Despite the claims of critics, the very term precludes the idea that "everything is constructed."'[57] Moreover, in discussing Kuhn's account of scientific discovery, Bloor writes: 'The scientist must come to realise *that* something is the case, and *what* is the case. There must be some generalised awareness of novelty and also a conceptualisation of the novelty.'[58] At first, the scientist only encounters nature as pure extantness, as an indeterminately existing thing, and hence as something which escapes conceptualisation. According to Bloor's theory of reference, when a thing cannot be conceptualised, it cannot be an object of reference

56   David Bloor (2004a), 'Sociology of Scientific Knowledge,' in *Handbook of Epistemology*, ed. by I. Niiniluoto, M. Sintonen and J. Woleński (Dordecht: Kluwer), pp. 919–62 (p. 942).

57   David Bloor (2003), 'Skepticism and the Social Construction of Science and Technology: The Case of the Boundary Layer,' in *The Skeptics: Contemporary Essays*, ed. by Steven Luper (Aldershot: Ashgate), pp. 249–65 (p. 263).

58   Bloor (2001), 'What Is a Social Construct?,' p. 150.

or intention. The scientist's encounter thus cannot be characterised as an epistemic one, as an act of knowing or believing, but only as a 'generalised awareness' of the thing's brute existence. Only by thematising the thing, only by interpreting its brute existence in the context of a world of pre-structured anticipations, can the scientist make sense of that thing, and only then can she begin to form concepts about the thing's essence, about what it is. Here, then, Bloor seems poised to limit the application of his theory of reference, and hence also his commitment to the subject-object distinction, so as to accommodate the phenomenological observation that we are able to experience nature in its brute state of indeterminate and undifferentiated existence. In other words, Bloor seems ready to accept the basic independence thesis of minimal realism.

Yet, Bloor then appears to lose his nerve. He writes that '[t]he pure "empiricist" encounter with the world corresponds roughly to the *that*.'[59] By modelling experience of nature in orthodox empiricist terms, Bloor appears to slip back into the problematic embrace of the subject-object distinction. As a consequence, he conflates a generalised awareness of nature in its brute existence for a conceptualisation of it in terms of an external world. He suggests that

> [t]he typical empiricist interrogation of a knowledge claim (to find out exactly what the claimants saw, heard, smelled, tasted, and touched) can be thought of as providing the raw materials out of which concepts are constructed. Notice that what is at issue here are sensory processes, that is, psychological and physiological causes of belief. [...] The causal story begins with observations not observation reports.[60]

Bloor assimilates the 'raw materials' of experience to objects of observation because he takes for granted the idea that the human being is a substance-subject which gains epistemic access to an external reality through 'sensory processes.' On this model, the subject is treated as an object distinguishable from other objects by its possession of a special power of perception. Observation, then, is meant to pierce the glass bulb separating the subject from the external world. The sceptic, of course, could not be happier with this particular arrangement.

---

59   Bloor (2001), 'What Is a Social Construct?,' p. 150.
60   Bloor (2001), 'What Is a Social Construct?,' p. 153.

The obstacle blocking SSK practitioners' unambiguous endorsement of the basic independence thesis is their tacit adherence to the glass-bulb model. Their reason for securing this schema at the centre of their social theory of knowledge is an admirable one: they wish to reconcile social constructivism with realism by holding fast the distinction between nature, on the one hand, and descriptions of nature, on the other. Unfortunately, they believe that this task entails a commitment to the fundamentality of the subject-object distinction. But such a commitment is unnecessary for, and indeed contrary to, their ultimate goal. As I hope to have shown thus far, once a distinction has been made between nature and world, the subject-object distinction can be set aside while still preserving the distinction between nature and its description. In other words, the basic independence thesis can be maintained without recourse to the orthodox subject-object schema. By accepting this thesis, and by thus trading in their residual realism for a minimal realism rooted in existential phenomenology, SSK practitioners would preserve their coveted distinction between nature and its description, and hence also be able to more effectively assert their credentials as both realists and social constructivists.

One implication of the present argument is that Heidegger's own early philosophy is compatible with the social constructivism favoured by SSK practitioners. This is a suggestion I am happy to accept. However, as we will see in the next section, the philosopher Joseph Rouse has presented a different interpretation of Heidegger, to the effect that the latter's early philosophy instead motivates a social constructivism in which no place at all can be found for a realist position, not even a minimal one. This presents a powerful challenge to the interpretation being elaborated here, and so we must give it careful consideration.

5. Rouse on Heidegger and Realism

Joseph Rouse is arguably the most prominent figure in science studies to have made positive use of Heidegger's early philosophy of science. In his 1987 book, *Knowledge and Power*, where he laid out the basic architecture of his reading of Heidegger, Rouse combines Heidegger's hermeneutics — the study of thinking as interpretation — with his phenomenology of practice in order to craft what Rouse calls a 'practical hermeneutics of science.'[61] This practical hermeneutics treats science as a collection of interrelated interpretative practices, rather than as an abstract system of concepts and theories. These interpretative practices are exemplified by material activities in the scientific laboratory. Rouse thus accepts Heidegger's existential conception of science, and attempts to further elaborate it by exploring the constitutive role played by material practice in relation to concrete modes of scientific existence.

Rouse argues that Heidegger's hermeneutics was motivated by the question of 'how it is that anything shows up at all.' According to him, Heidegger answered this question with the argument that the subjects for whom things 'show up' must have certain characteristics, the foremost among them being their membership in a 'self-adjudicating community.'[62] Members of a self-adjudicating community recognise one another on the basis of their shared ways of responding to a common environment. Rouse thus argues that Heidegger's word for the subject, Dasein, denotes the communal state of being 'socially and behaviorally self-adjudicating interpreters.'[63] In other words, the subjectivity of the subject is enmeshed in the intersubjective realm of a community of subjects. Heidegger calls this the subject's 'being-with,' and he argues that, like being-in-the-world, it is a fundamental aspect of the subjectivity of the subject.[64] On this construal of Heidegger, Rouse concludes that, if anything is to 'show up at all,' then it must show up for a self-adjudicating interpretive community. Furthermore, because this community is defined in terms of practical, as opposed to theoretical, acts of interpretation, it is on the basis of practical rather than theoretical interpretive acts that things 'show up.' Rouse locates these practical acts in the material practices of the sciences, with particular emphasis on

---

61   The present discussion draws, in part, from Jeff Kochan (2011a), 'Getting Real with Rouse and Heidegger,' *Perspectives on Science* 19(1), 81–115, which offers a more detailed critique of Rouse's practical hermeneutics of science in the context of his interpretation of Heidegger. Anna de Bruyckere and Maarten Van Dyck have tried to defend Rouse against this critique (Anna de Bruyckere and Maarten Van Dyck (2013), 'Being in or Getting at the Real: Kochan on Rouse, Heidegger and Minimal Realism,' *Perspectives of Science* 21(4), 453–62). However, their argument crucially depends on the false claim that I treat existence as the 'property' of a thing. My view is that a thing must exist in order to have a property. Properties constitute the essence (whatness), not the existence (thatness), of things.

62   Rouse (1987a), *Knowledge and Power*, p. 73.

63   Rouse (1987a), *Knowledge and Power*, p. 73.

64   Heidegger (1962a), *Being and Time*, p. 149 [114].

laboratory practices. For him, the laboratory is the principal site where things 'show up' in the sciences.

There is, however, a tension in Rouse's reading of Heidegger, a tension which does not exist in Heidegger's own work. On the one hand, Rouse is concerned with the conditions which make it possible for a thing to 'show up at all.' On the other hand, he describes those conditions as interpretive conditions, that is, as the social and behavioural conditions which enable a community to successfully interpret a thing. The tension is this. The conditions enabling something to 'show up at all' would seem to be existence conditions, that is, the conditions which enable a thing to exist at all. This is indeed how Rouse often presents them. Yet, conditions of existence are not the same as conditions of interpretation, for it seems clear that a thing must exist before it can be interpreted. Interpretation thus presupposes existence. Hence, to run existence and interpretation together, as Rouse does, is to court conceptual incoherence.

Rouse attempts to resolve this conflict, and so to escape the threat of incoherence, by assimilating existence to meaning. He writes that 'there is no fact of the matter about whether things that cannot intelligibly be encountered within a meaningful world exist or do not exist.'[65] And he endorses the 'invocation of meaning as the arbiter of [...] existence conditions for things.'[66] On this view, existence presupposes meaning. Hence, Rouse reverses the apparently common-sense claim that interpretation presupposes existence. For him, interpretation does not presuppose existence, it presupposes meaning. And meaning is, in his view, the condition of possibility for existence.

This position may allow Rouse to dodge the charge of incoherence, but at what cost? The claim that meaning precedes existence would seem to contradict common sense. Moreover, by making existence dependent on interpretation, Rouse effectively abandons the core realist doctrine of independent existence. Existence now means existence relative to an interpretive community. It must also be noted that Rouse's position fails to reflect Heidegger's own view of these matters. Indeed, as this chapter has already demonstrated, Heidegger offers a different way of resolving the conflict. Contrary to what Rouse claims, Heidegger's

---

65   Rouse (1987a), *Knowledge and Power*, p. 160.
66   Rouse (1987a), *Knowledge and Power*, p. 162.

hermeneutics of the subject is not driven by the question of how things show up *at all*, but rather by the question of how things show up as *what they are*. A key aspect of Heidegger's position, overlooked by Rouse, is his distinction between the that-being and what-being of a thing, that is, between its existence and its essence. Rouse is partly right that, for Heidegger, the being of a thing depends on its being meaningful, because Heidegger does argue that the what-being, or essence, of a thing depends on its being either practically or theoretically interpreted by a self-adjudicating community. But the that-being of the thing does not depend on its being so interpreted. A thing may exist without meaning anything at all, without being intelligible for a community.

Although Rouse shows no awareness of Heidegger's distinction between existence and essence, he does recognise that Heidegger's work includes elements which resist the use to which Rouse would like to put it. Principal among these is Heidegger's concept of 'change-over,' which, as discussed earlier in this chapter, describes the transformation in understanding of a thing from its being ready-to-hand to its being present-at-hand within-the-world. According to Rouse, the concept of change-over marks Heidegger's vestigial attachment to a theory-dominant account of the scientific enterprise. He thus rejects it as a retrograde move betraying Heidegger's otherwise laudable commitment to a practical hermeneutics of science which gives pride of place to material practice.[67] Indeed, according to Rouse, 'the theory-dominant perspective that Heidegger still retains [...] reduces experiment to a merely incidental practice in science.'[68] Even though Heidegger, as we saw above, offers some examples of the material practice of science, Rouse dismisses these as 'research practices that are only *associated* with theoretical cognition' rather than being constitutive of it.[69] He thus concludes that, on Heidegger's allegedly retrograde account, material

---

67   Rouse (1987a), *Knowledge and Power*, p. 74.
68   Rouse (1987a), *Knowledge and Power*, p. 79.
69   Rouse (1987a), *Knowledge and Power*, p. 76. I criticises this statement in more detail in Kochan (2011a), 'Getting Real with Rouse and Heidegger,' p. 105. Denis McManus also challenges the veracity of Rouse's statement (Denis McManus (2012), *Heidegger and the Measure of Truth: Themes from his Early Philosophy* (Oxford: Oxford University Press), p. 66 n. 60). Elsewhere, McManus has also carefully examined the complex relationship between the 'theoretical' and the 'practical' in Heidegger's early work (Denis McManus (2007), 'Heidegger, Measurement and the "Intelligibility" of Science,' *European Journal of Philosophy* 15(1), 82–105).

practice 'neither has "a life of its own" apart from theory nor makes a distinctive cognitive contribution.'[70]

Rouse views this alleged circumstance not only as an affront to his practical hermeneutics of science, he also asserts that the phenomenological concept of the change-over, which is meant to provide existential grounds for Heidegger's account of scientific theory, is unpersuasive. He has repeated this assertion several times over many years: 'Heidegger never does indicate what makes for this sudden leap to a new way of looking at things'; 'Heidegger is disturbingly vague about how this changeover can occur'; 'Heidegger does not describe how the practical tasks of science (experiment, instrumental manipulation, theoretical problem solving and calculation) are connected to the disclosure of things as present-at-hand'; 'Heidegger merely asserted such a changeover without an adequate phenomenological description of how it occurred'; 'Heidegger merely asserted such a changeover without adequately describing it.'[71] The problem with this repeated assertion is that Rouse has never developed it into a proper argument explaining why Heidegger's in fact not insignificant description of the change-over fails to meet Rouse's own standards of adequacy.[72] In fact, it turns out that Rouse's standards are ill-suited for measuring

70 Rouse (1987a), *Knowledge and Power*, p. 98.

71 Joseph Rouse (1985), 'Science and the Theoretical "Discovery" of the Present-at-Hand,' in *Descriptions*, ed. by Don Ihde and Hugh J. Silverman (Albany: SUNY Press), pp. 200–10 (p. 203); Rouse (1987a), *Knowledge and Power*, p. 75; Joseph Rouse (1998), 'Heideggerian Philosophy of Science,' in *Routledge Encyclopedia of Philosophy*, vol. 4, ed. by Edward Craig (London: Routledge), pp. 323–27 (p. 324); Joseph Rouse (2005a), 'Heidegger and Scientific Naturalism,' in *Continental Philosophy of Science*, ed. by Gary Gutting (Oxford: Blackwell), pp. 123–41 (p. 131); Joseph Rouse (2005b), 'Heidegger's Philosophy of Science,' in *A Companion to Heidegger*, ed. by Hubert L. Dreyfus and Harrison Hall (Oxford: Blackwell), pp. 173–89 (p. 181).

72 Nor has Rouse addressed any of the secondary literature which affirms the adequacy of Heidegger's description of the change-over and, in some cases, substantially elaborates on it. See, for example: Rainer A. Bast (1986), *Der Wissenschaftsbegriff Martin Heideggers im Zusammenhang seiner Philosophie* (Stuttgart-Bad Cannstatt: Frommann-Holzboog Verlag, pp. 139-62; Robert Brandom (1983), 'Heidegger's Categories in *Being and Time*,' *Monist* 6(3), 387–409 (pp. 403–04); Joseph J. Kockelmans (1985), *Heidegger and Science* (Lanham: The Center for Advanced Research in Phenomenology and The University Press of America), pp. 118-38; William McNeill (1999), *The Glance of the Eye: Heidegger, Aristotle, and the Ends of Theory* (Albany: SUNY Press), pp. 72–92; Tibor Schwendtner (2005), *Heideggers Wissenschaftsauffasung: Im Spiegel der Schriften 1919-29* (Frankfurt: Peter Lang), pp. 50-86; Hans Seigfried (1980), 'Scientific Realism and Phenomenology,' *Zeitschrift für philosophische Forschung* 34, 395–404 (passim).

the adequacy of Heidegger's description, because they are based on a misunderstanding of what role the concept of the change-over is meant to fulfill.

Rouse writes that 'it is not at all clear in [Heidegger's] account *how* one can get from a breakdown of practical involvement to the theoretical attitude. But once this happens, the ordinary functional contextuality of things gets *replaced* by the "mathematical projection of Nature."'[73] The standard of adequacy which Heidegger fails to meet thus demands an explanation of how a breakdown in our immersed involvement with a ready-to-hand nature can change over to a mathematical projection of a present-at-hand nature, a 'decontextualised theorising' which allegedly eliminates the ontological significance of material practice in the sciences. According to Rouse, Heidegger identifies the mathematical projection of nature with a decontextualised theorising and claims that '[w]hen we understand *theorizing*, we have understood what is essential about science.'[74] Rouse's Heidegger thus defines science as a theory-driven mathematical projection of nature in which thing show up as fully decontextualised and present-at-hand.

In his reading of Heidegger, Rouse appears to take for granted a fundamental distinction between theory and practice, an assumption he shares with more orthodox philosophers of science. However, whereas these philosophers usually seek to reduce the epistemic significance of practice to that of theory, Rouse argues for the reverse: he wants to reduce theory to practice. Furthermore, because he detects a resistance to his preferred direction of reduction in Heidegger's work, Rouse assumes that Heidegger must then belong to the orthodox camp, that he must be intent on reducing practice to theory. But there is an alternative possibility: namely, that Heidegger does not accept the distinction between theory and practice as a fundamental one. That Rouse has foisted a foreign distinction onto Heidegger is evinced by his elision of Heidegger's concept of the mathematical projection of nature with a theoretical stance towards nature. Because Heidegger seeks to explain science in terms of the mathematical projection, Rouse concludes that he must also be seeking to explain science reductively in terms of theory.

---

73   Rouse (1987a), *Knowledge and Power*, p. 75.
74   Rouse (1987a), *Knowledge and Power*, p. 96.

But Heidegger does not hold the distinction between theory and practice to be a fundamental one, and he does not identify the mathematical projection of nature with theory. In fact, as already discussed, Heidegger argues that the mathematical projection provides the existential condition of possibility for both scientific theory and practice. The distinction between the two is therefore not fundamental, but instead derives from their shared existential basis in the mathematical projection of nature. They are, so to speak, two sides of the same existential coin.[75]

Rouse's misunderstanding on this point appears to have arisen from his failure to recognise Heidegger's distinction between the existence and essence of a thing, and, more specifically, between a thing present-at-hand without-the-world and a thing present-at-hand within-the-world. This is the distinction between a thing which exists, but is not intelligible, and a thing which exists and is also intelligible. Hence, when Heidegger writes that, in the mathematical projection, 'something constantly present-at-hand (matter) is uncovered beforehand,' he means something present-at-hand in the first sense. When brute matter is uncovered in this projection, our basic understanding is directed towards 'those constitutive items in it which are quantitatively determinable (motion, force, location, and time).' The mathematical projection thus serves as a basic template which directs us to experience an independently existing nature as something essentially amenable to quantitative analysis. Heidegger writes that '[o]nly "in the light" of a Nature which has been projected in this fashion can anything like a "fact" be found and set up for an experiment regulated and delimited in terms of this projection.'[76] Hence, the mathematical projection is a condition of possibility for the scientific experiment as such. The readiness-to-hand of things in experimental practice presupposes the essential measurability of physical phenomena. Moreover, when there is a change-over in understanding, and those same ready-to-hand things become thematised as present-at-hand objects within an

---

75   Robert Crease also notes Rouse's failure to understand Heidegger on this point: 'In effect, what Rouse has done is taken the traditional priority of theory over praxis and stood it on its head, when what is needed is a rethinking of that relation' (Robert P. Crease (1993), *The Play of Nature: Experimentation as Performance* (Bloomington: Indiana University Press), p. 193 n. 43).

76   Heidegger (1962a), *Being and Time*, p. 414 [362].

experimental work-world, their essential determination as measurable does not change, but rather becomes articulated as the subject matter for theoretical representation. Hence, the mathematical projection is likewise a condition of possibility for theory as such.

The change-over thus marks a shift in experience within the range of possible understandings of nature opened up by the mathematical projection. The thing *with which* we were working now becomes a thing *about which* we concern ourselves. It shifts from being a resource for our activity to being the topic of our activity. This is, as we have seen, a process of objectification. Heidegger writes that the 'Being which Objectifies and which is alongside the present-at-hand within-the-world, is characterized by a *distinctive kind of making-present.*'[77] With this, Heidegger makes it clear that, whereas the mathematical projection correlates with independently existing nature — present-at-hand without-the-world — the objectifying, or thematising, process enabled by that projection, and which itself enables theory, correlates with things present-at-hand within-the-world. It is this crucial distinction which Rouse has failed to recognise. He mistakenly identifies the mathematical projection of nature with the scientific theorising which it enables, because he has not spotted the distinction Heidegger draws between a thing present-at-hand without-the-world and a thing present-at-hand within-the-world. He thus identifies the former with the latter, and thereby eliminates the conceptual space Heidegger had deliberately left open in his existential account of science for an independently existing nature.

As we have seen, Rouse assimilates existence to meaning in order to avoid the threat of incoherence facing his practical hermeneutics of science. He argues that interpretation is constitutive of existence, and thus rejects the common-sense belief that interpretation presupposes the existence of the thing interpreted. Giving up this common-sense belief is the price Rouse pays to protect his hermeneutics of scientific practice. Yet, as we have also seen, Heidegger allows for an alternative to Rouse's practical hermeneutics which does not sacrifice this belief, and which thus preserves the core realist doctrine of independent existence. But there is also a further reason to prefer Heidegger's account. It seems that

---

77   Heidegger (1962a), *Being and Time*, p. 414 [363].

Rouse's commitment to an abstract theory of universal hermeneutics has led him to ignore the concrete evidence which plays against that same theory. Indeed, this evidence can be found even in Rouse's own practice. His theoretical commitments conflict with the basic norms of intelligibility governing the very language he uses to articulate his theory. Avoided in theory, the threat of incoherence nevertheless re-emerges in practice.

Rouse's theoretical conviction that existence presupposes interpretation is clearly expressed in the following statement: 'what exists depends on the field of meaningful interaction and interpretation within which things can be encountered.'[78] Here, Rouse argues that interpretation is the condition of possibility for existence. A thing can only exist — can only 'show up at all' — within a field of meaning and interpretive practice. The relation of a thing to an interpretative practice is thus one of existential dependency. Yet Rouse immediately betrays this theoretical conviction in another statement, which only makes interpretation the condition of a thing's being present in a particular way: 'the possible ways a thing can be depends on the configuration of practices within which they become manifest.'[79] Now the configuration of practices, which are for Rouse constitutive of meaning, is no longer the condition of possibility for a thing's existence, as such, but instead for the range of possible ways in which it may show up as what it is. Rouse's emphasis has subtly shifted from identifying meaning with the that-being of a thing to identifying it with the what-being of that thing. The question being answered is no longer one of how things show up at all, but instead of how things show up as the bearers of the properties which manifest their essence, their *way* of being. This question is the same one as asked by Heidegger, and it can be answered without assimilating existence to meaning, and hence without threatening the doctrine of independent existence. This, I would suggest, is a more common-sense way of speaking about the relation between meaning and existence. Indeed, it is a way of speaking about that relation which natural language powerfully compels us to adopt, and so it is not surprising that Rouse quietly slides back into it.

---

78 Rouse (1987a), *Knowledge and Power*, p. 160.
79 Rouse (1987a), *Knowledge and Power*, p. 160–61.

Rouse nevertheless insists on tacking close to his theoretical commitments with the further claim that

> [b]elonging to the realm of possible determinations open within our practices is constitutive of a thing's being a thing at all. But this claim is just to say that having determinate properties, and interacting with other things in ways we must take account of, is a necessary condition for a thing to be.[80]

Rouse admits that '[t]his point is difficult to recognise,' but suggests that it can be made clear with an example of a case where 'thinghood' is in question.[81] He offers an example from the history of laboratory practice in biochemistry.

The case involves what biochemists eventually came to recognise as thyrotropin releasing hormone, or TRH.[82] Rouse writes that the name 'TRH' was originally used to designate 'whatever was physiologically active [...] in certain chromatographically isolated fractions of the hypothalami of sheep or pigs.' He notes that, at that early stage, biochemists did not know if TRH denoted 'a thing rather than an unstable artifact.' According to Rouse, the difference between an unstable artefact and a thing is that a chemical structure can be attributed to the latter. Once biochemists succeeded in attributing a chemical structure to what Rouse also refers to as 'the stuff in the fractions,' that stuff was no longer an unstable artefact but manifest itself as a genuine 'substance.' Hence, Rouse distinguishes between an 'unstable artifact' and 'stuff,' on the one hand, and a 'thing' and a 'substance' on the other, arguing that only the latter can be properly recognised as candidates for existence. What he refers to as 'the complex of practices that had developed over a hundred years of biochemistry' comprises the existence conditions for the thing called TRH. For Rouse, the 'crucial point' is this: 'not to show up in the ways that allow something to count as an x (in this case, as a chemical substance) is not to be a thing at all.' Yet it then becomes something of a puzzle what the terms 'unstable artifact' and 'stuff' are meant to refer to if not to something which exists. It would seem more coherent to say that the terms refer to a thing about which we can say *that* it is but not *what* it is, because *what* it is has not yet

---

80 Rouse (1987a), *Knowledge and Power*, p. 163.

81 Rouse (1987a), *Knowledge and Power*, p. 163

82 Rouse (1987a), *Knowledge and Power*, p. 163f.

been determined by the biochemists treating it as the subject matter for their investigation. It is thus not the existence of the thing which is in question, but its determinate properties. Contrary to what Rouse claims, the determination of those properties is not constitutive of that thing's being a thing at all, but rather of its being an object, a property-bearing substance, a thing present-at-hand within-the-world.

In this case, then, Heidegger's distinction between existence and essence provides a better resource for explaining what transpired than does Rouse's theoretical commitment to unfolding configurations of interpretive practice. On top of that, Heidegger's more moderate position also allows us to comfortably accommodate the core realist doctrine of independent existence. The threat posed to minimal realism by Rouse's practical hermeneutics of science has thus been defused. A practical hermeneutics of science will reveal its full worth only within the constraints of a phenomenology of scientific practice which recognises the independent existence of nature.

6. Minimal Realism and Scientific Practice

Rouse's critique of Heidegger's existential account of science attempts to separate the social constructivist elements in Heidegger's hermeneutics of subjectivity from the phenomenological elements supporting his minimal realism. Rouse adopted Heidegger's description of the subject in terms of 'socially and behaviorally self-adjudicating interpreters' — what Heidegger called the subject's elemental being-with-others — but failed to follow him in also accepting the doctrine of independent existence. This position was based on the reading of Heidegger which Rouse presented in his 1987 book, *Knowledge and Power*. Since the appearance of that book, however, Rouse has distanced himself not only from realism, but also from social constructivism, and especially from SSK. Indeed, he has more recently argued that '[s]ocial constructivism and realism are [...] vampires, the philosophical undead that still haunt our concepts and interpretations of nature, culture, and science.'[83] This chapter has already detected minimal life-signs in realism, enough to still count it among the living. In what follows, I will

---

83 Joseph Rouse (2002a), 'Vampires: Social Constructivism, Realism, and Other Philosophical Undead,' *History and Theory* 41, 60–78 (p. 63).

likewise argue that Rouse's report of the death of SSK is also greatly exaggerated.

Rouse's criticism of the social constructivist view of science is strikingly similar to his criticism of Heidegger's existential account of science. As he did with Heidegger, Rouse also charges SSK practitioners with espousing a theory-dominated view of science. From Rouse's perspective, then, SSK poses a threat to his practice-based philosophy of science. As we will see, SSK does indeed pose such a threat, not because it is theory-dominated (it need not be), but because it insists that scientific practices can be usefully studied in sociological terms.

A key entry point for Rouse in his critique of SSK is Richard Rorty's claim that '[n]atural science [is not] a natural kind.'[84] According to Rouse, natural science is not a natural kind because the products and norms of scientific investigation are historically variant, and also vary both across and within scientific disciplines. In other words, Rouse rejects the claim that there is 'an essence of science or a single essential aim to which all genuinely scientific work must aspire.'[85] The specific problem with SSK, he writes, is its insensitivity to the heterogeneity of the sciences. SSK practitioners act on the 'mistaken assumption [...] that scientific knowledge belongs to a single kind similar or distinguishable *in kind* in any interesting way from other kinds.'[86]

Rorty's argument that natural science is not a natural kind was meant to undercut essentialist solutions to the demarcation problem, that is, the problem of distinguishing genuine science from pseudoscience, astronomy from astrology, for example. The argument thus carries no weight against SSK, which holds that knowledge, as such, is a social and historical phenomenon and so any criterion demarcating scientific from other kinds of knowledge must itself be socially and historically contingent rather than essential. Indeed, Barnes, Bloor, and Henry write that 'demarcation criteria must be regarded as conventional, and their application in all cases as situated human action,' and they explicitly

---

84 Joseph Rouse (1996a), *Engaging Science: How to Understand Its Practices Philosophically* (Ithaca: Cornell University Press), p. 243; brackets original. Cf. Richard Rorty (1991), 'Is Natural Science a Natural Kind?,' in *Objectivity, Relativism, and Truth* (*Philosophical Papers*, vol. 1), by Richard Rorty (Cambridge: University of Cambridge Press), pp. 46–62.

85 Rouse (1996a), *Engaging Science*, p. 242.

86 Rouse (1996a), *Engaging Science*, p. 243.

treat the pseudoscientific status of astrology from this non-essentialist perspective.[87] Rouse's complaint seems to lie with the fact, not that SSK practitioners treat science as an essentially unique epistemic kind, but that they treat it as a kind at all. What Rouse disapproves of is SSK's claim that science is an epistemic kind which, like all other epistemic kinds, is amenable to sociological explanation: 'the vocabulary of social interaction (interests, negotiations, and so on) is supposed to hold the key to an adequate understanding of scientific work.'[88] Rouse thus joins other critics in attributing to SSK the doctrine that sociology 'can (potentially) account *fully* for the epistemic outcomes of scientific practices.'[89] And, along with those other critics, Rouse's attribution turns out, in crucial cases, to be a misattribution. For example, Rouse has David Bloor claiming, first, that the same kinds of causal explanation should be applied symmetrically to both true and false beliefs, and, second, that 'only sociological explanations could plausibly satisfy this demand.'[90] The first attribution is correct, and the second incorrect. The 'kinds' of causal explanation to be symmetrically applied may include sociological, psychological, biological, ecological, or any other naturalistic sort. As Bloor states in the *locus classicus* of his position, '[n]aturally there will be other types of causes apart from social ones which will cooperate in bringing about belief.'[91] The importance of sociological explanations attaches specifically to the normative conditions concerning belief formation, an issue central for understanding knowledge but certainly not the only such issue. Contrary to what Rouse alleges, then, social constructivists do not all claim that sociological explanation can account 'fully' for the epistemic outcomes of scientific practice. Some, like Bloor, argue only that a complete explanatory account of science must include sociological elements. Such elements are necessary, but not sufficient, for an explanation of science.

An initially more plausible criticism is Rouse's claim that SSK practitioners treat science as a 'theoretically coherent domain,' and 'mistakenly take the unity and theoretical integrity of "scientific

---

87　Barnes, Bloor and Henry (1996), *Scientific Knowledge*, pp. 142, 141.
88　Rouse (1996a), *Engaging Science*, p. 244.
89　Rouse (1996a), *Engaging Science*, p. 244; emphasis added.
90　Rouse (1996a), *Engaging Science*, p. 9.
91　David Bloor (1991), *Knowledge and Social Imagery*, 2nd edn (Chicago: University of Chicago Press), p. 7.

knowledge" for granted.'[92] This is the charge, mentioned above, that SSK practitioners deploy a theory-dominant view of science. It may well be that this criticism applies to some social constructivists. As before, I will here concentrate on crucial exceptions among proponents of SSK.

Barry Barnes has written that a 'theory is a cluster of accepted concrete applications.'[93] This looks like a definition of theory in terms of practice. The concrete applications Barnes refers to are 'problem solutions': 'particular concrete scientific achievement[s].'[94] Barnes, appropriating Kuhn, also calls them 'paradigms,' and writes that the 'most satisfactory way of describing scientific knowledge is simply as a repertoire of paradigms.' He furthermore argues that '[t]o speak instead of an abstract pattern of concepts and beliefs, or statements, can be seriously misleading.'[95] Nevertheless, Barnes goes on to do just that, albeit with clear qualification. He writes:

> It is always possible to reify the verbal component of a culture as a conceptual fabric, a structure made up of generalisations which connect concepts into a single integrated whole. It is true that something is lost by reducing linguistic activity to an abstract verbal pattern in this way. But the reification is irresistibly convenient, and harmless enough if its limitations are constantly borne in mind.[96]

It would be tedious to trawl through the works of Barnes and his SSK colleagues with the goal of judging whether they have 'constantly borne in mind' the limitations of this methodologically motivated reification. Suffice it to say here that Barnes's comments cast serious doubt on Rouse's blanket allegation that SSK practitioners uncritically view scientific knowledge as an integrated theoretical whole. Yet, those comments also suggest an explanation for how critics may come to think otherwise. Just because a writer has constantly borne in mind the limitations of their method, it need not follow that their readers are

---

92   Joseph Rouse (2002b), *How Scientific Practices Matter: Reclaiming Philosophical Naturalism* (Chicago: University of Chicago Press), p. 144 n. 7; Joseph Rouse (1999), 'Understanding Scientific Practices: Cultural Studies of Science as a Philosophical Program,' in *The Science Studies Reader*, ed. by Mario Biagioli (London: Routledge), pp. 442–56 (p. 451).

93   Barry Barnes (1982), *T. S. Kuhn and Social Science* (London: Macmillan), p. 124.

94   Barnes (1982), *T. S. Kuhn and Social Science*, p. xiv.

95   Barnes (1982), *T. S. Kuhn and Social Science*, p. 18.

96   Barnes (1982), *T. S. Kuhn and Social Science*, p. 71.

also constantly aware of, and hence manage to avoid, the incumbent dangers associated with the writer's own use of that method.

The problem is particularly acute in the case of Rouse's interpretation of Harry Collins, who, as we saw in Chapter One, rejects a role for external-world realism in SSK. Using a passage from Collins and Steven Yearley, Rouse tries to undermine the credibility of social constructivists by arguing that they reify, not just the practices of the natural sciences, but also the very sociological concepts they themselves use to explain the natural sciences. The passage in question reads: 'We provide a prescription: stand on social things — be social realists — in order to explain natural things. The world is an agonistic field (to borrow a phrase from Latour); others will be standing on natural things to explain social things.'[97] For Rouse, by distinguishing between 'social things' and 'natural things,' and then setting the two in opposition, Collins and Yearley erroneously 'presume the unity of each' when those unities, as well as their opposition, should better have been the subject of critical deconstruction.[98] He furthermore characterises the passage as 'presum[ing] that the social "world" and the natural "world" constitute relatively autonomous domains whose articulated descriptions then need to be brought into an appropriate relation to one another.'[99] As a consequence of this interpretation, we are now suddenly confronted with a version of the traditional epistemological problem, encountered in Chapter One, of how the world of the knower or knowers — the world of the subject — makes contact with the world of the knowable — the world of the object. Indeed, Rouse suggests that SSK practitioners harbour a 'vestigial commitment to epistemology.'[100]

Rouse is making two good points here, but if they are to be properly appreciated they must be separated from the misleading aspects of his argument. To that end, it is important to understand the broader context from out of which Rouse has plucked the above passage from

---

97 Rouse (1996a), *Engaging Science*, pp. 244–45. Cf. Harry M. Collins and Steven Yearley (1992), 'Journey into Space,' in *Science as Practice and Culture*, ed. by Andrew Pickering (Chicago: University of Chicago Press), pp. 369–89 (pp. 382–83).

98 Rouse (1996a), *Engaging Science*, p. 245.

99 Rouse (2002b), *How Scientific Practices Matter*, p. 136. A version of this criticism of SSK is repeated in Joseph Rouse (2015), *Articulating the World: Conceptual Understanding and the Scientific Image* (Chicago: University of Chicago Press), p. 323.

100 Rouse (2002b), *How Scientific Practices Matter*, p. 134.

Collins and Yearley. The two sentences immediately preceding it state: 'Our world is populated, we admit it, by philosophically insecure objects, such as states of society and participant's comprehension. But all worlds are built on shifting sands.' The third sentence following the passage states: 'We see the attractiveness of the idea of a comprehensive theory, but in its absence, life, although imperfect, is interesting.'[101] On this basis, Collins and Yearley can hardly be charged with an uncritical reification of the world into distinct and autonomous social and natural domains. Their appeal to the concept of 'social reality' is, as they clearly state in the passage quoted by Rouse, a 'prescription,' and they furthermore emphasise that this is an 'insecure' methodological recommendation made in full consciousness of the 'shifting sands' on which it rests. In further contrast to Rouse's claims about SSK, Collins and Yearley also clearly disavow the idea of a comprehensive theory, choosing instead to proceed imperfectly over shifting sands in a world which they nevertheless take to be of unceasing interest. This choice resonates with Barnes's own admission that the reification of concrete practices as an abstract conceptual fabric is irresistibly convenient from a methodological perspective, and harmless enough if its limitations are constantly borne in mind. Collins and Yearley explicitly acknowledge those limitations, and Rouse is wrong to suggest otherwise.[102]

On the other hand, Rouse is right to find some reason for worry here. With this we come to the first of his two good points. Rouse worries that SSK practitioners, by using particular social categories as resources in their explanations of science, will forego the opportunity to topicalise the contingencies of those resources. Within those contingencies, he fears, there may lie unresolved political tensions. There is no doubt that this is a genuine worry, and that the potential problems it responds to can be very great. There is, however, also no doubt that Barnes can be easily read as expressing this worry when he warns us of the limitations which accompany reification. There is also no doubt that the same worry may be fairly read into Collins and Yearley's observation that states of society and participant's comprehension are philosophically

---

101 Collins and Yearley (1992), 'Journey into Space,' p. 382.
102 Carol Steiner describes this passage from Collins and Yearley as an 'echo' of Heidegger's existential conception of science (Steiner (1999), 'Constructive Science and Technology Studies,' p. 602).

insecure. There thus appears to be little genuine disagreement between Rouse and SSK on this point. Accordingly, SSK practitioners should not object to having their explanatory resources topicalised and the social and historical contingencies of those resources exposed. They may sometimes find themselves surprised or even embarrassed in the process, but they will recognise such outcomes as consistent with their own methodology and so should accept them as contributions to the greater good. However, SSK practitioners may well object to being told by critics that they must perform this act directly on themselves if they wish to maintain the credibility of their field. For they will rightly suspect that this demand, when pushed too far, surreptitiously threatens the very possibility of their practice. As Bloor has observed: 'Nobody can turn every resource into a topic without finishing up with topics which they have no resources for tackling.'[103] If all the explanatory resources of SSK must be turned into topics before the critic will be satisfied, then clearly the critic's satisfaction depends on sociological explanation finally becoming impossible. It is hard to believe that Rouse would seriously want to place such a strong demand on SSK practitioners, not least because the same demand could easily be turned against his own attempt to explain science in terms of 'practice,' 'meaning,' and 'being,' categories the legitimacy of which he apparently takes for granted. For example, as we saw in the last section, Rouse neglects the distinction between essence and existence, which was prompted by Heidegger's 'question of being,' and so he mistakenly takes the assimilation of essence to meaning to also encompass existence. The point is thus a perfectly general one, applying to any explanatory enterprise, including Rouse's own practical hermeneutics.[104]

Let us turn now to Rouse's second good point. He argues that SSK practitioners still harbour a vestigial commitment to epistemology, and especially to an underlying presupposition that the problem of knowledge is one of explaining how a subject may acquire knowledge of

---

103 David Bloor (1999a), 'Anti-Latour,' *Studies in History and Philosophy of Science* 30(1), 81–112 (p. 92).

104 Rouse has more recently defended himself against this very kind of criticism, arguing that his method '*must* proceed from "in the thick of the human situation."' Hence: 'It is one thing to look at particular social practices [...] with "the cold eye of a stranger"; [...] It is another thing altogether to try doing so for social practices [...] generally' (Rouse (2015), *Articulating the World*, p. 168).

an independently existing external object. I naturally recognise this point as a good one, since I introduced it myself in Chapter One. I represented the problem with the glass-bulb model. Unlike Rouse, however, I view this problem as an opportunity, not to banish SSK practitioners into the wasteland of the philosophical undead, but to invite them into the verdant valley of existential phenomenology.

I have argued that the intelligibility of external-world scepticism depends on a prior, often tacit, acceptance of the glass-bulb model. To recognise external-world scepticism as a genuine epistemological problem, in need of some kind of solution, is to have already adopted an ontological image of the subject as a discrete and worldless substance-subject. I furthermore argued that SSK practitioners, despite their sometimes vigorous disagreements over how best to address this problem, are agreed with respect to the existence and intelligibility of the problem. This is because they all subscribe, to some significant degree or other, to the glass-bulb model which fuels that problem. I proposed not a solution, but a dissolution of this problem through the replacement of the glass-bulb model with Heidegger's existential concept of the subjectivity of the subject as being-in-the-world. Because being-in-the-world belongs to the basic existential structure of the subject, we never need to face the problem of how it gains epistemic access to that world. In my view, adopting Heidegger's concept adds philosophical strength to SSK, without significantly compromising its methodology or its goals. Indeed, as I have now argued in this chapter, the proposed combination is even compatible with the minimalist realism of many SSK practitioners because Heidegger, too, accepted the core realist doctrine of independent existence. This combination can succeed, in good part, because these social constructivists are able to accept, with dignity, Heidegger's claim that human existence is phenomenologically grounded in our immersed involvement with and alongside other entities in the world. Rouse is blind to the possibility of this combination, because he misattributes a theory-dominated account of science to both Heidegger and SSK. In this and the previous section, I have traced the errors in Rouse's interpretations, and hence shown that his criticisms against both parties do not succeed. The road thus remains clear for my proposed combination of SSK and Heidegger's existential conception of science. Furthermore, I have argued that this proposal, in contrast to Rouse's practical hermeneutics of science, can readily accommodate the core realist doctrine of independent existence. The proposed account provides the basis, in other words, for a minimally realist social constructivism about science.

7. Conclusion

I ended the introductory section of this chapter by noting Ginev's criticism of Rouse for not having paid adequate attention to the existential basis, and especially the cognitive specificity, of science. Ginev's criticism was inspired by Heidegger's existential conception of science. We have now seen that the core of this conception is the mathematical projection of nature, a term Heidegger used to denote the existential conditions which make science possible in both its practical and theoretical modalities. As a consequence of this projection, an independently existing nature comes to be understood *a priori* as something receptive to quantitative analysis. Given that the mathematical projection lies at the core of Heidegger's existential conception of science, it is ironic that Ginev follows Rouse in treating it as evidence for Heidegger's alleged dependency on a theory-dominant view of science. Indeed, Ginev even charges Heidegger with 'mathematical essentialism,' having apparently missed the point that Heidegger displaces the phenomenological priority of mathematical essence by explaining it in terms of the existential conditions which make it possible.[105] Heidegger's position might thus be better described as one of 'mathematical existentialism.'

As we have seen, Rouse, too, criticises essentialist accounts of science, drawing on Rorty's claim that 'natural science is not a natural kind'

---

105 Ginev (2011), *Tenets of Cognitive Existentialism*, p. 5. A more detailed criticism of Ginev, and Rouse, on this point may be found in Kochan (2015a), 'Scientific Practice and Modes of Epistemic Existence.' In a response to this criticism, Ginev has asserted that Heidegger never used existential conditions as an explanatory resource, and that his alleged mathematical essentialism was meant to be *sui generis*, 'untranslatable,' arising from 'something like a "mysterious act"' (Dimitri Ginev (2015), 'The Battle for Mathematical Existentialism and the War of the Heideggerian Succession: Rejoinder to Kochan,' in *Debating Cognitive Existentialism*, ed. by Dimitri Ginev (Leiden and Boston: Brill Rodopi), pp. 168–93 (p. 187). I disagree with Ginev, but, if he were right, then I would readily admit to demystifying Heidegger's existential conception of science so as to make it a more interesting and useful explanatory resource for science studies scholars.

to do so. Rorty's concern was to block attempts to demarcate science from pseudoscience — astronomy from astrology, for example — by attributing to science an absolute and unique essence. I have argued that this criticism fails when set against SSK practitioners, because they too reject the idea that science has an absolute essence. That is why they call themselves social constructivists. Rouse tries, however, to further press the case by criticising SSK for treating science as a 'kind' fully amenable to sociological explanation. I have shown that this argument is also not successful. SSK practitioners only argue that sociological categories are necessary, not that they are sufficient, for a complete explanatory account of science. In order to be sufficient, such an account may also have to include psychological, biological, and physical categories, and it should also include a place for an independently existing nature. SSK practitioners are thus realists in this minimal sense.

Rouse's argument is also unsuccessful against Heidegger. He alleges that Heidegger propounded an essentialist account of science in terms of theory: the genesis of scientific objects can be fully explained by reference to a mathematically structured conceptual scheme. This argument fails, first, because Heidegger did not view science primarily as a conceptual scheme but instead as a form of existence, and, second, because Heidegger did not claim to fully explain the genesis of scientific objects by reference to scientific modes of existence. Similar to SSK practitioners, Heidegger argued that these modes are responsible only for the essence of objects, not for their existence. Hence, Heidegger is also a realist in this minimal sense. Rouse misses this crucial moment in Heidegger's account of science, because he neglects Heidegger's distinction between the what-being and that-being, the essence and existence, of a thing.

As already mentioned, Ginev also criticises Rouse's effacement of the cognitive specificity of science, that is, his tendency to uncritically assimilate science to the broader sphere of cultural life in which it is necessarily embedded. This is the flip-side of the demarcation problem. Ginev's worry is not that non-scientific practices will penetrate the cognitive boundaries of scientific culture, thereby corrupting that culture. His worry is rather that science will permeate out into the broader culture, with the result that social life in general will become 'totally instrumentalized.'[106] According to Ginev, this threat follows from Rouse's mistaken assumption that anti-essentialism about science entails the rejection of its cognitive specificity. Yet, as Ginev points out, one may distinguish between scientific and non-scientific cultures, without recourse to essentialism, by recognising the existential specificity of scientific practice. The two spheres of culture are thus to be distinguished in existentialist rather than essentialist terms.[107] This move mirrors Heidegger's own displacement of an essentialist account with an existentialist account of science. It also mirrors Heidegger's identification, in some of his later work, of modern science with instrumental rationality. It is not clear, however, that this is an entirely apt characterisation of modern scientific practice. In any case, it remains, for present purposes, to consider how an account of the existential specificity of science may play against the distinction Rouse attempts to draw between SSK, on the one hand, and what he calls 'cultural studies of scientific knowledge,' on the other.[108]

Rouse includes feminist science studies within the realm of cultural studies of scientific knowledge, and this will be our focus here. In comparing feminist science studies with SSK, Rouse notes many important similarities between them, as well as some differences of both a technical and a fundamental nature. The main technical difference is that SSK practitioners have largely neglected issues of gender in their empirical and methodological work. This is true, and unfortunate. Yet, as Rouse argues, this neglect may only reflect an incomplete application of SSK's methods. In principle, that method may 'leave ample room for a full appreciation of the significance of gender relations as a social explanans for the content of scientific knowledge.'[109] Hence, Rouse

---

106 Ginev (2005), 'Against the Politics of Postmodern Philosophy of Science,' p. 198; Ginev (2011), *The Tenets of Cognitive Existentialism*, p. 102.

107 Ginev (2005), 'Against the Politics of Postmodern Philosophy of Science,' p. 202; Ginev (2011), *The Tenets of Cognitive Existentialism*, p. 108.

108 Rouse (1993), 'What Are Cultural Studies of Scientific Knowledge?,' *Configurations* 1(1), 57–94 (passim).

109 Joseph Rouse (1996b), 'Feminism and the Social Construction of Scientific Knowledge,' in *Feminism, Science, and the Philosophy of Science*, ed. by L. H. Nelson and J. Nelson (Dordecht: Kluwer), pp. 195–215 (p. 196).

moves to bracket this difference, turning his attention instead to what he considers the key fundamental difference between the two fields.

That difference hinges on Rouse's claim that SSK practitioners propound an 'epistemological' account of scientific knowledge. To view science in this way means to treat it as a definite 'kind,' or object of study, which may be surveyed as a 'totality,' for example, as a coherent and determinate system of representations, a conceptual scheme.[110] Hence, in distinguishing feminist science studies from SSK, Rouse writes that 'feminist scholars conceive of "knowing" as concretely situated, and as more interactive than representational. Knowledge is not merely a propositional attitude (belief or acceptance) toward some ideal or abstracted propositional content, but a *relationship* between knower and known.'[111] The nub of Rouse's argument, then, is his accusation that SSK trades on a reified picture of science which abstracts it from the concrete level of situated interaction favoured by feminist science studies. As I have argued in this chapter, this accusation flies wide of the mark. SSK practitioners may deliberately reify scientific practice in the interests of convenience, but their core methodological commitment is to a picture of science as a heterogeneous and shifting field of interaction which is nevertheless amenable to sociological analysis. Rouse's attempt to drive a substantive methodological wedge between feminist science studies and SSK is thus not successful. This is not to say that there do not remain important differences of orientation between these two fields, but these differences are of a technical rather than a fundamental nature. There is no reason, in principle, why the two fields cannot enter into greater cooperation with one another.

Indeed, Rouse's analysis even suggests, perhaps unwittingly, that feminist science studies, like SSK, is methodologically predisposed towards the minimal realist position I have commended in this chapter. He writes that '[f]eminist science studies have […] often been explicitly concerned with *different* ways in which knowers might interact with objects of knowledge.'[112] Although Rouse means to distinguish feminist

---

110 Joseph Rouse (1996b), 'Feminism and the Social Construction of Scientific Knowledge,' pp. 198, 199.

111 Joseph Rouse (1996b), 'Feminism and the Social Construction of Scientific Knowledge,' p. 203.

112 Joseph Rouse (1996b), 'Feminism and the Social Construction of Scientific Knowledge,' p. 204.

science studies from SSK with this observation, he could, in fact, have just as well made it of SSK. As we will see in Chapter Three, it is with explaining the difference between distinct ways of understanding a commonly encountered nature that SSK practitioners have often concerned themselves. Moreover, in a passage cited earlier in this chapter, Bloor writes that 'nature, in our ordinary way of thinking, is the object of knowledge, the thing that is known, while science is the knowledge we have of it, our theories about it and our description of it.'[113] I have already critiqued Bloor's phrase 'object of knowledge,' and those remarks could also be applied to the phrase 'objects of knowledge' in the passage from Rouse. More to the point is that, in both passages, science is presented as an activity distinct from, but directed towards, an independently existing nature. When epistemic groups disagree in their theoretical and practical attitudes towards nature, when they interact with nature in different ways, we can make sense of this disagreement, in part, by looking to the existential differences present between their distinct orientations towards nature. Such existential differences may be located within the cultures of science, but they may also mark a distinction between scientific and non-scientific cultural orientations towards nature. As Ginev reminds us, all science may be cultural, but not all culture is scientific. There would, of course, also be other ways of understanding such differences, ways which seek to prohibit all reference to an independently existing nature. As Rouse's own example illustrates, however, this kind of prohibition, while possible in theory, is difficult to maintain in practice. It is really much easier to simply accept minimal realism as the norm for science studies, and then to get on with one's research.

---

113 Bloor (2004a), 'Sociology of Scientific Knowledge,' p. 942.

Appendix

This Appendix supplements footnote 3 (p. 55) in this chapter.

In addition to works by Joseph Rouse, Trish Glazebrook, and Dimitri Ginev, discussed in this chapter, there are a significant number of other works also addressing the topic of Heidegger and realism. See:

William D. Blattner (1994), 'Is Heidegger a Kantian Idealist?,' *Inquiry* 37(2), 185–201.

William D. Blattner (2004), 'Heidegger's Kantian Idealism Revisited,' *Inquiry* 47(4), 321–37.

Patrick L. Bourgeois and Sandra B. Rosenthal (1988), 'Heidegger and Peirce: Beyond "Realism or Idealism,"' *Southwest Philosophy Review* 4(1), 103–10.

David R. Cerbone (1995), 'World, World-Entry, and Realism in Early Heidegger,' *Inquiry* 38(4), 401–21.

David R. Cerbone (2005), 'Realism and Truth,' in *A Companion to Heidegger*, ed. by Hubert L. Dreyfus and Mark Wrathall (Oxford: Blackwell), pp. 248–64.

Robert P. Crease (2009), 'Covariant Realism,' *Human Affairs* 19(2), 223–32.

Hubert L. Dreyfus (1991a), 'Heidegger's Hermeneutic Realism,' in *The Interpretive Turn: Philosophy, Science, Culture*, ed. by David R. Hiley, James F. Bohman and Richard Shusterman (Ithaca: Cornell University Press), pp. 25–41.

Hubert L. Dreyfus (2001), 'How Heidegger Defends the Possibility of a Correspondence Theory of Truth with Respect to the Entities of Natural Science,' in *The Practice Turn in Contemporary Theory*, ed. by Theodore R. Schatzki, Karin Knorr D. Cetina and Eike von Savigny (London: Routledge), pp. 151–62.

Hubert L. Dreyfus and Charles Spinosa (1999), 'Coping with Things-in-Themselves: A Practice-Based Phenomenological Argument for Realism,' *Inquiry* 42(1), 49–78.

Piotr Hoffman (2000), 'Heidegger and the Problem of Idealism,' *Inquiry* 43(4), 403–12.

Ka-wing Leung (2006), 'Heidegger on the Problem of Reality,' *The New Yearbook for Phenomenology and Phenomenological Philosophy* 6, 169–84.

Theodore R. Schatzki (1992), 'Early Heidegger on Being, the Clearing, and Realism,' in *Heidegger: A Critical Reader*, ed. by Hubert L. Dreyfus and Harrison Hall (Oxford: Blackwell), pp. 81–98.

Hans Seigfried (1980), 'Scientific Realism and Phenomenology,' *Zeitschrift für philosophische Forschung* 34, 395-404.

Lambert V. Stepanich (1991), 'Heidegger: Between Idealism and Realism,' *The Harvard Review of Philosophy* (Spring), 20–28.

Mark Basil Tanzer (1995), 'Heidegger's Critique of Realism,' *Southwest Philosophy Review* 11(2), 145–59.

Mark Basil Tanzer (1998), 'Heidegger on Realism and Idealism,' *Journal of Philosophical Research* 23, 95–111.

John Tiez (1993), 'Heidegger on Realism and the Correspondence Theory of Truth,' *Dialogue* 32(1), 59–75.

John Tiez (2005), 'Heidegger on Science, Realism, and the Transcendence of the World: *Being and Time*, Section 69,' *Idealistic Studies* 35(1), 1–20.

David J. Zoller (2012), 'Realism and Belief Attribution in Heidegger's Phenomenology of Religion,' *Continental Philosophy Review* 45(1), 101–20.

Trish Glazebrook brings much of this debate into conversation with recent work in philosophy of science:

Trish Glazebrook (2001a), 'Heidegger and Scientific Realism,' *Continental Philosophy Review* 34(4), 361–401.

Also in addition to works by Rouse, Glazebrook, and Ginev, discussed in this chapter, there is there is an extensive literature more generally addressing the topic of Heidegger and science. See:

Harold Alderman (1978), 'Heidegger's Critique of Science and Technology,' in *Heidegger and Modern Philosophy*, ed. by Michael Murray (New Haven: Yale University), pp. 35–50.

Babette E. Babich (1995), 'Heidegger's Philosophy of Science: Calculation, Thought, and *Gelassenheit*,' in *From Phenomenology to Thought, Errancy, and Desire: Essays in Honor of William J. Richardson, SJ*, ed. by Babette E. Babich (Dordecht: Kluwer Academic Publishers), pp. 589–99.

Giorgio T. Bagni (2010), 'Mathematics and Positive Sciences: A Reflection Following Heidegger,' *Education Studies in Mathematics* 73(1), 75–85.

Edward G. Ballard (1971), 'Heidegger's View and Evaluation of Nature and Natural Science,' in *Heidegger and the Path of Thinking*, ed. by John Sallis (Pittsburgh: Duquesne University Press), pp. 37–64.

Rainer A. Bast (1986), *Der Wissenschaftsbegriff Martin Heideggers im Zusammenhang seiner Philosophie* (Stuttgart-Bad Cannstatt: Frommann-Holzboog Verlag).

William Blattner (1995), 'Decontextualization, Standardization, and Deweyan Science,' *Man and World* 28, 321–39.

John D. Caputo (1995), 'Heidegger's Philosophy of Science: The Two Essences of Science,' in *From Phenomenology to Thought, Errancy, and Desire: Essays in Honor of William J. Richardson, SJ*, ed. by Babette E. Babich (Dordecht: Kluwer Academic Publishers), pp. 43–60.

John D. Caputo (2000), 'Hermeneutics and the Natural Sciences: Heidegger, Science, and Essentialism,' in *More Radical Hermeneutics*, by John D. Caputo (Bloomington: Indiana University Press), pp. 151–71, 281–84.

David R. Cerbone (2012), 'Lost Belongings: Heidegger, Naturalism, and Natural Science,' in *Heidegger on Science*, ed. by Trish Glazebrook (Albany: SUNY Press), pp. 131–55.

Catherine Chevalley (1992), 'Heidegger and the Physical Sciences,' in *Martin Heidegger: Critical Assessments*, vol. 4: *Reverberations*, ed. by Christopher Macaan (London: Routledge), pp. 342–64.

Robert P. Crease (1992), 'The Problem of Experimentation,' in *Phenomenology of Natural Science*, ed. by Lee Hardy and Lester Embree (Dordecht: Kluwer Academic Publishers), pp. 215–35.

Robert P. Crease (1993), *The Play of Nature: Experimentation as Performance* (Bloomington: Indiana University Press).

Shannon Dea (2009), 'Heidegger and Galileo's Slippery Slope,' *Dialogue: Canadian Philosophical Review* 48(1), 59–76.

Dennis Desroches (2003), 'Phenomenology, Science Studies, and the Question of Being,' *Configurations* 11(3), 383–416.

Dimitri Ginev (2012), 'Two Accounts of the Hermeneutic Fore-structure of Scientific Research,' *International Studies in the Philosophy of Science* 26(4), 423–45.

Trish Glazebrook (2000a), *Heidegger's Philosophy of Science* (New York: Fordham University Press).

Trish Glazebrook, ed. (2012a), *Heidegger and Science* (Albany: SUNY Press).

Alan G. Gross (2006), 'The Verbal and the Visual in Science: A Heideggerian Perspective,' *Science in Context* 19(4), 443–74.

Karlfried Gründer (1963), 'Heidegger's Critique of Science in Its Historical Background,' *Philosophy Today* 7(1), 15–32.

Patrick A. Heelan (1995), 'Heidegger's Longest Day: Twenty-Five Years Later,' in *From Phenomenology to Thought, Errancy, and Desire: Essays in Honor of William J. Richardson, SJ*, ed. by Babette E. Babich (Dordecht: Kluwer Academic Publishers), pp. 579–87.

Theodore J. Kisiel (1970), 'Science, Phenomenology, and the Thinking of Being,' in *Phenomenology and the Natural Sciences, Essays and Translations,* ed. by Joseph J. Kockelmans and Theodore J. Kisiel (Evanston: Northwestern University Press), pp. 167–83.

Theodore J. Kisiel (1992), 'Heidegger and the New Images of Science,' in *Martin Heidegger: Critical Assessments,* vol. 4: *Reverberations,* ed. by Christopher Macaan (London: Routledge), pp. 325–41.

Julian Kiverstein, ed. (2012), *Heidegger and Cognitive Science* (Basingstoke: Palgrave Macmillan).

Joseph J. Kockelmans (1970), 'Heidegger on the Essential Difference and Necessary Relationship between Philosophy and Science,' in *Phenomenology and the Natural Sciences,* ed. by Kockelmans and Kisiel, pp. 147–66.

Joseph J. Kockelmans (1985), *Heidegger and Science* (Lanham: The Center for Advanced Research in Phenomenology and The University Press of America).

David A. Kolb (1983), 'Heidegger and the Limits of Science,' *Journal of the British Society for Phenomenology* 14(1), 50–64.

George Kovacs (1990), 'Philosophy as Primordial Science (*Urwissenschaft*) in the Early Heidegger,' *Journal of the British Society for Phenomenology* 21(2), 121–35.

Lin Ma and Jaap van Brakel (2014), 'Heidegger's Thinking on the "Same" of Science and Technology,' *Continental Philosophy Review* 47(1), 19–43.

James E. McGuire and Barbara Tuchanska (2000), *Science Unfettered: A Philosophical Study in Sociohistorical Ontology* (Athens OH: Ohio University Press).

Denis McManus (2007), 'Heidegger, Measurement and the "Intelligibility" of Science,' *European Journal of Philosophy* 15(1), 82–105.

Graeme Nicholson (2008), 'Heidegger, Descartes and the Mathematical,' in *Descartes and the Modern,* ed. by Neil Robertson, Gordon McOuat and Tom Vinci (Newcastle: Cambridge Scholars Publishing), pp. 216–34.

Matthew Ratcliffe (2002), 'Heidegger's Attunement and the Neuropsychology of Emotion,' *Phenomenology and the Cognitive Sciences* 1(3), 287–312.

Hans-Jörg Rheinberger (2010a), *An Epistemology of the Concrete: Twentieth-Century Histories of Life* (Durham: Duke University Press).

Hans-Jörg Rheinberger (2010b), *On Historicizing Epistemology: An Essay* (Stanford: Stanford University Press).

William J. Richardson (1968), 'Heidegger's Critique of Science,' *New Scholasticism* 42(4), 511–36.

Michael Roubach (1997), 'Heidegger, Science, and the Mathematical Age,' *Science in Context* 10(1), 199–206.

John Sallis (1971), 'Toward the Movement of Reversal: Science, Technology, and the Language of Homecoming,' in *Heidegger and the Path of Thinking*, ed. by John Sallis (Pittsburgh: Duquesne University Press), pp. 138–68.

Tibor Schwendtner (2005), *Heideggers Wissenschaftsauffasung: Im Spiegel der Schriften 1919–29* (Frankfurt: Peter Lang).

Pablo Schyfter (2012), 'Standing Reserves of Function: A Heideggerian Reading of Synthetic Biology,' *Philosophy & Technology* 25(2), 199–219.

Robert Shaw (2013), 'The Implications for Science Education of Heidegger's Philosophy of Science,' *Education Philosophy and Theory* 45(5), 546–70.

Charles Sherover (1967), 'Heidegger's Ontology and the Copernican Revolution,' *Monist* 51(4), 559–73.

Carol J. Steiner (1999), 'Constructive Science and Technology Studies: On the Path to Being?,' *Social Studies of Science* 29(4), 583–616.

Chapter Three

Finitude, Humility, and the Bloor-Latour Debate

1. Introduction

Commenting on the minimal realism outlined in Chapter Two, some people have said that they find it very 'Kantian,' and expressed worry that it may be vulnerable to the dreaded Kantian 'two-world problem.' Minimal realism can indeed be justifiably described as 'Kantian,' but only in a highly qualified sense. In this chapter, I will present a detailed account of those qualifications. I will also defend minimal realism against the charge that it is vulnerable to the two-world problem. This defence will unfold as an intervention into the spirited 1999 debate between David Bloor and Bruno Latour, a debate which may well be one of the most dramatic dust-ups in the short history of science studies. The two-world problem was, I will argue, a central point around which this debate turned.

The two-world problem rests on the idea that there is, on the one hand, a world of appearance, and, on the other hand, a real world which underlies that appearance. The problem is enlivened by the further idea, attributed to Kant, that all we can know are appearances, that the real world underlying those appearances is something about

      http://dx.doi.org/10.11647/OBP.0129.03

which we can know nothing. The problem is: if we have no knowledge of the real world, then how can we know that it exists? The two-world problem is thus a version of the sceptical problem about the existence of the external world, to which we gave detailed attention in Chapter One. The Kantian version of this problem is usually articulated in terms of two distinct kinds of object, 'phenomena' and 'noumena.' The phenomena are the appearances, the things we can know. The noumena are the real things which underlie those appearances, the things we cannot know. Kant called a noumenon, the real thing about which we can know nothing, the 'thing-in-itself.' The problem thus becomes: if we cannot know anything about the thing-in-itself, then how can we know that it exists?

In Chapter One, I argued that external-world scepticism relies on a tacit acceptance of the glass-bulb model. The subject is trapped inside a glass bulb, and must find a way to burst through the barrier, thereby gaining access to the external world. We can now see that the glass-bulb model similarly grounds the two-world problem. Phenomena lie inside the bulb; things-in-themselves lie outside of it. How do we break out of the bulb, pushing past phenomenal appearances and grabbing hold of the thing-in-itself? How, indeed, when we have already been told that the thing-in-itself is something of which we can never grab hold, something which we can never know? One begins to wonder what the point ever was of positing the existence of the thing-in-itself. It seems like an idle wheel, spinning uselessly in imaginary space, never connecting with anything important or real.

But this is the wrong way to think about the thing-in-itself. The thing-in-itself is not a figment of philosophical imagination. It is real, and it does indeed connect with other things. In her commentary on Kant, Rae Langton has argued this point with clarity and force. The next section will summarise the most important parts of her argument. In a nutshell, there are not two worlds in Kant, but only one; so there are not two distinct kinds of object, but only one. As Langton reads Kant, this single object can be known in two different ways: either imperfectly, by humans, or absolutely, by God. The thing-in-itself is the object as it can be known only by God. It marks the limits of human epistemic power. For Langton, then, the thing-in-itself signifies the fact of our own inherent finitude vis-à-vis an infinitely powerful God. She suggests that, for Kant, the appropriate response to this fact of human finitude was to adopt an attitude of epistemic humility.

Langton's commentary provides a nice entry point into the Bloor-Latour debate. Latour first paints practitioners of the sociology of scientific knowledge (SSK) as beholden to a Kantian concept of the thing-in-itself, then sets this image up in terms of the two-world problem, and finally submits SSK to what are, at root, standard sceptical criticisms. Bloor attempts to deflect Latour's criticisms by arguing that acceptance of the concept of the thing-in-itself need not entail a further commitment to the two-world thesis. In making this argument, Bloor moves his account of the thing-in-itself closer to the one-world thesis, transforming Kant's transcendental and individualistic account into a naturalistic and sociological one. This naturalistic strategy leads to a potential problem, however, as Langton's commentary necessarily ties the concept of the thing-in-itself to the existence of a transcendent God, a supernatural and omniscient knower. Here Heidegger can come to Bloor's aid. In fact, Heidegger had much to say about the Kantian thing-in-itself, and he also firmly rejected the two-world thesis. However, in contrast to Langton, he conceptualised human finitude, not against the image of an omniscient God, but against the notion of a nature which necessarily slips free from our attempts to know it once and for all. As we will see, this move is compatible with Bloor's own account, and even provides further resources for strengthening his defence of SSK against Latour's challenge. The argument of this chapter thus reinforces and extends the combination of SSK and existential phenomenology pursued in Chapters One and Two. Furthermore, by recognising a role for an independently existing nature, Heidegger's account is also compatible with minimal realism. In the penultimate section of this chapter, I suggest an account of Heideggerian, in contrast to Kantian, humility: that is, an attitude of humility which arises, not in response to our finitude vis-à-vis an infinitely powerful God, but in response to the inevitable finitude of our epistemic powers when confronted with the insuperability of an independently existing nature. I commend this as a suitable attitude for a minimally realist approach in science studies.

Before moving into the argument of the chapter, it may be worth addressing a potential confusion on one point. It is important to carefully distinguish between the epistemic finitude to be discussed here and the doctrine of 'finitism' common in the SSK literature. They are not the same, though they are closely related. Bloor writes that 'finitism is probably the most important single idea in the sociological vision of knowledge.'[1] In my view, if SSK practitioners wish to protect their realist credentials, then finitude should also be counted among their most important ideas.

Finitism is the view that the correctness of an act of concept application is not determined by absolute standards. As Barry Barnes writes, every such act is 'open-ended and revisable': '[n]othing in the nature of things, or the nature of language, or the nature of past usage, determines how we employ, or correctly employ, our terms.'[2] What determines correct usage is, rather, the community of language users. Having been taught to which birds I should apply the word 'duck,' I may err by going on to apply it to a goose. I will then be corrected. The standard according to which I am corrected is a community standard; it is a social convention belonging to a tradition of language users. Hence, the standard is not absolute but historically contingent. It may change over time as the community changes. Other communities, in turn, may have different standards for making sense of the same thing. In short, our power to successfully create meaning is finite in scope, because it is bounded by the range of possibilities made available to us by the historical tradition in which we necessarily find ourselves.

It will be useful to place these considerations against the backdrop of Langton's argument that the Kantian thing-in-itself can be known in two distinct ways: one human and imperfect; the other divine and absolute. Here, the finitude of human thinking is defined by contrast to the infinite power of an omniscient God. In his naturalistic and sociological appropriation of Kant's thing-in-itself, Bloor replaces God with society. He argues, following Émile Durkheim, that 'God [is] really the social collectivity.'[3] Hence, the final arbiter of right and wrong in acts of thinking and naming is not God, but *us*.

---

1   David Bloor (1991 [1976]), *Knowledge and Social Imagery*, 2nd edn (Chicago: University of Chicago Press), p. 165.

2   Barry Barnes (1982), *T. S. Kuhn and Social Science* (London: Macmillan), p. 30.

3   David Bloor (1983), *Wittgenstein: A Social Theory of Knowledge* (London: Macmillan), p. 20.

So much, I think, agrees with Heidegger's own account of the thing-in-itself. But now we come to an ambiguity in SSK's presentation of finitism, one which Heidegger's notion of finitude can help us to clear up. Recall, from Chapter Two, Heidegger's distinction between the existence and the essence of a thing. Heidegger traced the roots of this distinction back to the Biblical view that all things owe their existence to a divine Creator. Medieval Christian metaphysicians transformed this doctrine into the claim that God first establishes the essence of a thing, and then actualises that essence by bringing the thing into existence. On this model, there is, strictly speaking, no independently existing thing, because the thing depends on God for its existence. However, because God does not depend on us, neither does the thing. We are thus justified in being realists about the thing. Now turn back to the Kantian thing-in-itself. The ambiguity in the finitist account is this: if God is the source of both the essence and the existence of the thing, and if God is really the social collectivity, then does it not follow that society is the source of both the essence and the existence of the thing? If this is so, then the finitist account would seem incompatible with the core realist doctrine of independent existence. The finitism of SSK thus seems to conflict with its realism. As we will see, Latour forcefully exploits this apparent conflict in his dispute with Bloor.

Heidegger's notion of finitude may be used to resolve this conflict by bringing into sharper focus the metaphysical presuppositions which have been left largely unexamined by the proponents of finitism. In doing so, we can fully complete the naturalisation of Kant's transcendental account of the thing-in-itself. On this fully naturalised account, the social collectivity does not construct the thing *ex nihilo*; it only constructs the categories by which its members may make sense of that thing. The thing is thus a thing-in-itself. As such, it marks the extreme limit of our constructive power. It is against the independent existence of this thing that we come to recognise our basic condition as finite beings. Residual supernatural notions of an infinitely constructive power have no place in this recognition. With these considerations in mind, let us now proceed into the chapter.

2. Kantian Humility and the Thing-in-Itself

Two distinct theses figure prominently in Kant's *Critique of Pure Reason*: an epistemological thesis; and a metaphysical thesis. The epistemological thesis endorses an empiricist theory of knowledge, which states that our physical senses provide the only means by which to gain knowledge of the world. Furthermore, Kant specifies that our senses are causally affected by things, and describes this in terms of our 'receptivity' towards those things. The metaphysical thesis, in turn, asserts the existence of what Kant variously calls a 'substance' or 'absolute substance,' a 'transcendental object,' a 'noumenon,' or a 'thing-in-itself.' Notoriously, Kant argued that, although things-in-themselves exist, we cannot know them. We can only know what affects our senses, and things-in-themselves can do no such thing. As we saw above, Kant has thus been widely interpreted as positing the existence of two distinct and separate worlds, one which touches our senses, a world of appearances, and the other comprised of things-in-themselves. This two-worlds view has received a considerable amount of criticism since it was first attributed to Kant.

In her 1998 book, *Kantian Humility: Our Ignorance of Things in Themselves*, Rae Langton rejects this two-world interpretation, and offers an alternative interpretation of Kant's position. She argues that, for Kant, there was only one world: the world of things. These things, however, feature two distinct, non-overlapping sets of properties: relational and intrinsic.[4] Relational properties causally affect our senses, while intrinsic properties do not. Hence, employing Kant's epistemological, or empirical, thesis, we can conclude that we do indeed know the things themselves, but only through their relational properties. Kant's term 'thing-in-itself' refers to the thing as it is on its own, independently of any causal relation to a subject. The thing-in-itself is the thing as it exists *in itself* in contrast to as it exists *in relation to us*. In a lucid turn of phrase, Langton thus describes the autonomous thing-in-itself as 'lonely.'[5] The term 'thing-in-itself' refers not to one kind of thing rather than another — to a noumenal rather than a phenomenal thing, a

---

4   Rae Langton (1998), *Kantian Humility: Our Ignorance of Things in Themselves* (Oxford: Clarendon Press), p. 12.
5   Langton (1998), *Kantian Humility*, p. 19.

transcendental rather than an empirical object — but to two distinct ways in which one single thing may exist: either in itself or in relation to a knower. According to Langton, it is precisely the thing as it exists in itself, in isolation from us, which Kant says we cannot know. When Kant says that we cannot know the thing-in-itself, he means specifically that we cannot know it because it is autonomous and lonely, because it has no relation to anything beyond itself. Existing in this way, the thing has only intrinsic properties, properties which extend only to itself, and thus cannot affect our senses. Hence, although we can know that the thing-in-itself exists, because it is identical with the thing as it relates to our senses, we can never know what it is in respect of its intrinsic properties. As Langton writes, '[w]e can know *that* there are things that have intrinsic properties without knowing *what* those properties are.'[6] Of course, this sounds a lot like the minimal realism I drew from both Heidegger and SSK in Chapter Two. However, as we shall see in this chapter, both Heidegger and SSK depart from Kant in a number of important and interesting ways. For the moment, let us focus on what they all have in common.

There are three things shared by Kant, Heidegger, and SSK relevant to the present discussion. First, all are committed to receptivity, that is, to the view that knowledge necessarily depends on the way things causally affect our senses. They all thus endorse, in one way or another, the empirical basis of all knowledge. Second, they are all committed to epistemic finitude. In other words, they accept that we are finite beings, and hence that there is a necessary limit to what we can know. For example, they all agree that we cannot gain knowledge about things-in-themselves. Third, all three are committed, one way or another, to an attitude described by Langton as 'humility.' This is a direct result of their acceptance of epistemic finitude. In the case of Kant, according to Langton, the move from finitude to humility is mediated by Kant's distinction between sensible intuition, on the one hand, and 'intellectual intuition,' on the other. Sensible intuition is a characteristic of finite humans, and it allows them to perceive the relational properties of things. Intellectual intuition, by contrast, belongs only to God, and it allows God to pick out the intrinsic properties of things, among which

---

6   Langton (1998), *Kantian Humility*, p. 13.

Kant includes things-in-themselves.[7] Kantian humility must thus be understood as humility in the face of the absolute knowledge of a divine being. In the case of both Heidegger and SSK, so I argue, the move from finitude to humility is mediated by a rejection of absolute knowledge, as such, and the endorsement instead of an account of knowledge as an inherently contextual phenomenon, the scope of which is determined, in significant part, by the social and historical conditions under which the subject necessarily finds itself. We thus express humility, not before the omnipotence and omniscience of an absolute God, but before a natural world which constantly outstrips our best efforts to know it.

A crucial claim in the argument of this chapter is that the concept of an 'intrinsic property' is only accidentally related to the concept of a 'thing-in-itself.' In other words, we can relieve the thing-in-itself of its intrinsic properties without threatening our belief in its independent existence. In my view, this claim is implicit in the positions taken up by both SSK practitioners and Heidegger, and it lies at the core of their consequential departure from Kant's own position.

Langton attributes two fundamental claims to Kant: first, that the thing-in-itself must have an independent existence, that it must exist and be lonely; second, and most crucial for Kant, that the thing-in-itself must possess intrinsic properties. Langton argues that the possession by the thing-in-itself of intrinsic properties is consistent with its autonomous existence and its being lonely. Let us allow that this was Kant's view. Notice, however, that the fact that these two claims are consistent with one another does not entail that their connection is also a necessary one. While the independent existence of the thing-in-itself may be consistent with its possessing intrinsic properties, it is also consistent with its not possessing any such properties. This possibility is, in my view, the one best suited to understanding the respective positions of Heidegger and SSK practitioners when it comes to the thing-in-itself.

As it turns out, Langton also entertains this as a possible view held by Kant, one in which the thing-in-itself is a 'bare substratum' without intrinsic properties.[8] The idea is that if we cannot know the intrinsic properties of the thing-in-itself, then we cannot assert that those properties exist. Indeed, we may just as well assert that they do not exist.

---

7   Langton (1998), *Kantian Humility*, p. 45.
8   Langton (1998), *Kantian Humility*, p. 29.

The fact that such properties do not exist can then be used to explain why we cannot know, indeed why not even a divine agent could know, that they exist. Langton thinks that if Kant had held such a view, then '[h]e would be guilty of no contradiction.'[9] However, she then goes on to argue that, although this position has 'some prima facie plausibility,' there is strong evidence that Kant rejected it.[10] On her reading, Kant was fully committed to the views that, first, '[i]f a substance can exist on its own, it must have properties that are compatible with its existing on its own,' and, second, that we cannot know the intrinsic properties of this independently existing substance, or thing-in-itself.[11] Langton cites the following passage from Kant's *Critique of Pure Reason* in order to support her reading: 'Substances in general must have some intrinsic nature, which is therefore free from all external relations.'[12]

According to Langton, in this passage Kant is not saying that the intrinsic properties of the thing-in-itself cannot be known at all. He is, rather, saying that we cannot know them given our current cognitive endowment. Indeed, she goes on to cite another passage, wherein Kant refers to the thing-in-itself as something 'of which we know, and *with the present constitution of our understanding can know*, nothing whatsoever.'[13] Langton thus concludes that, for Kant, the thing-in-itself does possess intrinsic, knowable, properties, but that the limited nature of our current cognitive abilities prevents us from gaining any knowledge of those properties. It would seem, then, that Kant's commitment to the thesis that things-in-themselves possess intrinsic properties, unknowable by humans, implies a further commitment to the corresponding thesis that there exists a superhuman agent who is able to know such properties. It is difficult to imagine what else might have motivated Kant's commitment to the first thesis if not his commitment to the second. It thus looks like Kant's doctrine that the thing-in-itself possesses intrinsic properties may, at base, have been motivated by a theological doctrine

---

9   Langton (1998), *Kantian Humility*, p. 31.
10   Langton (1998), *Kantian Humility*, p. 32.
11   Langton (1998), *Kantian Humility*, p. 19.
12   Langton (1998), *Kantian Humility*, p. 32; cf. Immanuel Kant (1998 [1781/1787]), *Critique of Pure Reason*, ed. and trans. by Paul Guyer and Allen W. Wood (Cambridge: Cambridge University Press), p. 374 (A274/B330).
13   Langton (1998), *Kantian Humility*, p. 32; cf. Kant (1998), *Critique of Pure Reason*, p. 348 (A250).

attesting the existence of a subject with absolute cognitive faculties, in a word, God.

In what follows, I shall argue that Heidegger modifies Kant's doctrine of humility by rejecting the possibility of absolute, or divine, knowledge of the intrinsic properties of the thing-in-itself, and so also the possibility of the existence of such properties. For the time being, we should bear in mind the interrelatedness of Kant's respective beliefs in the existence of intrinsic properties and the possibility of absolute knowledge, as this will prove crucial for our understanding of what is at stake in the debate between Bloor and Latour. Let us turn, then, to this important disagreement in the science studies literature.

3. Latour's Attack on Social Constructivism

Latour's condemnation of the alleged Kantianism of SSK practitioners is of a piece with his allegedly radical appropriation and revision of Bloor's symmetry principle.[14] This principle famously stipulates that the sociology of knowledge should be symmetrical in its style of explanation, that the same types of cause should be used to explain both true and false, rational and irrational, or successful and unsuccessful beliefs.[15] Latour argues that Bloor's symmetry principle is, in fact, profoundly asymmetrical. At the root of this purported asymmetry is Bloor's allegedly uncritical adoption of the Kantian subject-object distinction, where the 'subject pole' is occupied by the transcendental Ego and the 'object pole' is occupied by the thing-in-itself. According to Latour, Kant gathered all explanatory resources around the transcendental subject, thereby reducing the thing-in-itself to absolute passivity.[16] The result of this is a striking asymmetry in the way Kant treated the subject and object poles. In Latour's view, by reducing the thing-in-itself to absolute passivity, Kant robbed it of any role in explanations of the genesis of scientific knowledge. All the epistemic action takes place around the subject pole. Latour claims that, in the field of science studies, Bloor's

---

14   Bruno Latour (1992), 'One More Turn after the Social Turn…,' in *The Social Dimensions of Science*, ed. by Ernan McMullin (Notre Dame: University of Notre Dame Press), pp. 272–94.

15   Bloor (1991), *Knowledge and Social Imagery*, p. 7.

16   Latour (1992), 'One More Turn after the Social Turn…,' p. 278.

1976 book, *Knowledge and Social Imagery*, stands as the 'high-tide mark' of Kantian asymmetry.[17] According to Latour, Bloor has simply replaced the transcendental Ego at Kant's subject pole with Durkheim's 'macro-Society.'[18] In this sociological version of Kant, all explanatory resources are now gathered around society while the thing-in-itself, or 'Nature,' remains absolutely passive.

Latour describes Bloor's Durkheimian appropriation of Kant as the 'social turn' in science studies, and he calls for its rejection and replacement through 'one more turn after the social turn.' This further turn is meant to finally result in the symmetry Bloor had sought after but failed to achieve. In formulating this new, allegedly more radical, symmetry principle, Latour takes the explanatory resources away, not just from the object pole, but also from the subject pole. Like nature, society too will no longer serve as a resource in explanations of the genesis of scientific knowledge, but will instead stand, like nature, as a topic in need of explanation. Latour claims that it would be more in keeping with the empirical findings of science studies if both nature and society were to be viewed as constructed, and thus as explicable in terms of their constructedness. For Latour, then, the thing-in-itself is not an independently existing thing which necessarily rebuffs all of our attempts to know it. Instead, it is a wholly constructed thing, and we may thus come to know it by studying the processes by which it has come into existence. It is thus not really something existing in itself at all, but something which exists only in relation to us. Note that, in his critique of SSK, Latour has shifted the focus of analysis from the construction of knowledge about nature to the construction of nature as such. The distinction between knowledge and its object has been erased along with Kant's distinction between subject and object. As we will see later in this chapter, this shift in emphasis marks the weak point in Latour's criticism of SSK.[19]

Latour's new symmetry seems to be accompanied by an implicit epistemological assumption, namely, that we know only what we

---

17   Latour (1992), 'One More Turn after the Social Turn…,' p. 278.

18   Latour (1992), 'One More Turn after the Social Turn…,' p. 277.

19   Olga Amsterdamska had early on identified and criticised Latour's tendency to elide descriptions of nature with the nature being described (Olga Amsterdamska (1990), 'Surely You Are Joking, Monsieur Latour!' *Science, Technology, and Human Values* 15(4), 495–504).

make. He holds that we can know the thing-in-itself because we have participated in its production. For Latour, then, the Kantian thing-in-itself is not really an independently existing thing. It needs us in order to exist, and thus is neither lonely nor autonomous. As a result, Latour's new symmetry principle would appear to jettison the Kantian doctrine of humility, since there appears to be nothing to whose existence we have not contributed, nothing which lies outside the scope of our constructive influence, and hence nothing about which we cannot gain knowledge. Since the notion of the thing-in-itself just is, by definition, a name for what lies beyond our epistemic reach, the unrestrained constructivity of Latour's proposal leads him to reject this notion, along with the finitude and humility it implies.

4. Bloor's Defence of Social Constructivism

Seven years after Latour's attack appeared in print, Bloor responded with a strongly worded defence, in which he rejects Latour's claim that SSK practitioners are little more than 'unreconstructed Kantians.'[20] Indeed, he argues that Durkheim's reworking of Kantian themes allows for a viable naturalistic and sociological reading of Kant's subject-object distinction. For Bloor, a naturalistic reading of that schema must be distinguished from an individualistic and transcendental reading. He argues that the rejection of this latter reading of the distinction does not necessitate rejecting it in all cases. The subject-object distinction may still prove useful for anti-individualistic and naturalistic analyses of scientific knowledge production. Understood naturalistically, the subject-object distinction is, according to Bloor, a 'biological given.'[21] As we will see, Bloor's proposed naturalisation of the distinction marks a substantial departure from Kant's original position, including Kant's notion of the thing-in-itself.

While acknowledging that the thing-in-itself may refer to the noumenal basis of Kant's subject-object distinction, Bloor suggests that it can also be employed in reference to 'more common-sense ideas about the independence of the objects of nature from our ideas about

---

20   David Bloor (1999a), 'Anti-Latour,' *Studies in History and Philosophy of Science* 30(1), 81–112 (p. 106).

21   Bloor (1999a), 'Anti-Latour,' p. 107.

them.'[22] Introducing a simplified, individualistic model, he briefly describes a naturalistic setting in which an organism learns about its environment by causally interacting with it. This process, he suggests, involves varying degrees of active engagement and disengagement with the environment. These are causal and biological processes which do not derive from culture, but are instead presupposed by it.[23] Hence, according to Bloor, a naturalised subject-object distinction can be used as a conceptual tool for explaining the way in which one part of nature (the organism, or subject) interacts with another part of nature (the organism's environment, or object).[24] Unfortunately, Bloor does not develop his naturalistic description beyond the individual level. However, if we were to extend this simplified, individualistic model to the more complex social level, his general idea would seem to be that organisms depend on one another as they engage with, disengage with, and otherwise learn about their environment. On Bloor's account, then, there is a biologically given level of fundamental sociality which will be presupposed in any naturalistic account of the emergence of culture, in general, and science in particular.[25]

Bloor thus charges Latour with having failed to sufficiently distinguish between the historically contingent ways of formally articulating the subject-object distinction, on the one hand, and the biological and causal phenomenon which constitute the natural ground for that distinction, on the other. Bloor argues that Kant's individualistic and transcendental presentation of the distinction may be both socialised and naturalised in a way which brings it more firmly in line with its original biological and causal foundation. On this basis, Bloor furthermore rejects Latour's claim that SSK reduces the object pole of the distinction, that is, nature, to absolute passivity. According to the naturalistic construal of the subject-object distinction, the object pole, the thing-in-itself, is conceived as active because it exercises causal

---

22   Bloor (1999a), 'Anti-Latour,' p. 86.

23   Bloor (1999a), 'Anti-Latour,' p. 86.

24   Bloor (1999a), 'Anti-Latour,' p. 106–07.

25   For recent work on this topic, see: Michael Tomasello (2014), *A Natural History of Human Thinking* (Cambridge, MA: Harvard University Press), especially chpt. 4, 'Collective Intentionality'; and Kim Sterelny (2012), 'Language, Gesture, Skill: The Co-Evolutionary Foundations of Language,' in *Philosophical Transactions of the Royal Society B* 367, 2141–51. My thanks to Andrew Buskell for bringing this literature to my attention.

agency.[26] That nature is causally active is entailed by Kant's claim that we can only know a thing if it can affect us. As Langton argues, knowledge entails receptivity. The basic idea is that we form beliefs about nature on the basis of our causal interaction with it. However, although nature plays a necessary causal role in the formation of our beliefs about it, it is not a sufficient cause for those beliefs. On Bloor's account, the claim that a necessary but insufficient causal role must be played by nature is methodologically crucial, for without it we could not make sense of the fact that two scientists may form contradictory beliefs about the same natural phenomenon. This claim is, of course, also consistent with minimal realism.

Bloor elaborates on the necessary but insufficient causal role played by the thing-in-itself in the formation of scientific knowledge by discussing the contradictory interpretations of Robert Millikan and Felix Ehrenhaft in respect of the natural effects of what we now know as electrons.[27] On the basis of the experimental data, Millikan believed that he had secured evidence confirming Rutherford's electron theory. In contrast, Ehrenhaft, also on the basis of the experimental data, believed that he had secured evidence falsifying Rutherford's theory. Bloor argues that, because both interpretations were based on data produced by natural causes, that data alone cannot explain their divergence.

> If we believe, as most of us do believe, that Millikan got it basically right, it will follow that we also believe that electrons, as part of the world Millikan described, did play a causal role in making him believe in, and talk about electrons. But then we have to remember that (on such a scenario) electrons will *also* have played their part in making sure that Millikan's contemporary and opponent, Felix Ehrenhaft, *didn't* believe in electrons. Once we realise this, then there is a sense in which the electron 'itself' drops out of the story because it is a common factor behind two different responses, and it is the cause of the difference that interests us.[28]

In the cases of both Millikan and Ehrenhaft, a complete causal explanation for their respective interpretations must refer to something beyond the thing-in-itself, the independently existing natural thing, to which they were equally exposed. For SSK practitioners, this 'something beyond'

---

26   Bloor (1999a), 'Anti-Latour,' p. 91.
27   Bloor (1999a), 'Anti-Latour,' p. 93.
28   Bloor (1999a), 'Anti-Latour,' p. 93.

is social causation. Only by citing both natural and social causes can the sociologist uncover the necessary and sufficient conditions for the formation of the beliefs in question. Because Millikan and Ehrenhaft were similarly affected by nature, the difference in their respective interpretations must therefore be explained by a divergence in the social conditions influencing the formation of their respective interpretations of the data. This line of argument follows from what is commonly called the 'underdetermination thesis.' This thesis states that the data resulting from natural causes underdetermines the interpretations which arise in response to that data. This is just another way of saying that such data is necessary but not sufficient for explaining the way in which it is interpreted. The underdetermination thesis plays a fundamental role in the methodology of SSK, in particular, and of social constructivism, in general. Let us now turn to Latour's response to Bloor's defence of social constructivism.

5. Where the Dust Settles in the Debate

In responding to Bloor's defence, Latour no longer argues that SSK treats the thing-in-itself, the object pole of the subject-object distinction, as absolutely passive. Instead, he argues that SSK's underdetermination thesis is premised on an unacceptably impoverished view of the role played by objects. He criticizes Bloor for allegedly making the claim that, in the disagreement between Millikan and Ehrenhaft over the existence of electrons, the electron 'itself' makes 'no difference': 'Now, I want someone to explain to me what it is for an object to play a role *if it makes no difference.*'[29] Latour then goes on to argue that, for SSK, 'electrons "themselves" are not allowed to cause our interpretations of them, no matter how much scientists engage in making them have a bearing, a causality, on what they (the scientists) say about them (the electrons).'[30] I will comment on Latour's complaint in the next but one section. Note for now, however, that, as will be clear from the long passage from Bloor quoted at the end of the last section, Bloor does not claim that the electron 'makes no difference' to what Millikan and Ehrenhaft believe

---

29   Bruno Latour (1999a), 'For David Bloor ... and Beyond: A Reply to David Bloor's "Anti-Latour,"' *Studies in History and Philosophy of Science* 30(1), 113–29 (p. 117).

30   Latour (1999a), 'For David Bloor ... and Beyond,' p. 119.

about it. His claim is, rather, that its causal effects cannot explain the difference between the scientists' respective beliefs. As Bloor observes, Latour has 'confused different "differences."'[31]

In any case, on the basis of his interpretation of this passage, Latour concludes that the underdetermination thesis is an 'absurd position' not worth defending even when it is being attacked by 'even more stupid enemies.'[32] Latour then proceeds to rephrase Bloor's term 'electron "itself"' as 'electron in itself,' allowing him to more easily identify it with the Kantian thing-in-itself.[33] On this basis, he argues that for the SSK practitioner, just as for Kant, the thing-in-itself plays no other role than to allow one to distinguish between competing philosophical schools. More specifically, he claims that the thing-in-itself serves only to protect SSK against charges of idealism. This, in Latour's view, is the only real difference it makes.

Yet, as Langton's reading of Kant suggests, the notion of the thing-in-itself was motivated not by Kant's desire to deflect charges of idealism, but rather by his recognition that our ability to acquire knowledge of nature is inescapably finite. Latour thus seems to have misunderstood the motivation behind Kant's position. Indeed, Latour characterises the thing-in-itself, not as the mark of an independently existing nature, not as a sign of human finitude and a reason for humility, but rather as a symptom of an unseemly absolutism and hubris: 'It is through nature that the whole history of absolutism has been developed.'[34] He thus proposes that the concept of nature be topicalised and deconstructed. Insofar as the term 'nature' here stands for the Kantian transcendental

---

31   David Bloor (1999b), 'Reply to Bruno Latour,' *Studies in History and Philosophy of Science* 30(1), 131–36, p. 134.

32   Latour (1999a), 'For David Bloor ... and Beyond,' p. 117.

33   Latour (1999a), 'For David Bloor ... and Beyond,' p. 118. Bloor's electron passage has also attracted strong criticism from the philosophers of science Tim Lewens and Nick Tosh, though for different reasons (Tim Lewens (2005), 'Realism and the Strong Program,' *British Journal for the Philosophy of Science* 56(3), 559–77; Nick Tosh (2007), 'Science, Truth and History, Part II. Metaphysical Bolt-holes for the Sociology of Scientific Knowledge?,' *Studies in History and Philosophy of Science* 38(1), 185–209). I elsewhere defend Bloor against these particular criticisms (Jeff Kochan (2010a), 'Contrastive Explanation and the "Strong Programme" in the Sociology of Scientific Knowledge,' *Social Studies of Science* 40(1), 127–44). Martin Kusch offers a critical, if not always accurate, overview of this dispute (Martin Kusch (2018), 'Scientific Realism and Social Epistemology,' in *Routledge Handbook of Scientific Realism*, ed. by Juha Saatsi (London: Routledge), pp. 261–275).

34   Latour (1999a), 'For David Bloor ... and Beyond,' p. 127.

object, a thing-in-itself possessing intrinsic but unknowable properties, this may well be good advice. However, this is not what Bloor means by the term 'nature' and, as we will see in the next section, Heidegger provided just the sort of topicalisation and deconstruction called for by Latour. I will argue that Heidegger's conclusions tend to support Bloor rather than Latour.

As we have seen, Bloor maintains that the thing-in-itself, which he takes to be an independently existing 'natural object,' does not on its own provide sufficient grounds for explaining the interpretive disagreements which may arise between scientists regarding the experimental data related to that object. Central to Bloor's position is his conviction that the resulting interpretations are underdetermined by that data. This conviction traces its conceptual roots, in significant part, back to Kant. However, in his response to Latour's rejection of the Kantian thing-in-itself, Bloor furthermore argues that the underdetermination thesis does not depend on, nor does it push us towards, viewing natural objects in the way Kant did.[35] Bloor appears to both move towards and pull away from Kant.

The apparent conflict in Bloor's attitude may be resolved once we recognise that he does not reject the basic common-sense impulse behind Kant's notion of the thing-in-itself — belief in an independently existing nature — but instead follows Durkheim in formally developing that impulse differently than did Kant. Where Kant viewed the thing-in-itself as a transcendental object standing opposite an individual subject, Bloor views it a natural object standing opposite a society of subjects. Bloor then argues that both nature and society make a necessary contribution to the development of scientific knowledge. Latour misreads Bloor as denying a significant causal role to nature, and, on that basis, he concludes that the underdetermination thesis is absurd. But, for Bloor, the underdetermination thesis is not absurd, because the causal efficacy of an independently existing nature is beyond doubt. Against Latour's misplaced criticism, Bloor reasserts that 'the richness of the natural world, and the complexity of the scientist's engagement with it, is central to the thesis of underdetermination when properly understood, and hence to the Strong Program' in SSK.[36]

---

35 Bloor (1999b), 'Reply to Bruno Latour,' p. 134.
36 Bloor (1999b), 'Reply to Bruno Latour,' p. 134.

This is where the dust settles in the exchange between Bloor and Latour. I have argued that SSK's endorsement of a modified version of Kant's notion of the thing-in-itself fits hand in glove with its realist commitment to the independent existence of nature. Latour's attack on this modified notion of the thing-in-itself may thus be viewed as a simultaneous attack on SSK's realism. In Chapter Two, I promoted a minimal realist reading of Heidegger's work, and suggested that this minimal realism bears much in common with SSK's residual realism. In fact, it turns out that the realism on both sides is based, in significant part, on a critical appropriation of Kant's notion of the thing-in-itself. Indeed, Heidegger had much to say about the Kantian thing-in-itself. I want to now show that attention to Heidegger on this point will help us to more fully illuminate some of the key points in the dispute between Bloor and Latour, as well as to more fully understand the ways in which Bloor's notion of the thing-in-itself departs from that of Kant.

6. Heidegger and the Thing-in-Itself

In *Being and Time* and other writings from the late 1920s, Heidegger offers numerous critical comments on Kant's thing-in-itself, in particular, and his subject-object distinction, more generally. Heidegger is not so much concerned with the distinction between subject and object as he is with the presuppositions Kant relies on in schematising that distinction as a formal structure. In Heidegger's view, Kant grounds the distinction uncritically in an ontology which construes both subject and object in terms of substance. Heidegger's criticism focuses especially on Kant's assumption that the subject is to be understood, first and foremost, as a thinking substance.[37] As we saw in Chapter One, Heidegger's existential analysis of the subject is meant to dig beneath the orthodox notion of the

---

37   Heidegger argues that Kant uncritically adopted his ontology of the subject from Descartes. Hence, Kant 'failed to provide an ontology with Dasein as its theme or (to put this in Kantian language) to give a preliminary ontological analytic of the subjectivity of the subject' (Martin Heidegger (1962a [1927]), *Being and Time*, trans. by John Macquarrie and Edward Robinson (Oxford: Blackwell), p. 45 [24]; following scholarly convention, page numbers in square brackets refer to the original 1927 German edition of *Being and Time*). In the same period, Heidegger also comments that 'Kant is still working with a very crude psychology' (Martin Heidegger (1982a [1975]), *Basic Problems of Phenomenology*, trans. by Albert Hofstadter (Bloomington: Indiana University Press), p. 50).

subject as a thinking substance which seeks access to the world, bringing out instead the more fundamental existential state of the subject as already existing in the world alongside and along with other entities. Hence, Heidegger does not reject the subject-object distinction as such; he instead challenges its Kantian formulation which presupposes a theory-laden conceptualisation of the subject in terms of substance.

We can draw on Heidegger's critique of Kant in order to challenge Langton's Kantian claim that the thing-in-itself is necessarily an object with intrinsic properties. Note that this critical move does not force us to also reject epistemic humility, or the finitude which motivates that humility. Heidegger argues that the problem presented by Kant's thing-in-itself is not an epistemological but a metaphysical one. In Kant, the notion of the thing-in-itself correlates with the existence of an absolute knower, and so a rejection of Kant's notion entails the corresponding rejection of an absolute knower.[38] The move here is to repudiate the thing-in-itself just insofar as it correlates with an absolute thinking substance-subject, but not to repudiate the concept as such. The problem Heidegger locates in Kant is not based on the epistemic question of whether one could gain knowledge of the intrinsic properties of things-in-themselves, but on the metaphysical question of whether a knower with the requisite absolute epistemic powers could at all exist. For Kant, the answer to this metaphysical question was yes, but the absolute knower is God not the finite human. The finite human should thus express humility in the face of God's infinite epistemic power.

According to Heidegger, Kant makes a distinction between the thing-in-itself as an object grasped by absolute understanding, on the one hand, and the thing-in-itself as an object grasped by finite understanding, on the other. The former grasps the thing-in-itself absolutely, in terms of the thing's own intrinsic properties, while the latter only grasps it as an 'appearance' and remains ignorant of those intrinsic properties. Heidegger thus reads Kant in a similar way to Langton. Both scholars reject the 'two-world' thesis. For Heidegger, as for Langton, there are not two different kinds of thing in Kant, the phenomenal and noumenal, but rather one kind of thing understood in two different ways, either

---

38   Martin Heidegger (1984a [1978]), *The Metaphysical Foundations of Logic*, trans. by Michael Heim (Bloomington: Indiana University Press), p. 164.

finitely or infinitely.[39] In his commentary on Kant, Heidegger wrote that 'the entity "in appearance" is the same entity as the entity in itself, and this alone. As an entity, it alone can become an object, although only for a finite [act] of knowledge.' Later he emphasised, in a marginal note alongside this passage, that this is 'not the sameness of the What, but rather the That of the X!'[40] The X is the thing-in-itself, an entity about which we can say *that* it exists, but not *what* it is. Heidegger's conception of the thing-in-itself thus draws on his distinction between existence and essence, that-being and what-being, which was discussed at length in Chapter Two.

Heidegger furthermore argues that, because Kant took the fundamentality of substance ontology for granted, he uncritically conceptualised the subject-object distinction as one between a discrete substance-subject, on the one hand, and a discrete substance-object, on the other. As a consequence, Kant was forced to address the problem of how the subject-substance crosses over, or transcends, the barrier separating it from the world conceived as object-substance, a world containing independently existing things. This is the problem of the glass-bulb model, introduced in Chapter One. For Kant, only an absolute or infinite subject is capable of this kind of transcendence, that is, of grasping independently existing things as they are in themselves. Because human beings are finite creatures without absolute epistemic powers, this kind of transcendence is closed off to us. Kant thus concludes that we grasp things only as they appear to us, not as they are in themselves.

Heidegger challenges Kant's treatment of the subject-object distinction by re-interpreting Kant's notion of transcendence. As we saw in Chapter One, Heidegger argues that the subject is not a substance condemned to an inner realm from which it must win its freedom in order to achieve knowledge of an external world. On the contrary, the subject already exists alongside and along with other

39   Martin Heidegger, Martin (1997 [1929]), *Kant and the Problem of Metaphysics*, 5th edn, enlarged, trans. by Richard Taft (Bloomington: Indiana University Press), p. 22; Heidegger (1984a), *The Metaphysical Foundations of Logic*, p. 164.

40   Heidegger (1997), *Kant and the Problem of Metaphysics*, p. 22; translation modified, brackets original. For the marginal note, see Heidegger (1997), *Kant and the Problem of Metaphysics*, p. 22 n. 1.

entities in the world. Hence, argues Heidegger, the subject is not in search of transcendence; it is transcendence itself. The subject does not transcend its own finite limits in order to achieve contact with an independently existing thing-in-itself. Rather, as being-in-the-world, the subject already exists alongside independently existing things. Hence, its transcendence carries it, not towards those things, but away from them, towards a recognition of the projected world which provides the existential conditions structuring its possibilities for understanding and engaging with those independent things.

> [W]hat Dasein surpasses in its transcendence is not a gap or barrier 'between' itself and objects. But entities, among which Dasein also factically is, get surpassed by Dasein. Objects are surpassed in advance; more exactly, entities are surpassed and can subsequently become objects. [...] [A]s transcending, Dasein is beyond nature, although, as factical, it remains environed by nature. [...] That towards which the subject transcends is what we call *world*. [...] [W]e characterize the basic phenomenon of Dasein's transcendence with the expression *being-in-the-world*.[41]

Against Kant's notion of the subject as a substance, Heidegger offers an account of the subject in terms of being-in-the-world. In this way, he dissolves the epistemological problem which so exercised Kant, namely, the problem of how an internal thinking substance may cross over to, so as to then grasp, an external object. This external object is Kant's thing-in-itself, which, according to Langton, Kant construed as a substance possessed of intrinsic properties.

In abandoning the Kantian construal of the thing-in-itself, Heidegger also abandons the corresponding notion of absolute understanding. He describes his rejection of this latter notion as 'ontic atheism,' that is, the repudiation of the idea that God is a substance with an infinite power to absolutely grasp the objects of the world in their intrinsic features.[42] By abandoning the notion of an infinitely powerful subject-substance, against which Kant measured the finitude of the human being, Heidegger suggests a reconceptualisation of the meaning of human finitude. He still understands finitude in terms of receptivity, but receptivity is no

---

41   Heidegger (1984a), *The Metaphysical Foundations of Logic*, p. 166; translation modified.
42   Heidegger (1984a), *The Metaphysical Foundations of Logic*, p. 165 n. 9.

longer a sign of our lamentable inability to achieve absolute knowledge. It is no longer a deprived state in which we find ourselves permanently condemned to being affected by objects whose intrinsic properties we can never know. In contrast to Kant, Heidegger argues that the subject's finitude is not an 'ontic' condition, that is, not a consequence of its being a finite substance. It is, rather, an ontological condition, that is, a consequence of the finite range of possibilities available to the subject as a form of existence. This finite range of possibilities opens up a horizon in which the subject may encounter things as what they are. Their whatness, their essence, is circumscribed by the horizon of existential possibilities within which the subject is able to make sense of them. However, to make sense of a thing entails that that thing exists as something of which sense may be made. Furthermore, if we are to make sense of a thing, it must first affect us. We must, in other words, be receptive to its influence. Receptivity is thus our capacity to be affected by the things-in-themselves alongside which we exist in the world.

Unlike Kant, Heidegger argues that the things towards which we are receptive, the things which affect us, are not objects with intrinsic properties which lie forever beyond our ken. Indeed, for Heidegger objecthood is a projection rather than an affect. The objecthood of the object, its fundamental essence as object, is something we construct rather than receive. This is why, in the long quote two paragraphs above, Heidegger writes that only in transcending things do we come to perceive them as objects. A thing affects us, but our understanding of that thing as an object with determinate properties, our making sense of its whatness in terms of objecthood, is something we project onto the thing rather than something we receive from it. One may say that the whatness of a thing is underdetermined by the way in which it affects our senses. Heidegger's attention to the projective element in our understanding highlights an aspect of Kant's philosophy which carries us beyond the exposition of Langton with which this chapter began. Langton's commentary focuses almost exclusively on Kant's notion of receptivity, but we must now broaden the scope of our attention to include Kant's notion of spontaneity. Heidegger called this projective spontaneity 'construction,' writing that 'the explicit execution of the projecting, and even what is grasped in the ontological, must necessarily be construction.'[43] For Heidegger, then, objecthood lies on the constructive side of Kant's distinction between receptivity and constructivity. In critical response to Kant and neo-Kantianism, Heidegger writes that '[an] entity is without a subject, but objects exist only for a subject that does the objectifying.'[44] In other words, a thing becomes an object for us only when we constructively thematise it as such. Heidegger is making the general point that what affects our senses is a thing about which we may know that it is but not what it is. What affects us, in other words, is the thing described in Chapter Two, a thing which exists but lacks determinate properties: the thing-in-itself as Heidegger now construes it. On this reading of Heidegger, we cannot know what the thing-in-itself is, we cannot grasp its intrinsic properties, not because we are finite, but because the thing-in-itself has no such properties. The issue here is thus not, as Langton says it was for Kant, the finitude of our receptivity, for even an infinitely receptive knower will fail to be affected by what was never there in the first place. The issue here is instead the finitude of our constructivity. Unlike an infinitely constructive knower, we do not construct the thing-in-itself in an act of knowing; we instead only construct the categories through which we are able to know it. Whereas Kantian humility is prompted by the finitude of our receptivity, Heideggerian humility is prompted by the finitude of our constructivity.

To sum up this section, the categories which enable understanding are projected through our constructive power. The finite number of categories available to us determines the limited number of ways in which we may make sense of things-in-themselves. This finite constructive project provides the basis for metaphysics, that is, the study of the basic ontological categories by which we come to know what things are. Heidegger wrote that metaphysics is grounded in 'the humanness of reason, i.e., its finitude.' Metaphysical knowledge is, according to Heidegger, a direct consequence of our finitude, our inescapable mortality, rather than of our presumed ability to transcend that finitude, to reach, infinitely, for heaven. Because the finitude of our

---

43   Heidegger (1997), *Kant and the Problem of Metaphysics*, p. 163.
44   Heidegger (1982a), *Basic Problems of Phenomenology*, p. 157; translation modified.

constructive power makes impossible a transcendent grasp of the thing-in-itself — leaving us to be only affected by it in its brute, independent existence — our attention is instead pushed away from the thing-in-itself and towards the constructive categories we must employ in order to make sense of it as a thing present-at-hand within-the-world. For Heidegger, metaphysics is nothing other than the study of these categories and their relations to one another. Orthodox metaphysics, in contrast, treats these existential categories as ontic, that is, as extant mental things referring to the intrinsic properties of the things we seek to know, rather than as ontological, that is, as the existential structures of being-in-the-world which enable us to know those things.

For Kant the problem of finitude springs from our failure to grasp things in terms of their own autonomous categories. These categories, the intrinsic properties of things, lie beyond the reach of our finite powers of construction. Not so for Kant's absolute thinker. This divine subject possesses infinite powers of construction, and hence there is nothing which can exist beyond its reach, nothing which is autonomous and lonely. The absolute subject has no need for receptivity, because it absolutely affects things rather than being affected by them. It requires nothing to exist beyond itself, because when it creates, it does so like God, *ex nihilo*. On this reading, then, orthodox metaphysics lacks humility because it effaces the problem of finitude by seeking to grasp things absolutely. On such a view, the thing-in-itself is an affront to the infinite constructivity of an absolute subject. There is thus no room in orthodox metaphysics for the thing-in-itself. Heidegger argues that if this orthodoxy were to abandon its 'presumption,' by giving up its 'pride' and accepting the basic existential fact of its own finitude, indeed, if it were to recognise ontology as springing from the very essence of finitude, then metaphysics will have finally found its true meaning.[45] According to him, 'the struggle against the "thing in itself,"' the origins of which he locates in German Idealism, springs from a failure to understand the way in which the 'humanness of reason,' that is, reason's finitude, forms the essential core of Kant's problematic.[46]

---

45   Heidegger (1997), *Kant and the Problem of Metaphysics*, p. 88.
46   Heidegger (1997), *Kant and the Problem of Metaphysics*, pp. 171, 15.

7. Putting the Bloor-Latour Debate to Rest

We saw earlier that Bloor attempts to preserve the common-sense impulse behind Kant's version of the subject-object distinction, but he departs dramatically from Kant by naturalizing and socialising that distinction. Bloor's reformulation of the distinction is compatible not only with the view that things exist independently of our knowledge of them, but also with the view that such things possess intrinsic properties existing independently of our knowledge of them. However, Bloor rejects the metaphysical claim that our descriptions of things, that is, our specification of their properties by applying concepts or categories to them, strictly correspond to the independent nature of those things. Indeed, Bloor has not endorsed the view that the thing-in-itself possesses intrinsic properties. His position is thus also compatible with the view that the thing-in-itself possesses no intrinsic properties at all. This view, like that of Heidegger, departs significantly from the Kantian position as described by Langton.

Bloor's departure from Kant is, however, left somewhat unclear by the fact that he continues to refer to things as 'objects' without spelling out in sufficient detail how his use of this term differs from Kant's original usage. Latour seizes on Bloor's terminology and submits it to strong criticism. He dismisses Bloor's use of the term 'object,' apparently without attempting to properly understand what Bloor means by that term. Indeed, he reads Bloor as having meant a Kantian substance possessed of intrinsic properties. Latour seems to reason that, if we are receptive to objects, then, since objects have intrinsic properties, we must also be receptive to those properties. Hence, making reference to the Millikan-Ehrenhaft controversy over the existence of the electron, Latour criticises Bloor for allegedly claiming that 'electrons "themselves" are not allowed to cause our interpretations of them.'[47] Now, as we saw earlier, Latour knows that Bloor grants electrons a necessary but insufficient causal role in the formation of our beliefs about them. Latour complains, however, that in this capacity 'they don't do very much.'[48] Latour wants an account where electrons do more. Indeed, he appears to want an account where

---

47   Latour (1999a), 'For David Bloor ... and Beyond,' p. 119.
48   Latour (1999a), 'For David Bloor ... and Beyond,' p. 117.

electrons possess intrinsic properties, and where those properties both necessarily and sufficiently determine our interpretations of them, an account, in short, where electrons *make us* know them as what they are. Only such an account could effectively short-circuit the underdetermination thesis central to the methodology of SSK.

By suggesting that electrons not only necessarily but also sufficiently determine our interpretations of them, Latour rolls back the symmetry principle introduced by Bloor, reintroducing an old and familiar asymmetry into explanations of the truth and falsity of scientific beliefs and descriptions. Scientific descriptions are true when they correspond to the independently existing properties of the things they describe. False scientific descriptions, in contrast, are false because they do not correspond to the independent properties of the things they describe. As is common with Latour's rhetorical style, he obscures this regressive move behind the claim that he is extending the symmetry principle in a radically new way, by claiming to introduce 'one more turn after the social turn.'[49] But it seems that Latour has made not so much a critical advance on Bloor's symmetry principle as he has an obfuscating retreat into a more orthodox position, albeit one wrapped up in unorthodox terminology. It is for this reason that Bloor, in step with Harry Collins and Steven Yearley, has argued that 'something remarkably like direct or naive realism turns up in Latour's methodology.'[50] Yet, as I will argue shortly, this cannot be the full story.

In the meantime, it must be acknowledged that Latour's criticism is motivated by a genuine, if misplaced, worry. In Latour's view, Bloor appears to place the object on the side of receptivity and its intrinsic properties on the side of constructivity, and, on this basis, he rightly wonders how an object could be separated from its intrinsic properties in this way. Indeed, such a position may well not even be coherent. But this is not Bloor's position, because he does not require what he

---

49   Elsewhere, I characterise Latour's rhetorical strategy as one of dissimulation (Jeff Kochan (2010b), 'Latour's Heidegger,' *Social Studies of Science* 40(4), 579–98). In still another place, I call (tongue in cheek) for yet one more turn after the Latourian turn, a turn which delivers us to the position being outlined here (Jeff Kochan (2015b), 'Putting a Spin on Circulating Reference, or How to Rediscover the Scientific Subject,' *Studies in History and Philosophy of Science* 49, 103–07).

50   Bloor (1999a), 'Anti-Latour,' p. 94. Cf. similar criticisms of Latour in Harry M. Collins and Steven Yearley (1992), 'Journey into Space,' in *Science as Practice and Culture*, ed. by Andrew Pickering (Chicago: University of Chicago Press), pp. 369–89.

calls 'objects' to have intrinsic properties. His is, admittedly, a rather unconventional use of the term 'object,' and so it is perhaps not surprising that Latour failed to properly understand it. According to Bloor's usage, 'object' denotes an indeterminate material thing, one which exists independently of our beliefs about it, or involvements with it. It is the thing-in-itself, and our knowledge of its existence is a consequence of our receptive rather than our constructive relation to it. Hence, Bloor's position does not require that objects possess intrinsic properties, and it is compatible with the claim that they do not. Indeed, as Latour's criticism nicely brings out, Bloor's position is best understood as requiring that the thing-in-itself does not possess any intrinsic properties at all. On this interpretation, Bloor is closer to Heidegger than to Kant. When Heidegger argues that we project objecthood in our understanding of things as objects, he means that our relation to objects is a constructive one. However, in contrast to Bloor, by 'object' Heidegger means a substance with intrinsic properties. This is the same meaning employed by Latour, and attributed to Kant by both Langton and Heidegger.

Heidegger's criticism of Kant may help to throw further light on Latour's own position. Indeed, despite being charged with naive realism, there is evidence suggesting that for Latour, too, our relation to objects is a constructive, or projective, one. This evidence is, however, obscured by the fact that Latour also espouses the view that our relation to objects is receptive rather than constructive. The root of the problem here may lie in Latour's failure to properly distinguish between the that-being and the what-being of a thing, between a thing's existence and its essence. An elision of these two aspects of the being of a thing may explain his disinclination, noted earlier, to distinguish between our constructive knowledge of nature and the nature we constructively know, or, put otherwise, between our interpretations of nature and the nature we interpret. This puts Latour into a similar camp to Joseph Rouse, whose practical hermeneutics was discussed in Chapter Two. If this diagnosis of the problem is correct, then it may help to resolve an apparent contradiction in Latour's work. He has, for example, asserted that things, including those he calls 'nonhumans,' determine our interpretations of them, that our epistemic relation to them is a receptive one. Latour thus laments Bloor's failure to allow the electron-nonhuman to play a sufficient role in determining the difference between Millikan's and Ehrenhaft's respective interpretations of their data. This looks like a strong stance in favour of a robust realism. And yet, Latour also argues that his 'new active nonhumans are utterly different from the boring inactive things-in-themselves of the realist's plot.'[51] So he appears to also reject realism. We need to pull the various tangled threads apart in order to understand what is going on here.

On the basis of a number of Latour's statements, it would be natural to interpret him as affirming the view that the things we call electrons causally determine the categories by which we now know them. This suggests that the electron is an independently existing substance with determinate properties, that we are receptive to those properties, and that those properties cause our knowledge of them. We encountered this interpretation in Chapter One, labelling it the 'natural attitude' which both SSK and existential phenomenology treat as a topic for investigation. Latour appears to adopt the natural attitude as a resource when he asserts that electrons cause our interpretations of them. This was the basis for his rejection of the underdetermination thesis. Yet consider, more fully now, what Latour writes in his characterisation of Bloor's position: 'electrons "themselves" are not allowed to cause our interpretations of them, no matter how much scientists engage in making them have a bearing, a causality, on what they (the scientists) say about them (the electrons).'[52] This cannot be a straightforward description of scientists' receptivity towards electrons. Although he argues, on the one hand, that electrons cause our interpretations of them, he also argues, on the other, that scientists *make them* exercise that causation. Electrons make us know them as what they are, because we make them make us know them thus. It looks, then, like Latour's 'new active nonhumans' owe much of their activity to humans. Electrons do not, after all, sufficiently determine our interpretations of them. Our epistemic relation to them is not sufficiently determined by our receptivity towards them, but only by a combination of both receptivity and constructivity. Because Latour's account does not eliminate constructivity, it does not threaten the underdetermination thesis.

---

51 Latour (1992), 'One More Turn after the Social Turn…,' p. 284.
52 Latour (1992), 'One More Turn after the Social Turn…,' p. 119.

On Latour's account, we receive what we construct. Our interpretations are based on our reception of physical effects whose causal conditions we have played a necessary role in constructing. There appears to be no room here for an independently existing nature. In abandoning the 'boring inactive things-in-themselves of the realist's plot,' Latour appears to have given up on realism entirely. As I argued in Chapter Two, this is the price paid for failing to recognise the distinction between the existence and essence of a thing. Like Rouse, Latour elides the construction of the essence of a thing with the construction of its existence. One may thus describe Latour's position as a kind of 'pragmatic idealism.' Here the governing idea is that no thing can exist independently of our practical activities, both linguistic and otherwise. The independently existing thing-in-itself disappears in an endless cycle of interpretation, or what Latour has elsewhere called 'circulating reference.'[53] Latour's rejection of independent existence thus seems to undermine Bloor, Collins, and Yearley's suggestion that Latour is a naive realist. It would be more accurate to say that naive realism is expressed in Latour's rhetoric, but that his methodology pushes him more towards idealism. The crucial point, for the present argument, is that Latour's abandonment of realism pulls the rug out from under his argument against underdetermination. For once one allows that no thing can exist independently of our relations to it, one can no longer intelligibly assert that an independently existing thing can sufficiently determine our interpretations of it. Notwithstanding agile rhetorical performances to the contrary, one cannot have one's cake and eat it too.[54]

---

53   Bruno Latour (1999b), *Pandora's Hope: Essays on the Reality of Science Studies* (Cambridge, MA: Harvard University Press), p. 24. For a critique of Latour on 'circulating reference,' which also addresses his rejection of Kantian epistemology, see Kochan (2015b), 'Putting a Spin on Circulating Reference.'

54   One can, however, have one's cake at one moment, and then eat it at another. In 1987, Latour described science as 'two-faced': on the one side, 'science in the making,' on the other, 'made science.' In a controversy, scientists speak a constructivist language, but, once the controversy has been settled, they speak a realist language. There is no contradiction, because the respective contexts of the languages are different (Bruno Latour (1987), *Science in Action: How to Follow Scientists and Engineers through Society* (Cambridge, MA: Harvard University Press), p. 4). Fifteen years later, Latour claimed to be justified in speaking both languages, but he also seemed to elide the contextual difference between them (Bruno Latour (2002), 'The Science Wars: A Dialogue,' *Common Knowledge* 8(1), 71–79 (p. 77)). The Millikan-Ehrenhaft controversy is a case of science in the making, and so one would expect Latour to use a constructivist language when referring to it. Indeed, although he may have

One final point deserves mention before we finally lay the Bloor-Latour debate to rest. The fact that Latour treats electrons as thoroughly constructed things would seem to support my earlier observation that his position offends against the Kantian doctrine of humility. For Latour, there exists nothing about which we cannot have knowledge, because we know only what we make, and all existent things depend on our constructive power. Here, Latour may respond that we only partially construct things, and so we know them only partially. But this will not deflect the criticism. The core realist doctrine is that of independent existence. Even if the thing-in-itself is only partially the result of our constructive power, then it does not exist independently of that power, and hence it cannot provide the grounds for a genuinely realist position. There are no 'lonely' things in Latour's ontology, nothing to mark the finite limits of our constructive power. Latour's argument against social constructivism nicely demonstrates how, when minimal realism is rejected, then so too is epistemic humility. I have recommended minimal realism as a suitable position for science studies. It follows from this that a minimally realist science studies should also adopt an attitude of epistemic humility.

8. The Humility of Science Studies

If finitude is best met with an attitude of humility, then social constructivism should adopt an attitude of humility. Indeed, insofar as minimal realism presupposes the finitude of our indigenous constructive powers, this realism suggests a methodological commitment to humility. Resisting the temptation to believe that we can leap beyond the natural limits of our understanding is no small matter. As we saw in the case of Latour, metaphysically fuelled ambition may override humility and derail realism. But, as we also saw in Chapter One, even the more restrained SSK practitioners sometimes overstep the boundaries of their methodological commitments and thus threaten their realist credentials. David Bloor, for example, observes that the

---

found it rhetorically expedient to also use realist language in debating this case with Bloor, the inertia of his own established methodology, and the logical weight of his own earlier distinction between the two contexts, ultimately returns him to a constructivist language.

ways in which we conceptualise our experience will always involve a simplification of that experience, and infers from this that nature is enormously complex.[55] This inference seems to presuppose that the enormous diversity of possible ways in which human beings come to understand their experience of nature must somehow correspond to the enormous complexity of nature itself. But this attribution of a specific, intrinsic property to nature — complexity — seems to contradict Bloor's claim that our categories of understanding do not map onto nature itself in this way. The trouble here is that any claim to know nature in itself seems to violate epistemic finitude. We are finite knowers because we must be affected by nature in order to gain knowledge of it. But the concepts we apply in making sense of our experience are constructions projected onto nature rather than affections received from it. This goes too for the concept of complexity. It seems that the only attribution we may make with respect to nature itself is a purely privative one. Nature provides no ready-made categories by which we could know it, because it has no determinate properties of its own. Hence, it would seem more accurate to describe nature as incomprehensible rather than as enormously complex. The tremendously diverse ways in which we come to understand nature is not indicative of the inherent complexity of nature itself, but rather of the immense richness of our nevertheless finite constructive power. This point would appear to agree with Bloor's comment that '[t]here is much that has been achieved with our finite and contingent resources.'[56]

Heidegger argues that autonomous things, things left to themselves, 'lonely' things, as Langton puts it, are 'essentially devoid of any meaning at all.'[57] From a phenomenological perspective, in an unconstructed experience of things we encounter those things as incomprehensible. We fail to make sense of them within the constructive field of finite interpretative possibilities available to us. What is more, Heidegger also suggests that things may directly assault and disrupt this constructive field of possibilities. In such cases, things are not just without meaning, but they also act against meaning, that is to say, against our ability to

---

55 Bloor (1999a), 'Anti-Latour,' p. 90.
56 David Bloor (2007), 'Epistemic Grace: Antirelativism as Theology in Disguise,' *Common Knowledge* 13(2–3), 250–80 (p. 250).
57 Heidegger (1962a), *Being and Time*, p. 193 [152].

comprehend them. He writes, for example, that 'natural events [...] can break in upon us and destroy us.'[58] This observation suggests that the things of nature may startle or shock us in a way which disrupts, or perhaps even destroys, our constructive power of understanding. In such moments, we are reduced to pure receptivity. Nature affects us we know not how.

Heidegger uses the German word *Befindlichkeit* to name our receptivity, our ability to be affected by an independently existing nature. Hence, Kant's distinction between receptivity and constructivity may be viewed as resurfacing in Heidegger as a distinction between affectivity and constructivity. *Befindlichkeit* denotes the situation or state in which one finds oneself, as in 'I found myself increasingly worried about the future' or 'I found myself suddenly cheered by the passing festivities.' In Heidegger scholarship, the standard translation for *Befindlichkeit* is 'state of mind,' but Hubert Dreyfus has also translated it as 'affectedness.' I prefer to translate it as 'affectivity.'[59] *Befindlichkeit* derives from the reflexive verb *sich befinden*, which means 'to be here' or 'to be located here.' It thus has a tight connection with Heidegger's word for the subject, Dasein, or 'being here.' Recall from Chapter One that, according to Heidegger, the subject always finds itself *in* the world; one of its fundamental existential features is being-in-the-world. As we also noted, in Chapter Two, Heidegger gives an equally fundamental role to being-with-others in his account of the subject. To already be in the world means to also already be together with other persons in that world. Being-in-the-world is, in other words, a fundamentally social phenomenon. In constructively understanding the entities — persons and things — with whom and alongside which it exists, the subject also already finds itself receptively oriented towards those entities. Hence, the affectivity of the subject is, for Heidegger, another fundamental aspect of its existence. Indeed, altogether Heidegger notes at least four basic existential elements of subjectivity: being-in-the-world; being-with-others; affectivity; and constructivity (or projective understanding).

---

58   Heidegger (1962a), *Being and Time*, p. 193 [152].
59   Hubert L. Dreyfus (1991b), *Being-in-the-World: A Commentary on Heidegger's* Being and Time, *Division I* (Cambridge, MA: The MIT Press), chpt. 10. 'Affectivity' better captures the connotation of activity in *Befindlichkeit*.

Let us now consider the way in which worldly things may disrupt the subject's constructive attempts to make sense of them. Following Heidegger on this point will help us to also better understand how his views relate to the notion of epistemic humility. One particular shape taken by affectivity, one specific state of mind, to which Heidegger gives considerable attention, is anxiety. Heidegger writes that '[t]hat in the face of which one is anxious is completely indefinite.'[60] It is not, however, the indefiniteness of things themselves which causes our anxiety, but the indefiniteness of our existence as being-in-the-world. Heidegger suggests that it is in the face of our own indefinite existence that we feel anxious. Normally, we make sense of our own existence through our dealings with the things and persons with which and with whom we share our world. According to Heidegger, when those dealings break down, things 'slip away': 'We can get no hold on things. In the slipping away of beings only this "no hold on things" comes over us and remains.'[61] The idea seems to be that, in such situations, which Heidegger notes happen 'rarely enough and only for a moment,' our constructive power fails to get a hold on things and determine their meaning, which is to say, their essence or whatness.[62] As a consequence, we lose our ability to give meaning to the world and to our place in it. In situations like these, our constructive power is deflected back from things, and our relation to those things thus becomes entirely determined by our receptivity towards them.[63] Heidegger observes that, in this forcing back of our understanding, things suddenly reveal themselves as strange and 'radically other.'[64]

This returns us to the claim made by Joseph Fell, discussed in Chapter Two, that Heidegger uses the term 'present-at-hand' in a handful of distinct ways. In particular, we saw that Heidegger uses 'present-at-hand' not just to denote a thing which has been thematised as an object, but also a thing which has suddenly become 'unhandy' through a local breakdown in a global context of practical involvements which have

---

60   Heidegger (1962a), *Being and Time*, p. 231 [186].

61   Martin Heidegger (1993a [1978]), 'What Is Metaphysics?,' trans. by David Farrell Krell, in *Basic Writings*, revised and expanded edn, by Martin Heidegger, ed. by David Farrell Krell (New York: HarperCollins), pp. 93–110 (p. 100).

62   Heidegger (1993a), 'What Is Metaphysics?,' p. 100.

63   Cf. Heidegger (1962a), *Being and Time*, p. 232 [187].

64   Heidegger (1993a), 'What Is Metaphysics?,' p. 103.

otherwise remained undisturbed. As Fell writes, 'this is the experience of a presentness at hand that was *already there* but which in my practical preoccupation I was simply not attending to.'[65] Yet Fell also argues that the term 'present-at-hand' may refer to things in cases where there is a *global* breakdown of the significance relations which give meaning to things in the world.[66] In such cases, the subject finds itself in a state of anxiety before the unintelligibility of that world, an anxiety which reveals the brute contingency of the categories it normally takes for granted when making sense of both things and other people. These things now lose their taken-for-granted meaning, becoming strange. I would like to suggest that this strange and alien thing is the thing-in-itself as construed by Heidegger. As such, it marks the epistemic limit of our constructive power, and hence of our ability to make sense of, much less to know, nature. The thing-in-itself thus marks the boundaries of our finitude, and hence moves us towards humility. A breakdown in intelligibility, especially at the global level, thus reveals the sheer contingency of our categories of understanding. They no longer fix themselves onto things, but are instead driven back from them, an event which disrupts our ability to make sense of nature, which unsettles our taken-for-granted assumptions about what nature in itself is, and which may be experienced as a state of anxiety. In the last but one section, we saw that Heidegger describes transcendence as the surpassing of an independently existing nature. We can now add that, for Heidegger, our recognition of this transcendence, for example, our recognition that the objecthood of a thing is not intrinsic to it but something we project onto it, may be accompanied by a feeling of anxiety. The anxiety which may follow on our realisation of the contingency of the categories by which we make sense of the world is an affective recognition of our epistemic finitude, and hence a reason for humility.

The main point to draw out of this is that the notion of the thing-in-itself is not, for Heidegger, a merely theoretical one. It is also the notion of a thing alongside which we live in the world, a thing which we may, if only rarely and momentarily, experience in its bare existence when

---

65   Joseph P. Fell (1989), 'The Familiar and the Strange: On the Limits of Praxis in the Early Heidegger,' in *Heidegger and Praxis*, ed. by Thomas J. Nenon (The Southern Journal of Philosophy 28, Spindel Conference Supplement), pp. 23–41 (p. 31).

66   Fell (1989), 'The Familiar and the Strange,' p. 30.

our customary patterns of sense-making are momentarily disturbed, or, more dramatically, when we find ourselves in a state of global existential disturbance. If Heidegger's phenomenological analysis of anxiety is correct, then it would seem to lend some empirical weight to the notion of the thing-in-itself as lacking intrinsic properties of its own, and hence also to the more general notion of a nature which exists independently of the categories we normally use to understand it.

In fact, there is clinical evidence suggesting that Heidegger's analysis of anxiety offers a credible description of a genuine human experience which may, in some rare cases, become pathological. Russell Nieli, for example, argues that Heidegger is describing a radical kind of alienation experience, falling under the psychiatric labels of 'derealization' and 'depersonalization,' which attracted the attention of psychiatrists and psychologist around the turn of the last century.[67] As illustrations of this kind of experience, he cites representative patients' statements from William James's 1890 book, *The Principles of Psychology* — 'I looked about me with terror and astonishment: *the world was escaping from me*' — and Karl Jaspers's 1913 study, *General Psychopathology* — 'All objects appear so new and startling, I say their names over to myself and touch them several times to convince myself they are real.'[68]

For Heidegger, such unsettling experiences of the world or of objects are experiences of our own existence, because the meaning of things is a product of our constructive power, and the world, as we saw in Chapter One, is internally related to our existence as being-in-the-world. Hence, Heidegger wrote that, with the alienation experienced in anxiety, '[b]eing-in enters into the existential "mode" of the "*not-at-home* [*das Un-zuhause*],"' and, more to the point, that '[i]n anxiety one feels "*uncanny*" [*unheimlich*].'[69] This recalls Sigmund Freud's 1919 essay, 'The "Uncanny,"' where he wrote that '[t]he German word "*unheimlich*" is obviously the opposite of "*heimlich*" ["homely"], "*heimisch*" ["native"] — the opposite of what is familiar [...]. The better orientated in his environment a person is, the less readily will he get the impression of something uncanny in regard to the objects and events

67   Russell Nieli (1987), *Wittgenstein: From Mysticism to Ordinary Language* (Albany: SUNY Press), p. 17.
68   Nieli (1987), *Wittgenstein*, p. 24.
69   Heidegger (1962a), *Being and Time*, p. 233 [188–89].

in it.'[70] Anxiety, then, is an affective state in which we experience the things around us as slipping free from the taken-for-granted categories by which we, in the normal course of life, constructively make sense of them. By marking the limit of our constructive power, anxiety brings us face-to-face with our own inherent finitude.

More recently, the American Psychiatric Association, while tacitly invoking an ontology rejected by Heidegger, has described derealisation as 'the sense that the external world is strange or unreal,' and they designate it as a common symptom of post-traumatic stress disorder (PTSD).[71] The Heidegger-influenced psychiatrist Patrick Bracken, in turn, argues that traumatic experiences may lead to 'ruptured meanings' which signal the disintegration of one's world: 'The experience of very frightening events can have the effect of shattering any sense of living in an orderly world that has inherent structures of meaning and order.'[72] Global manifestations of this phenomenon have been noted especially among combat veterans. Russian psychologist Madrudin Magomed-Eminov writes that veterans of the Soviet war in Afghanistan suffered a 'loss of meaning to life' precipitating a general 'existential crisis.'[73] Bracken, for his part, cites part of the following passage from US Vietnam War veteran Tim O'Brien's 1990 short story collection *The Things They Carried*.[74] O'Brien writes that 'war has the feel — the spiritual texture — of a great ghostly fog, thick and permanent.' He continues:

> There is no clarity. Everything swirls. The old rules are no longer binding; the old truths are no longer true. Right spills into wrong. Order blends into chaos, love into hate, ugliness into beauty, law into anarchy, civility into savagery. The vapor sucks you in. You can't tell where you are, or why you're there, and the only certainty is overwhelming ambiguity. [...] In war you lose your sense of the definite [...].[75]

---

70   Sigmund Freud (1985), 'The "Uncanny,"' in *Art and Literature*, vol. 14 of *The Penguin Freud Library*, ed. by Albert Dickson (London: Penguin Books), pp. 339–76 (p. 341).

71   DSM-IV-TR (2000), *Diagnostic and Statistical Manual of Mental Disorders*, 4th edn, text revision (Arlington: American Psychiatric Association), p. 530.

72   Patrick J. Bracken (2002), *Trauma: Culture, Meaning and Philosophy* (London: Whurr), pp. 147, 142.

73   Magomed-Eminov (1997), 'Post-Traumatic Stress Disorders as a Loss of the Meaning of Life,' in *States of Mind: American and Post-Soviet Perspectives on Contemporary Issues in Psychology*, ed. by Diane F. Halpern and Alexander E. Voiskounsky (Oxford: Oxford University Press), pp. 238–50 (p. 239).

74   Bracken (2002), *Trauma*, p. 142.

75   Tim O'Brien (1990), *The Things They Carried* (New York: Broadway Books), p. 88.

Similar experiences were common among soldiers during the First World War (1914–1918). According to the German Army Medical Service, 613,047 German soldiers were treated during the war for 'diseases of the nervous system.'[76] In the decade following the war, the continued suffering of traumatised veterans became an issue dominating national debate. The affected veterans were overwhelmingly from the lower economic strata, and they saw themselves increasingly medicalised and blamed for their condition.[77] Pressured from the right, the government's labour ministry gradually cut pensions and health care to psychologically disabled veterans, causing widespread resentment.[78] It would be extraordinary if Heidegger had been unaffected by these events, during which period he was developing his existential account of anxiety. I do not mean to suggest that he deliberately shaped his account in response to these events, but that these events may have provided the social conditions in which anxiety could emerge as a compelling resource in the development of his phenomenology of human existence and his views on the independent existence of nature. I know of no evidence that Heidegger, who was himself exempted from combat duty on medical grounds, sympathised with the plight of the traumatised veterans, and, as Michael Zimmerman has shown, there are reasons to think that he emphatically did not.[79] Nevertheless, in Heidegger's hands, anxiety was stripped of its psychiatric meaning and reconceptualised as

---

76 Doris Kaufman (1999), 'Science as Cultural Practice: Psychiatry in the First World War and Weimar Germany,' *Journal of Contemporary History* 34(1), 125–44 (p. 125).

77 George L. Mosse (2000), 'Shell-Shock as a Social Disease,' *Journal of Contemporary History* 35(1), 101–08 (pp. 103–04).

78 Jason Crouthamel (2002), 'War Neurosis versus Saving Psychosis: Working-Class Politics and Psychological Trauma in Weimar Germany,' *Journal of Contemporary History* 37(2), 163–82 (p. 165).

79 Michael Zimmerman discusses the influence exercised on Heidegger by the popular writings of the war enthusiast and decorated combatant Ernst Jünger (Michael E. Zimmerman (1990), *Heidegger's Confrontation with Modernity: Technology, Politics, Art* (Bloomington: University of Indiana Press), pp. 66–76). Even Jünger's celebrated steely nerves had their limits, however. Recounting one battlefield experience, he writes: 'after a moment's blank horror I took to my heels like the rest and ran aimlessly into the night.' Later, 'I threw myself on the ground and broke into convulsive sobs' (Ernst Jünger (1929), *The Storm of Steel*, trans. by Basil Creighton (London: Chatto & Windus), pp. 245, 246). Those whose front-line combat experience is merely vicarious have the privilege of celebrating the heroics and ignoring the horrors.

an existential relation to an independent nature which slips free from the categories by which we attempt to make sense of it.

The social psychologist, James Averill, describes emotions in general as social constructions, and he suggests that they are thus a legitimate topic for the sociology of knowledge. Specifically, Averill argues that most standard emotions are 'institutionalized patterns of response,' which presuppose 'highly structured cognitive systems.'[80] He contrasts these standardised emotional responses to the responses symptomatic of anxiety, of which a cardinal feature is 'cognitive disintegration.' From the viewpoint of SSK, however, Averill has the relation between cognitive order and institutional order backwards. As Bracken rightly argues, it would, in fact, be better to say that highly structured cognitive systems presuppose institutionalised patterns of emotional response. Hence, when there is a significant disturbance in these institutionalised patterns, the result may be the kind of cognitive disintegration marked by a feeling of anxiety. A broken mind does not result in a broken world. The causal relation runs the other way around.

Heidegger viewed anxiety as a characteristic emotional response to disruptions in the coherence relations which normally obtain between ourselves and the things and persons with which and with whom we inhabit and share a world. This disturbance correlates with a breakdown in the constructive power by which we make sense of that world. When that failure is catastrophic, this power loses its grip on things in general and is thrown back onto itself. The result is a global breakdown in meaning, an experience wherein we encounter things-in-themselves in what Heidegger called their 'empty mercilessness.'[81] I have argued that Heidegger's phenomenological description of such failures of meaning, his description of anxiety, offers a potential account of the way in which we may experience nature as existing independently of the categories by which we normally come to know and productively interact with it. On this account, anxiety emerges as a direct consequence

80 James R. Averill (1980a), 'Emotion and Anxiety: Sociocultural, Biological, and Psychological Determinates,' in *Explaining Emotions*, ed. by Amélie O. Rorty (Berkley: University of California Press), pp. 37–72 (p. 68); see also James R. Averill (1980b), 'A Constructivist View of Emotion,' in *Theories of Emotion*, ed. by Robert Plutchik and Henry Kellerman (New York: Academic Press), pp. 305–39.

81 Heidegger (1962a), *Being and Time*, p. 393 [343].

of our epistemic finitude; it is the unsettling realisation that the basic categories structuring our understanding of nature do not pick out anything intrinsic to nature itself. Anxiety is the state in which we find ourselves when confronted with the existential fact of our finitude. The appropriate response to this inescapable existential fact is to adopt an attitude of epistemic humility.

9. Conclusion

In this chapter, I have argued that the minimal realism proposed in Chapter Two presupposes the finitude of human reason. That an independent nature always exists beyond the reach of our constructive power, that nature itself must always slip free from all attempts to determine its intrinsic properties, is a basic presupposition motivating the core realist doctrine that things exist independently of any practical and theoretical interactions we may have with them. I have furthermore suggested that recognition of our own inherent limitations vis-à-vis knowledge of nature is best met with an attitude of epistemic humility.

This argument was played out in the context of the well-known debate between David Bloor and Bruno Latour. I have presented the disagreement between Bloor and Latour as a debate over the appropriate attitude science studies should take towards the Kantian thing-in-itself. The thing-in-itself stands for the independent existence of nature. Latour dismisses the thing-in-itself as irrelevant to explanations of scientific knowledge. Bloor, by contrast, consequentially modifies Kant's original concept, replacing Kant's transcendental and individualistic formulation with a naturalistic and sociological one. As a consequence, Latour's attempt to dismiss Bloor's social constructivism as being in hock to the Kantian notion of the thing-in-itself, with all of its incumbent difficulties, largely fails. Bloor's treatment of the thing-in-itself is naturalistic and causal, and, as such, it is compatible with minimal realism. I have furthermore argued that Latour, for all his rhetorical affirmations of realism, is most coherently read as methodologically committed to a position of pragmatic idealism: things are constructed and only exist within fields of practice. On Latour's account, minimal realism evaporates along with the independently existing thing-in-itself entailed by such realism.

According to Rae Langton, Kant introduced the notion of the thing-in-itself in recognition of the finitude of human knowledge. Because our knowledge is finite, nature will always exist independently of our attempts to know it. Langton suggests that this finitude provides good grounds for adopting an attitude of epistemic humility. I have developed an account of epistemic humility through a discussion of Heidegger's own appropriation of Kant's notion of the thing-in-itself. Heidegger reconstrues the thing-in-itself in terms of an independently existing and indeterminate nature which may, on occasion, deflect our attempts to determine what it is according to our own indigenous constructive powers. Heidegger's treatment of Kant, I have suggested, is compatible with that of Bloor. This treatment also reveals the way in which Latour's rejection of the thing-in-itself fits together with his enthusiasm for an unrestrained constructivism which oversteps finitude, and so undercuts humility, in its denial of an independently existing nature. This failure of humility is, perhaps, most strongly exemplified in Latour's conviction that he has successfully abandoned the subject-object distinction and 'headed off in a different direction.'[82] But this distinction is not like a suitcase to be dropped off at the hotel before one dashes out to see the sights of a new and exciting city. Neither is it just a few words to be summarily excised from language. It is a structure in our thinking which has developed over centuries and with which we must live as a part of our cultural inheritance. This is our condition as finite, historical, and social beings. The only way to gain a free relation to the subject-object distinction, and the broader existential and conceptual structures which sustain it, is to trace the historical threads which weave it into the taken-for-granted patterns of our thinking. As we will see in Chapter Four, this was a task towards which Heidegger put much effort.

---

82   Latour (1999b), *Pandora's Hope*, p. 295.

Chapter Four

Things, Thinking, and the Social Foundations of Logic

1. Introduction

In Chapter Three, I argued that the doctrine of the independent existence of things, as the basis for a minimal realism, is inextricably bound up with the fact that human knowledge is inherently limited. The idea that things possess an independent existence follows from the recognition that our epistemic capabilities are irremediably finite. Chapter Three was primarily concerned with the independent existence of things. This chapter will be oriented more towards epistemic finitude, or more specifically, the finitude of what I will call 'thinking.' Thinking was a fundamental concept for Heidegger, and he meant it in the broadest possible sense to include not just mental activity, conventionally construed, but all practical acts expressing the possession of knowledge. Thinking is thus present not only in deliberative, propositionally structured actions, but also in actions which are non-deliberative and non-propositional in nature. On this account, 'thinking' means 'cognitive activity,' that is, the activity of knowing. When we say that someone knows how to ride a bicycle in traffic, we need not claim that she possesses propositionally structured knowledge of bike riding, knowledge which she may deliberatively apply in the performance itself. Heidegger's concept of thinking thus also encompasses skill. In

                    http://dx.doi.org/10.11647/OBP.0129.04

his view, skilled performance entails a mode of thinking which need not be articulable in propositional form. His concept of thinking thus appears similar to philosopher of science Ian Hacking's concept of reasoning. Hacking also attempts to stretch the term 'reasoning' beyond its conventional usage, so as to also include the embodied aspects of practical action, and he explicitly acknowledges the tension which arises with such stretching: 'Even my word "reasoning" has too much to do with mind and mouth and keyboard; it does not, I regret, sufficiently invoke the manipulative hand and attentive eye.'[1] Hacking has more recently rebranded his concept of reasoning as one of 'thinking and doing' in an attempt to relieve some of this tension.[2] According to Heidegger, thinking is always dependent on doing of some kind, and doing, if it manifests the possession of knowledge, must also always involve thinking. Theory and practice thus go hand in hand, together with mind and body.

This view sits at the core of Heidegger's phenomenological project of explaining the logical structure of scientific thinking in terms of its existential foundations. As we saw in Chapter Two, Heidegger argued that science, as a coherent body of logically interconnected propositions, derives from a specific mode of existence, namely, the specific ways in which scientists involve themselves with things and one another, and the specific ways in which they come to understand their collective involvement with those things. In fact, Heidegger conceived of human existence as being fundamentally structured by the relationship between things and thinking, with scientific existence exemplifying a special mode of that relation. As we saw in Chapter One, this move provided him with the means by which to deflect scepticism about the existence of an external world. Such scepticism presupposes an image of the subject as contained within a glass bulb, and asserts that thinking will never penetrate the wall of that bulb and thus never achieve epistemic access to the things from which it is separated. Heidegger, argued, in response, that thinking is always already in relation to things, because being-in-the-world is a fundamental structure of human existence. As we will see

---

1   Ian Hacking (1992), '"Style" for Historians and Philosophers,' *Studies in History and Philosophy of Science* 23, 1–20 (p. 3).

2   Ian Hacking (2012), '"Language, Truth and Reason" 30 Years Later,' *Studies in History and Philosophy of Science* 43(4), 599–609 (p. 601).

in this chapter, Heidegger furthermore argued that this fundamental relation between things and thinking is always marked by a *directedness* of thinking towards things. This just means that thinking is a necessarily intentional phenomenon, that it is always a thinking *about*.

Heidegger's attempt to elucidate the existential genesis of science as a body of logically interrelated propositions was an attempt to delineate the way in which the relational phenomenon of intentionality came to be specified as a relation between thinking, construed as the propositionally structured act of a mental substance, and a thing, construed as a property-bearing substance. Heidegger thus described the existential genesis of science as a historical process by which intentionality became increasingly specified according to the model of the proposition.[3] According to Heidegger, this was to have a profound influence on the way both things and thinking came to be understood in the philosophical tradition. In particular, thinking itself came to be identified with logic, and all legitimate forms of thinking, including scientific thinking, were then viewed as ultimately grounded in logic. Heidegger noted that this conclusion leads to the circular argument that science, as a logically structured body of knowledge, is itself grounded in logic. Logic grounds logic. Heidegger's alternative argument that science, and hence also logic, is grounded in the informal, pre-propositional structures of existence was his attempt to soften this circularity.

In this chapter, we will first consider Heidegger's account of the historical process by which scientific thinking came to be viewed as ultimately governed by self-validating rules of logic, and then link this account to more recent work in the sociology of scientific knowledge (SSK). By historicising the prevailing logical picture of scientific thinking, Heidegger sought to loosen up intuitions about its apparent necessity, and thus to prepare readers for his own phenomenological alternative. It must be emphasised that, in doing so, Heidegger was not promoting an irrationalist or anti-logical theory of science. Indeed, Heidegger was

---

3   In an interpretation otherwise quite different from my own, Hans-Jörg Rheinberger also emphasises the deeply historical nature of Heidegger's account of science, with particular attention to Heidegger's preoccupation with the material aspect of modern science (Hans-Jörg Rheinberger (2010b), *On Historicizing Epistemology: An Essay* (Stanford: Stanford University Press)). This material aspect will take centre stage in Chapter Six.

instead motivated by a sense of distress at the failure of the orthodox account to provide a foundation for science as a cultural enterprise. In this, he was of one mind with many of his Central European contemporaries during the interwar period. Describing the situation in Weimar Germany, Paul Forman has written that such feelings of distress were 'widespread among the educated middle classes, but especially oppressive in academia.' These were unsettling feelings of 'moral and intellectual crisis, a crisis of culture, a crisis of science and scholarship.'[4] Forman argues that Weimar intellectuals felt compelled to address the perceived crisis in order to maintain their own credibility, and this often led them to 'repudiate the traditional methods and doctrines of [their] discipline.'[5]

A striking example of this circumstance was Heidegger's mentor Edmund Husserl. Husserl sought to address what he too called 'the crisis of science' through the methods of transcendental phenomenology.[6] For him, phenomenology provided the methodological means by which to finally establish philosophy as a 'rigorous science.'[7] This rigorously scientific philosophy was meant to ground all the other sciences, including the scientific philosophy of the mathematical logicians, whose own attempts to ground science in a self-sufficient logic Husserl dismissed as 'nothing but naïveté.'[8] Contrasting his own declaredly more radical phenomenological science to the scientific project of mathematical logicians, Husserl contended that '[o]nly when this radical, fundamental science exists can such a logic itself become a science.'[9] Science was thus not to be grounded in a self-sufficient logic, but rather in the pre-logical phenomena which Husserl sought to expose through the methods of his transcendental phenomenology.[10]

---

4   Paul Forman (1971), 'Weimar Culture, Causality, and Quantum Theory, 1918–1927: Adaptation by German Physicists and Mathematicians to a Hostile Intellectual Environment,' *Historical Studies in the Physical Sciences* 3, 1–115 (p. 26).

5   Forman (1971), 'Weimar Culture, Causality, and Quantum Theory,' p. 28.

6   Edmund Husserl (1970), *The Crisis of European Sciences and Transcendental Phenomenology,* trans. by David Carr (Evanston: Northwestern University Press), p. 3.

7   Edmund Husserl (1965), 'Philosophy as Rigorous Science,' trans. by Quentin Lauer, in *Phenomenology and the Crisis of Philosophy,* by Edmund Husserl (New York: HarperCollins), pp. 71–147.

8   Husserl (1970), *Crisis of European Sciences,* p. 141.

9   Husserl (1970), *Crisis of European Sciences,* p. 141.

10 For discussions of Husserl's philosophy of science see: Patrick Heelan (1987), 'Husserl's Later Philosophy of Natural Science,' *Philosophy of Science* 54 (3), 368–90;

It is clear that Heidegger's ambitions closely tracked those of his former mentor. Indeed, Heidegger too argued that phenomenological research represents 'nothing less than the more explicit and more radical understanding of the idea of scientific philosophy.'[11] Furthermore, Heidegger also described his phenomenological method as a 'transcendental science.'[12] Yet Heidegger's concept of the transcendental differed profoundly from that of Husserl. As discussed in Chapter Three, Heidegger rejected the Kantian notion of the transcendental subject, replacing it instead with the finitude of human existence as being-in-the-world. Because the subject is already in the world, it does not need to transcend its indigenous condition in order to make contact with that world. Indeed, on Heidegger's account, transcendence is not transcendence towards things in the world, but instead away from them and towards the existential possibilities which inevitably structure our everyday projective understanding of the things as we typically encounter them. For Heidegger, then, phenomenology as transcendental science meant a scientific investigation of the structures of possibility giving shape to actual acts of thinking. As we will see in this chapter, Heidegger located the conditions of possibility for thinking, including logical thinking, in the finite, historical existence of human beings. Husserl, in contrast, urged a conception of transcendental subjectivity which saw the human being escaping the finite conditions of worldly existence on the basis of an 'immortal' human spirit.[13] Not only did

---

David Hyder and Hans-Jörg Rheinberger, eds. (2010), *Science and the Life-World: Essays on Husserl's* Crisis of European Sciences (Stanford: University of Stanford Press); Jeff Kochan (2011b), 'Husserl and the Phenomenology of Science,' *Studies in History and Philosophy of Science* 42 (3), 467–71; Joseph Rouse (1987b), 'Husserlian Phenomenology and Scientific Realism,' *Philosophy of Science* 54 (2), 222–32; Robert Sokolowski (1979), 'Exact Science and the World in which We Live,' in *Lebenswelt und Wissenschaft in der Philosophie Edmund Husserls*, ed. by Elisabeth Ströker (Frankfurt: Vittorio Klostermann), pp. 92–106; and Elisabeth Ströker (1997), *The Husserlian Foundations of Science* (Dordecht: Kluwer Academic Publishers). For a brief, and only partial, introduction to phenomenological philosophy of science, spotlighting the works of Husserl, Heidegger, Patrick Heelan, and Joseph J. Kockelmans, see: Jeff Kochan and Hans Bernhard Schmid (2011), 'Philosophy of Science,' in *The Routledge Companion to Phenomenology*, ed. by Sebastian Luft and Søren Overgaard (London: Routledge), pp. 461–72.

11   Martin Heidegger (1982a [1975]), *Basic Problems of Phenomenology*, trans. by Albert Hofstadter (Bloomington: Indiana University Press), p. 3.

12   Heidegger (1982a), *Basic Problems of Phenomenology*, p. 17.

13   Husserl (1970), *Crisis of European Sciences*, p. 299.

Heidegger reject the idea of a pure and boundless reason implied in Husserl's appeal to immortality, he furthermore critiqued this notion as a historical possibility actualised in the early-modern period only because thinking had already begun to view itself in propositional terms. In Heidegger's view, overcoming both the philosophical doctrine of immortal, or infinite, reason, as well as the propositional model of thinking on which it is partly based, entails a deconstruction of the philosophical orthodoxy back to its origins in Plato. This chapter recounts some of the key moments in Heidegger's deconstruction of that orthodoxy, framing it as an attempt to ground logic, and thus science in general, in the pre-propositional and ineluctably finite structures of human existence.

Heidegger was careful not to commit the self-defeating error of claiming a transcendental (in the orthodox sense) viewpoint from which to declare the historical contingency of thinking as such. In fact, he openly admitted that the 'investigation which we are now conducting is determined by its historical situation [...] and by the preceding philosophical tradition.'[14] However, his reaction to this predicament was not a studied complacency regarding the historical origins of his own basic concepts. Rather, he sought to articulate a historical account of the contingency of those concepts by deconstructing them 'down to the sources from which they were drawn.'[15] By analysing the conventionalised concepts of modern philosophy as the historical actualisation of a tradition construed in terms of possibilities, Heidegger aimed not only to demonstrate the contingency of the basic concepts of modern formalised logic, but also the legitimacy of his own existential phenomenology both as an expression of possibilities latent in the philosophical tradition and as being better equipped than formal logic to provide a defensible foundation for the sciences.

The explicit reflexivity of Heidegger's method, his recognition that the legitimacy of his own concepts was also historically contingent, strongly resonates with SSK's reflexivity tenet. This tenet states that SSK's 'patterns of explanation would have to be applicable to sociology itself [...] otherwise sociology would be a standing refutation of

---

14 Heidegger (1982a), *Basic Problems of Phenomenology*, p. 22.
15 Heidegger (1982a), *Basic Problems of Phenomenology*, p. 23.

itself.'[16] However, as we already well know from earlier chapters, the similarity between the two methods does not end there. Indeed, like Heidegger, SSK practitioners have applied their method extensively in an investigation of the foundations of logic. In the latter part of this chapter, the methods of each will be compared. Both parties embrace a doctrine of finitude, and hence reject the contrary notions of an immortal spirit, an unbounded reason, an infinite faculty of thinking, and the like. Both also grant priority to informal over formalised modes of thinking. The benefits of this comparison run in both directions. On the one hand, Heidegger had little to offer by way of detailed empirical illustrations of the contingency and informal basis of logical thinking. SSK can thus help to fill out Heidegger's theoretical account with empirical studies. On the other hand, Heidegger's work can help to untangle some conceptual knots in the sociology of logic. In particular, Heidegger's phenomenological description of different modes of intentionality can put into SSK practitioners' hands a non-propositional account of intentionality which is compatible with their own naturalistic and causal account of knowledge. This promises to save SSK practitioners from the difficulties which threaten to follow from their outright rejection of intentionality as a legitimate explanatory resource. Before plunging into this comparative work, however, let us first take an extended tour through Heidegger's phenomenological history of logic.

2. Heidegger on the Unity of Things and Thinking

Heidegger argues that our concept of the thing as a property-bearing substance is necessarily related to our concept of thinking as possessing a propositional structure. The property-bearing substance, on the one side, and proposition-based thinking, on the other, are 'mirror images' of one another, and they share a 'deeper lying root.'[17]

As discussed in earlier chapters, for Heidegger the concepts of a property-bearing substance and a proposition-based thinking are not foundational concepts, but derive from the more fundamental existential

16   David Bloor (1991), *Knowledge and Social Imagery*, 2nd edn (Chicago: University of Chicago Press), p. 7.

17   Martin Heidegger (1967 [1962]), *What Is a Thing?*, trans. by William B. Barton, Jr., and Vera Deutsch (Chicago: Henry Regnery), p. 47.

structures of our subjectivity. In Chapter Two, we reviewed Heidegger's phenomenological analysis of propositional thinking as arising from an interruption or breakdown in the smooth, unreflective mode of thinking characteristic of our normal, everyday dealings in the world. As we saw, this change-over of thinking from unreflective immersion to propositional reflection was accompanied by a corresponding transition in our experience of things within the world from things ready-to-hand to things present-at-hand, that is, to property-bearing substances, or objects. In Chapter Three, we discussed Heidegger's interpretation of the way Kant articulated this dialectical relationship between things and thinking. In particular, Heidegger credits Kant with the fundamental insight that human thinking, as a finite faculty, entails the independent existence of the things to which that thinking is directed. Kant called this aspect of thinking, which follows from the fact of human finitude, 'receptivity.' Heidegger rephrased this as *Befindlichkeit*, which I translate as 'affectivity.' The receptivity of thinking means that thinking is always a response, in one way or another, to things. In Heidegger's terminology, thinking is a basic feature of Dasein's existence, and that existence necessarily takes place in a world of things. There can, in short, be no thinking, much less any knowledge, in the absence of experience. On the other hand, as was argued in Chapter Two, although we are, as thinking beings, necessarily related to things, things do not in turn depend on us for their own existence. Things outstrip our ability to understand them, and so demonstrate our finitude. This insight provides the basis for minimal realism.

Heidegger analyses the relation between things and thinking in terms of four components, with two on each side. The two components on the side of the thing are its existence and its essence, and the two components on the side of thinking are its receptivity and its constructivity. In the case of the first component of each, the relation runs from the thing, as existent, to thinking, as receptive. In the case of the second component of each, the relation runs in the opposite direction, from thinking as constructive, to the thing as possessed of determinate properties, as having an essence. As we have seen, Heidegger describes this second relation as one of 'projection.' The idea is that, while the thing itself can exist independently of thinking, its essence, articulated in terms of its properties, cannot. The thing has no properties in the absence of projective thinking.

We are now in a better position to understand what Heidegger means when he says that our concept of the thing as a property-bearing substance is necessarily related to our concept of thinking as possessing a propositional structure. His argument is that the constructive component of thinking must first be brought into a propositional form before we can begin to speak intelligibly about things as substances with properties. In Chapter Two, we examined the stages Heidegger identified in his phenomenological analysis of how thinking takes on propositional form in the act of thematising a ready-to-hand thing as a present-at-hand object. The two crucial points we may draw from this analysis, for present purposes, are that thinking is not fundamentally propositional in form, and that logic, as the science of thinking, provides us with an only derivative account of what thinking is. In other words, the formal, propositional structure of thinking is not a fundamental structure discovered through logical enquiry. It is rather a derivative structure which we construct in the course of thematising thinking as an object of investigation. One important implication of this is that truth — as correspondence between a proposition and the independently existing property of a substance — is a derivative form of truth. Truth, as correspondence, depends on a thematising project which simultaneously constructs thinking as a propositional act, on the one hand, and the thing, towards which that act is directed, as a property-bearing substance, on the other.

Heidegger argues that the orthodox attitude in philosophy, which takes formal logic as the foundation of thinking and substance ontology as revealing the fundamental structure of things, is not absolutely valid, but instead based on historically contingent presuppositions. We will give more detailed attention to Heidegger's historical argument in the following sections. Note for the time being, however, that Heidegger calls the basic constructive relation of thinking to things, the inherent directedness of the former to the latter, 'intentionality.'[18] Hence, the phenomenological study of constructivity may be viewed more generally as the study of intentionality. The basic phenomenological feature of intentionality is its directedness towards something. An intentional act is a directed act. Heidegger often describes intentionality as the way in which the subject 'comports' itself towards things. In his

---

18   Heidegger (1982a), *Basic Problems of Phenomenology*, p. 58.

view, the orthodox attitude in philosophy takes for granted, and relies on, a historically specific mode of intentionality. In Kant's philosophy, this historical mode was conceptualised in terms of constructivity. Heidegger also refers to it as 'productive comportment.'[19]

Much of Heidegger's philosophy may be viewed as an exploration of the limits and the latent possibilities in the subject's productive comportment towards things, or, put another way, an exploration of the limits and possibilities of a philosophical tradition which understands intentionality by analogy to production. He argues that productive comportment, as a fundamental but often unacknowledged concept in the philosophical tradition, is the source of the distinction between existence and essence.[20] The concept of production entails the prior existence of material: 'If we bring to mind productive comportment in the scope of its full structure we see that it always makes use of what we call *material*, for instance, material for building a house.'[21] Furthermore, given the intimate relation between the existence and essence of things, on the one hand, and the receptivity and constructivity of thinking, on the other, we can conclude that productive comportment also provides a conceptual root for the relatedness of receptivity and constructivity, and thus for the relation between things and thinking in general. This, then, provides the background for Heidegger's more narrow argument that the property-bearing substance and proposition-based thinking are mirror images, sharing with one another an underlying root. Intentionality provides that underlying root. It serves to unify things and thinking, and has been traditionally construed by philosophers on the model of production.

The key point here is that intentionality, as productive comportment, plays a unifying role in Heidegger's explanation of the relation between things and thinking. Another point is that the analogy to production specifies the meaning of intentionality as more than mere directedness. On this construal, intentionality is directedness guided by a pre-existing standard. As Heidegger writes, '[a]ll forming of shaped products is effected by using an image, in the sense of a model, as guide and

---

19   Heidegger (1982a), *Basic Problems of Phenomenology*, p. 105.
20   Heidegger (1982a), *Basic Problems of Phenomenology*, p. 105.
21   Heidegger (1982a), *Basic Problems of Phenomenology*, p. 115.

standard.'[22] Putting these two points together, Heidegger's overall claim is that the philosophical tradition interprets the phenomenon of intentionality as the experience of being guided by a pre-existing standard, or image, such that one's thinking will come into proper relationship with the things. As we will see in the next four sections, Heidegger traces the historical course of this model of intentionality from Plato's doctrine of the good, through Aristotle's categorial analysis of the proposition and Descartes's emphasis on the propositionally structured subject 'I,' to Kant's phenomenological investigation of the imagination. This history of the concept of intentionality will, in turn, prepare the way for a detailed consideration, in Chapters Five and Six, of the emergence of early-modern mathematical and experimental science. But for now, let us take a look at Heidegger's phenomenological history of thinking as logic.[23]

3. Heidegger's Phenomenological History of Logic: Plato

Heidegger addresses the question of the historical relation between things and thinking in the context of the development of logic as the scientific study of thinking. By taking this approach, he aims to challenge the orthodox view of logic as a free-floating and ultimate form of thinking, one which provides the grounds for all of the other sciences. For Heidegger, then, the attempt to unearth the foundations of

---

22 Heidegger (1982a), *Basic Problems of Phenomenology*, p. 106.
23 For work addressing Heidegger's views on logic in the context of late-nineteenth and early-twentieth century developments, see: Albert Borgmann (1978), 'Heidegger and Symbolic Logic,' in *Heidegger and Modern Philosophy*, ed. by Michael Murray (New Haven: Yale University Press), pp. 3–22; Steven Galt Crowell (1992), 'Lask, Heidegger, and the Homelessness of Logic,' *Journal of the British Society for Phenomenology* 23(3), 222–39; Steven Galt Crowell (1994), 'Making Logic Philosophical Again (1912–1916),' in *Reading Heidegger from the Start: Essays in his Early Thought*, ed. by Theodore Kisiel and John van Buren (Albany: SUNY Press), pp. 55–72; Thomas A. Fay (1977), *Heidegger: The Critique of Logic* (The Hague: Martinus Nijhoff); Stephan Käufer (2001), 'On Heidegger on Logic,' *Continental Philosophy Review* 34(4), 455–76; Stephan Käufer (2005), 'Logic,' in *A Companion to Heidegger*, ed. by Hubert L Dreyfus and Mark A. Wrathall (Oxford: Blackwell), pp. 141–55; Jitendranath Mohanty (1988), 'Heidegger on Logic,' *Journal of the History of Philosophy* 26(1), 107–35; Greg Shirley (2010), *Heidegger and Logic: The Place of* Lógos *in* Being and Time (London: Continuum).

logic is simultaneously an attempt to expose the fundamental historical presuppositions of science as such. He argues that the essence of thinking, of 'judgement,' has been determined by logic, and more specifically by the proposition, since ancient times.[24] By excavating logic down to its foundations, Heidegger seeks to challenge the perceived self-evidence of this determination, to expose to the light of critical reflection what ancient philosophers had themselves found continually disturbing and obscure.[25]

Heidegger's historical analysis is scattered across a number of works. Here, I will only gather together the highlights, which should suffice to capture the overall trajectory of his considerations. Heidegger organises his history of thinking into three chapters: first, the recognition of a mutual relation between the thing and the propositionally structured thought, the latter guiding the categorial determinations of the former; second, the mathematical interpretation of the proposition, which in turn provided the basic principles of pure thinking; and third, the emergence of a critique of pure thinking, which follows from things having been determined on the basis of a propositionally structured thinking.[26] Heidegger elaborates these three chapters of history through discussions of the philosophies of Aristotle, Descartes, and Kant, respectively. The prologue to this history, however, belongs to Plato.

Recall that Heidegger focuses his attention on the way in which intentionality unifies things and thinking. It does this, he says, in accordance with a pre-existing standard of some kind. Heidegger's historical analysis traces the ways in which this pre-existing, unifying standard, as an implicit and inherent feature of subjectivity, has been recognised and articulated over the course of the philosophical tradition, beginning with Plato. With Plato, argues Heidegger, this standard was conceptualised as an image or model, the 'look' a thing has in the imagination of its producer before it is produced. This look underpins the philosophical meaning of Plato's concept of the idea: 'It

---

24　Heidegger (1967), *What Is a Thing?*, p. 149.

25　Martin Heidegger (1962a [1927]), *Being and Time*, trans. by John Macquarrie and Edward Robinson (Oxford: Blackwell), p. 21 [2]. (Following scholarly convention, page numbers in square brackets refer to the original 1927 German edition of *Being and Time*.)

26　Heidegger (1967), *What Is a Thing?*, p. 108.

is this anticipated look of the thing, sighted beforehand, that the Greeks mean ontologically by *eidos*, idea.'[27]

This ancient analogy between looking and thinking carries with it a connotation of illumination, for looking entails the presence of light — above all, the sun. Plato thus drew a comparison between visible things and thinkable things, arguing that sunlight is to vision what the idea of the good is to scientific thinking.[28] According to Heidegger, Plato believed that the good provides an illumination by which to distinguish between 'a shadow' and 'the real.'[29] The idea of the good thus provides the guidance we need in order to bring our thinking into proper contact with the real. It is what prevents us from wandering aimlessly among the shadows, without hope of ever discovering truth. The key point is that achieving such knowledge is not simply a matter of observing an enormous number of things, as a naive empiricism might suggest. One must also be able to distinguish the epistemically good things from the epistemically bad ones, that is, the things which contribute to knowledge from those which do not.[30] The good provides the standard by which such distinctions are made. Plato's idea of the good thus represents the condition of possibility for scientific thinking. It is the *a priori* element in cognition which makes scientific knowledge, as such, possible.

On Heidegger's reading, Plato's account of the good, as a unifying standard combining things and thinking in the experience of knowing, is modelled on an account of production. Clearly, then, the idea of the good is not just one idea among many, but rather the first, or primary, idea. It serves to organise all the secondary ideas, the categories or concepts which give specific content to our understanding, into a unified whole. The idea of the good, writes Heidegger, lies beyond all other ideas, giving them the 'form of wholeness,' or 'communality.'[31] He suggests that, because it plays this fundamental organising role, of productively forming all other ideas into a unified whole, the idea of the good 'is nothing but the demiourgos, the producer pure and simple.'[32]

---

27   Heidegger (1982a), *Basic Problems of Phenomenology*, p. 106.
28   Heidegger (1982a), *Basic Problems of Phenomenology*, p. 283.
29   Heidegger (1982a), *Basic Problems of Phenomenology*, p. 285.
30   Heidegger (1982a), *Basic Problems of Phenomenology*, p. 285.
31   Martin Heidegger (1984a [1978]), *The Metaphysical Foundations of Logic*, trans. by Michael Heim (Bloomington: Indiana University Press), p. 184.
32   Heidegger (1982a), *Basic Problems of Phenomenology*, p. 286.

This suggests that Plato's account of knowledge should be read as a causal one. Just as the craftsperson forms matter and ideas into artefacts, so the idea of the good forms things and thinking into knowledge.

According to Heidegger, Plato's importance lies in the fact that his conceptualisation of thinking on the model of production has provided the basis for all subsequent philosophical enquiry. However, the philosophical nature of this analogy remains obscure in his work. Heidegger describes Plato's account as having a 'mythic' quality in which the philosophical point fails to fully present itself.[33] The claim that genuine philosophical, or more generally scientific, insight may arise from mythical imagery would seem to challenge the orthodox view that science and mythology are diametrically opposed to one another. Indeed, Heidegger insists that mythology, like science, 'has its basis in specific experiences and is anything but pure fiction or invention.'[34] The important point here is that Heidegger viewed Plato's mythical account of the foundations of knowledge as a proto-scientific account from out of which a more precisely determined philosophical theory could then be developed. According to Heidegger, it was Aristotle who would take the first significant step along this path.

4. Heidegger's Phenomenological History of Logic: Aristotle

On more than one occasion, Heidegger calls Aristotle 'the father of logic.'[35] However, the fact that Aristotle has gained a pre-eminently authoritative place in the philosophical canon did not, writes Heidegger, happen as a matter of course but 'only after arduous struggles and controversies' which finally concluded in the thirteenth century.[36] Remarkably, medieval Christian theology and ancient Greek ontology

---

33 Heidegger (1984a), *The Metaphysical Foundations of Logic*, p. 184.
34 Heidegger (1982a), *Basic Problems of Phenomenology*, p. 234.
35 Heidegger (1982a), *Basic Problems of Phenomenology*, p. 179; Heidegger (1962a), *Being and Time*, p. 257 [214]; cf. Martin Heidegger (2009 [1998]), *Logic as the Question Concerning the Essence of Language*, trans. by Wanda Torres Gregory and Yvonne Unna (Albany: SUNY Press), p. 5.
36 Heidegger (1982a), *Basic Problems of Phenomenology*, p. 118.

shared a common interpretation of creation in terms of production, but the former understood creation as production from out of nothing, while the latter understood it as production from 'out of a material that is already found on hand.'[37] Whereas for medieval theologians both the that-being and the what-being of a thing are produced, for ancient philosophers only the what-being is produced. For the Greeks, in other words, existence itself was not subject to creation. Consequently, with its assimilation into medieval theology, ancient ontology became reformulated in a way which obscured its original problematic, a circumstance which began to find correction only in the eighteenth century.[38]

According to Heidegger, Aristotle's signal achievement was the introduction of a more precise formulation of the whatness of the thing.[39] In developing this doctrine, Aristotle used as his guideline the Greek concept of *logos*.[40] Heidegger writes that *logos* has been variously interpreted to mean 'reason,' 'judgement,' 'concept,' 'definition,' 'ground,' and 'relationship.'[41] He notes, however, that the word *logos* is derived from the same root as *legein*, which means 'to talk' or 'to hold discourse.'[42] He thus suggests as a basic signification of *logos* the German word *Rede*, which may be translated as 'speech.'[43] The connection to thinking is clear. Speech is one central way in which thoughts are expressed or communicated. Heidegger argues that *legein*, discourse, or *Rede* 'is the clue for arriving at those structures of Being which belong to the entities we encounter in addressing ourselves to anything or speaking about it.'[44]

One important connotation of *logos*, as speech, is that it functions to 'let something be seen,' which means that it points something out to the

---

37 Heidegger (1982a), *Basic Problems of Phenomenology*, p. 118

38 Heidegger (1982a), *Basic Problems of Phenomenology*, p. 118

39 Heidegger (1982a), *Basic Problems of Phenomenology*, p. 85.

40 Heidegger (1967), *What Is a Thing?*, p. 106.

41 Heidegger (1962a), *Being and Time*, p. 55 [32].

42 Heidegger (1962a), *Being and Time*, p. 47 [25]; see translator's note 3, p. 47.

43 Heidegger (1962a), *Being and Time*, p. 55 [32]; cf. Heidegger (1984a), *The Metaphysical Foundations of Logic*, p. 1.

44 Heidegger (1962a), *Being and Time*, p. 47 [25].

listener.[45] Another important connotation of *logos* is that of 'relation' or 'relationship.'[46] Plato's mythological image of the good as demiurge, or producer, is thus an example of *logos* in that it points out the createdness of the thing, as well as the relation of unity between things and thinking. In contrast to Plato, however, Aristotle focussed on *logos* as 'statement' or 'proposition.' The crucial move here is from a conception of thinking as a collection of images to a conception of thinking as a collection of propositions, as speech. Heidegger describes this move as 'decisive': 'Thinking is here conceived in the sense of talking and speaking.'[47] Moreover, speaking is further specified in terms of propositions: 'Aristotle is the first to give the clearer metaphysical interpretation of the *logos* in the sense of the propositional statement.'[48] In his attempt to treat the thing by analogy to the proposition, Aristotle thus takes a dramatic step beyond Plato. Nevertheless, it should be emphasised that Aristotle still adopts, without question, Plato's image of the thing on the model of the product, or artefact. The proposition provided Aristotle with a useful model by which to more precisely determine and formalise the ontological structure of the thing conceived in this way.

It is because Aristotle undertook a systematic analysis of the proposition that Heidegger names him the father of logic. Logic, in this sense, is the 'science of *logos*,' the formal analysis of speech, reason, or thinking made possible through the instrument of the proposition.[49] *Logos*, as proposition, does not just point something out — as in 'Look, a bird!' — it points something out about something — as in 'The bird flies.' *Logos* as proposition is, in other words, a composite; it combines two or more things. Heidegger argues that the proposition, as a determination of something as something, is an expression of thinking. Hence, logic, as the science of *logos*, is also the science of thinking.[50] Furthermore, a proposition combines things in a particular way: 'Flies bird the' is not a proposition, not *logos*. Logic thus originates in the

45   Heidegger (1962a), *Being and Time*, p. 56 [32].
46   Heidegger (1962a), *Being and Time*, p. 58 [34].
47   Heidegger (2009), *Logic as the Question Concerning the Essence of Language*, p. 17.
48   Martin Heidegger (2000 [1953]), *Introduction to Metaphysics*, trans. by Gregory Fried and Richard Polt (New Haven: Yale University Press), p. 61.
49   Heidegger (1984a), *Metaphysical Foundations of Logic*, p. 22.
50   Heidegger (1984a), *Metaphysical Foundations of Logic*, p. 1.

attempt to make explicit whatever it is that governs the intelligibility of combinations of terms within a proposition. According to Heidegger, Aristotle recognised that the implicit organiser in the proposition was being, which is rendered explicit in the term 'to be' as well as its cognate 'is.'[51] In order to be properly analysed, then, the implicit presence of being in the proposition 'The bird flies' must be made explicit, by reformulating the proposition as 'The bird is flying.' The general form of the proposition thus becomes 'S is P,' where the 'is' combines two distinct terms. However, Heidegger writes that Aristotle furthermore observed that the 'is' separates as well combines. We cannot understand the combination of S and P unless we have first understood them as each signifying ontologically distinct, but potentially combinable, things.[52] The proposition 'The bird is flying' points out that two distinct things, one material and the other motive, are combined in a single intelligible event.

The term 'is' does not signify a distinct thing in the way that the terms 'bird' and 'flying' do. Heidegger interprets Aristotle as arguing that the 'is' signifies *nothing*, that is, no distinct thing. Rather, it *con*signifies a relation between things, 'a certain combining, which cannot be thought unless what is already combined or combinable has been or is being thought.'[53] The 'is' does not point to a thing existing among other things, but instead to an aspect of thinking: 'the being-combined of what is thought in thinking.'[54] If we think of the 'is' as pointing out a relation, then we immediately see that it presupposes the existence of two or more relata, or things related. On the account Heidegger attributes to Aristotle, these relata are not mutually distinct and separate things existing beyond thinking, but rather the significations in thinking of those things. The 'is' in the proposition 'The bird is flying' does not combine the bird and the motion of flying. It combines the concept 'bird' and the concept 'flying.' Hence, Heidegger writes that in Aristotle '*logos* is conceived as a *connecting of notions*, as a conjoining of meanings, as a

---

51   Heidegger (1982a), *Basic Problems of Phenomenology*, p. 180.
52   Heidegger (1982a), *Basic Problems of Phenomenology*, p. 182.
53   Heidegger (1982a), *Basic Problems of Phenomenology*, p. 181: cf. Aristotle (1941a), *De Interpretatione*, trans. by E. M. Edghill, in *The Basic Works of Aristotle*, ed. by Richard KcKeon (New York: Random House), pp. 38–61 (p. 41 [lines 16b19–25]).
54   Heidegger (1982a), *Basic Problems of Phenomenology*, p. 182.

binding together of concepts.'[55] From this, he concludes that the starting point of Aristotle's science of *logos* is a clarification of the precise concepts from out of which composite *logoi* of the form 'S is P' come to be organised. The starting point, in other words, is the definition of concepts. Aristotle's doctrine of the concept, including its definition, must therefore precede his doctrine of *logos* as proposition.[56]

Aristotle thus sets about differentiating and defining types of concepts in order to prepare the ground for a study of the ways in which those concepts may be combined in a proposition. In order to give a precise account of the proposition, he recognised that one must first give a precise specification of the concepts it includes. Recall that for the ancient Greeks thinking is grounded in experience. Hence, for Aristotle the definition of a concept was closely linked to a clarification of the nature of the experienced thing which that concept is meant to signify. A more precise determination of the concept thus went hand in hand with a more precise determination of the whatness of the thing. Ontology proceeds from grammar. The science of *logos* provides an entry point for a science of being.

Aristotle's key philosophical contribution was to narrow the focus of enquiry down to one particular kind of *logos*, the proposition. As a result, the subsequent course taken by ontology was powerfully conditioned by the range of possibilities made available by the composite linguistic form 'S is P.' Consider one consequence of this. In the *Categories*, Aristotle observes that one particular kind of concept could be represented only in the subject position of the proposition, never in the predicate position. This is the kind of concept signifying concrete individuals. Aristotle called such individuals 'primary substances.'[57] By further studying the constraints imposed by the proposition, Aristotle then identified another kind of concept which could, unlike the concept of primary substance, also be represented in the predicate position of the proposition. One example of a proposition in which these two kinds of concept are represented is 'The shirt is white.' Aristotle wrote that

---

55   Heidegger (1984a), *Metaphysical Foundations of Logic*, p. 23.
56   Heidegger (1984a), *Metaphysical Foundations of Logic*, p. 23
57   Aristotle (1941b), *Categoriae*, trans. by E. M. Edghill, in *The Basic Works of Aristotle*, ed. by Richard KcKeon (New York: Random House), pp. 3–37 (p. 9 [lines 2a11–13]).

the predicate 'white' represents the category of 'quality,' in this case, the shirt's quality of being white.[58] In saying of the shirt that it is white, the proposition signifies the fact that a particular whiteness is present in the shirt, or, more generally, that a particular quality is present in a particular substance. Aristotle emphasised that by 'present in' he does not mean that the whiteness is in the shirt as a part is in a whole. He means rather that this particular whiteness could not exist independently of this particular shirt.[59] More generally, although a particular quality is ontologically distinct from the concrete substance in which it is present, it would not exist if it were not present in that substance.

To us, Aristotle's point may seem rather trivial. He is simply telling us that the thing is a substance with qualities, or, put more broadly, with properties. The point Heidegger wants to make, however, is that our self-evident understanding of the thing as a property-bearing substance was the outcome of considerable intellectual effort and philosophical ingenuity. It is not obvious that the thing is best understood by analogy to the proposition. When we see a white shirt we do not see a shirt, on the one hand, and its being white, on the other. We see a white shirt. Separating the shirt into substance and property may strike one, after all, as a rather odd thing to do.

By analysing the thing through the composite *logos* of the proposition, Aristotle determined that the thing too is composite in its structure, and, more specifically, that it is a substance with distinct properties. Just as the 'is' of the proposition separates, so as to then intelligibly combine, concepts in accordance with specific grammatical principles, so too are the thing and its properties distinguished and then combined in accordance with specific ontological principles. In his systematic analysis of *logos* as proposition, Aristotle not only laid the foundations for the subsequent field of logic, he also rendered a more precise formulation of Plato's obscure and largely informal notion of a unifying standard which serves to organise things and thinking into knowledge. The mythical image of the demiurge was thus replaced by the more systematically tractable logic of the proposition.

---

58   Aristotle (1941b), *Categoriae*, p. 9 (line 1b29).
59   Aristotle (1941b), *Categoriae*, p. 7 (lines 1a23–24).

5. Heidegger's Phenomenological History of Logic: Descartes

The second chapter in Heidegger's history of things and thinking begins with the emergence of modern natural science.[60] With the arrival of the early-modern period, Aristotle's narrow focus on *logos* as proposition not only continues to guide formal enquiry into the nature of the thing, the philosophical consequences of this orientation are further strengthened and become radicalised in a profound way. Heidegger argues that the fundamental foundation distinguishing modern science from its predecessors can be located in that which 'rules and determines' the basic activities of science.[61] This foundation is twofold. First comes what he describes as the 'work experiences' of modern science. This designates, above all, the modern scientist's distinctive 'direction and mode of mastering and using what is,' or, put more prosaically, the scientist's 'manner of working with the things.'[62] It is important to emphasise that Heidegger is here offering a phenomenological description. He seeks to pick out the distinctive and fundamental features of the way the modern natural scientist *experiences* her work with things. The second aspect of the foundation of modern science is the range of possibilities within which the scientist may meaningfully understand her experience of working with things. As was discussed in Chapter Three, the study of these possibilities in distinction from their concrete actualisation as the whatness of things is what Heidegger means by the term 'metaphysics.' When we transcend the things which normally preoccupy us, and reflect instead on the existential conditions determining the whatness of those things, then we are doing metaphysics in Heidegger's sense. Hence, Heidegger calls this second aspect the 'metaphysical projection of the thingness of things.'[63] The idea is that, with this projection, the scientist's experience of her work with things is guided by a metaphysical thinking which determines the thingness of those things, that is, the specific kind of whatness rendering them amenable to scientific understanding.

---

60   Heidegger (1967), *What Is a Thing?*, p. 65.
61   Heidegger (1967), *What Is a Thing?*, p. 68.
62   Heidegger (1967), *What Is a Thing?*, pp. 66, 68.
63   Heidegger (1967), *What Is a Thing?*, p. 68.

In the last section, we considered the way in which Aristotle's focus on the proposition facilitated a specific metaphysical projection of the thingness of things as property-bearing substances. This was the first chapter in Heidegger's phenomenological history of logical thinking. The second chapter addresses how, with the rise of modern natural science, the proposition becomes mathematically interpreted in a way which then establishes the basic principles of pure thinking, or pure reason. For Heidegger, a key early-modern figure exemplifying this next stage is René Descartes. Descartes uncritically accepted Aristotle's articulation of the thing as substance, but then proceeded to radicalise it. He took Aristotle's claim that substance exists independently of all else, that it is autonomous, and argued that God is the only substance which truly meets this criterion. Indeed, drawing from the medieval Christian notion of God as an uncreated creator who produces the cosmos from nothing, Descartes argued in *Principles of Philosophy* that God is the basis for all other entities: 'We perceive that all other things can exist only by the help of the concourse of God.'[64] Hence, the notion of production lying behind Aristotle's ontological treatment of the proposition is transformed in Descartes from creation out of material already on hand to creation from out of nothing. The guiding principle which unifies things and thinking, represented by Plato with the mythic image of the demiurge, and by Aristotle in the 'is' of the proposition, is transplanted by Descartes into the subject position of the proposition, where it comes to signify the most perfect substance and ultimate ground of all other things. With this move, argues Heidegger, the conditions determining the thingness of the thing, and, more immediately, the substantiality of the substance, get buried beneath the dogmatic assumption that substance provides the only and ultimate basis for any knowledge of things.[65] The nature of substance can thus no longer be clarified in terms of its substantiality, because questions of substantiality are now immediately referred back to substance. An ontology of substance, an enquiry into the conditions determining substance in its fundamental

---

64   Heidegger (1962a), *Being and Time*, p. 125 [92]; René Descartes (1969a), *The Principles of Philosophy*, trans. by Elizabeth S. Haldane and G. R. T. Ross, in *The Philosophical Works of Descartes*, vol. 1, by René Descartes (Cambridge: Cambridge University Press), pp. 201–302 (p. 239 [Principle LI]).

65   Heidegger (1962a), *Being and Time*, p. 127 [94]; Heidegger (1982a), *Basic Problems of Phenomenology*, p. 124.

whatness, thus passes outside the scope of analysis. As a consequence, the nature of a substance can, for Descartes, become known only indirectly through the study of its most enduring properties: 'there is always one principal property of substance which constitutes its nature and essence, and on which all the others depend.'[66] On the other hand, because substance can still provide an absolute basis for knowledge, it supplies Descartes with a secure means by which to build up an account of knowledge in terms of certainty.

In a parenthetical comment, Heidegger suggests that the priority given by Descartes to certainty had its historical roots in the Christian doctrine of salvation, especially as regards the security of the individual.[67] According to this doctrine, among created things humans are distinctive because their 'eternal salvation is in question.'[68] However, Heidegger is more concerned with the ascendency of mathematics immediately before and during Descartes's lifetime. He argues that mathematics presented an ideal tool for those hoping to fulfill a growing cultural desire for certain, or absolute, knowledge.[69] In particular, mathematics provided a template for philosophers seeking to ground knowledge in indubitable first principles and axioms. For example, in *Rules for the Direction of the Mind*, likely his first philosophical work, Descartes wrote that 'in our search for the direct road towards truth we should busy ourselves with no object about which we cannot attain a certitude equal to that of the demonstrations of Arithmetic and Geometry.'[70]

Drawing from the philosophical tradition, Descartes set out to discover axiomatic certainty, of the kind exemplified in mathematics, through an analysis of the proposition. His goal was to construct unimpeachable axioms on the secure bedrock of absolute substance. For Descartes, God, being neither created nor sustained by anything beyond itself, was the first, 'absolutely perfect,' substance.[71] However, he furthermore contended that there exist two kinds of created

---

66 Descartes (1969a), *Principles of Philosophy*, p. 240 (Principle LIII).

67 Heidegger (1967), *What Is a Thing?*, p. 99.

68 Heidegger (1967), *What Is a Thing?*, p. 109.

69 Heidegger (1967), *What Is a Thing?*, p. 100.

70 René Descartes (1969b), *Rules for the Direction of the Mind*, trans. by Elizabeth S. Haldane and G. R. T. Ross, in *Philosophical Works of Descartes*, vol. 1, by René Descartes (Cambridge: Cambridge University Press), pp. 1–77 (p. 5 [Rule II]).

71 Descartes (1969a), *Principles of Philosophy*, p. 241 (Principle LIV).

substance, which each exist independently of any other kind of created substance, and which may themselves provide a solid basis for indubitable, axiomatic knowledge of God's creation.[72] Descartes called these two kinds of substance *res extensa* and *res cogitans*, the 'extended thing' and the 'thinking thing,' or what are more commonly known as 'body' and 'mind.' These two attributes, 'extendedness' and 'thinking,' are, according to Descartes, the principal properties constituting the nature of body and mind, respectively, and on which all other possible attributes of body and mind are themselves dependent. There was, for Descartes, an especially close relationship between the human mind and God, because both, in his view, are examples of a thinking substance.[73] He was, however, careful to maintain a strict qualitative difference between the two.[74]

Taking thinking substance as the foundational principle in his architectonic of certainty, Descartes determined that the fundamental axiom of a secure system of knowledge was the proposition 'I think.' As is well known, Descartes claimed that the indubitable truth of the proposition 'I think' entails the truth of the further proposition 'I am.' He argued that the proposition 'I am' is already present, implicitly, in the assertion 'I think.' According to Heidegger, an absolute mathematical principle must exclude anything which may have been given beforehand; it cannot have anything in front of it.[75] Taking mathematics as his model, Descartes thus sought a proposition which refers only to itself, excluding all possible traces of prior experience. This absolute proposition turned out to be the proposition in general, *as such*, the pure positing of a thinking which asserts.[76] Just as the perfectly general proposition posits only itself, so thinking in general, construed as absolutely mathematical, takes note only of what it already has. What Descartes discovered, argues Heidegger, is that pure thinking of this kind is always an 'I think,' *ego cogito*. Wherever pure thinking posits only itself, it must encounter the ego, the 'I.' Thus, in Heidegger's words, Descartes concludes that '[i]n "I posit" the "I" as the positer is co- and

---

72   Heidegger (1962a), *Being and Time*, pp. 125–26 [92].
73   Descartes (1969a), *Principles of Philosophy*, p. 241 (Principle LIV).
74   Descartes (1969a), *Principles of Philosophy*, p. 239–40 (Principle LI).
75   Heidegger (1967), *What Is a Thing?*, p. 104.
76   Heidegger (1967), *What Is a Thing?*, p. 104.

pre-posited as that which is already present, as what is. The being of what is determined out of the "I am" as the certainty of the positing.'[77] The subject of the most fundamental proposition — the 'I' of the 'I think' — thus becomes the absolute substance which must 'stand under' everything else. With Descartes, argues Heidegger, the individual 'I' becomes the foundational subject, 'that with regard to which all the remaining things first determine themselves as such.'[78] These things, which stand in relation to and are determined by the individual subject, become 'objects.' Through Descartes's intervention, the grammatical and logical distinction between subject and object thus becomes interpreted as a metaphysical distinction between mind and body. This is, argues Heidegger, a radical change in the way thinking, and subjectivity more generally, is understood to relate to things. The whatness of things now becomes illuminated on the basis of a mathematical impulse towards absolute first principles, which have themselves been grounded in the thinking subject construed as the foundational and individual 'I.'

With Descartes, the 'I' principle thus becomes the fundamental axiom of all knowledge.[79] Although he adopted without criticism Aristotle's focus on the proposition, as well as his account of the thing as a property-bearing substance, Descartes displaced the organising principle in the proposition from the 'is' to the subject position. The ontological guideline distinguishing and combining the categories in general thus gets buried in the ultimate category of substance, where it no longer presents itself for investigation. Hence, with Descartes, the conditions determining the substantiality of substance, the thingness of the thing, get concealed behind a dogmatic appeal to the ultimacy of substance. As the absolute ground of all enquiry, the 'I' resists any further explication. It becomes self-grounding. However, Heidegger argues that the 'I' principle, while fundamental, is not the only fundamental axiom of knowledge which emerges from Descartes's intervention. Indeed, because every propositional assertion necessarily implicates the 'I' principle, such assertions must always posit only what lies in the 'I' as the original subject. Hence, what is posited in

77 Heidegger (1967), *What Is a Thing?*, p. 104.
78 Heidegger (1967), *What Is a Thing?*, p. 105.
79 Heidegger (1967), *What Is a Thing?*, p. 107.

the predicate must not speak against the subject. From this, Heidegger concludes that every proposition co-posits, along with the 'I' principle, an equally fundamental principle of non-contradiction.[80] The idea is that the content of the 'I' as the foundational subject must be perfectly consistent with itself. As a consequence, reason in Descartes becomes formulated in terms of purity. The 'I think' provides the basis for reason, and non-contradiction ensures the purity, in the sense of logical consistency, of that reason. The fundamental mathematical axioms of the 'I' principle and the principle of non-contradiction thus combine to determine thinking as pure reason. This, in turn, provides the standard governing all determinations of the thingness of things: 'The question about the thing is now anchored in pure reason, i.e., in the mathematical unfolding of its principles.'[81] The Cartesian notion of 'pure reason' thus traces its roots back to Aristotle's narrow construal of *logos* in terms of the proposition. However, the purity in the notion results more immediately from Descartes's appropriation of mathematical techniques in order to establish the indubitable certainty of knowledge.

Descartes conceptualised thinking by analogy to mathematical practice, so as to then initiate a radically new formulation of the thingness, or whatness, of the thing as determined by the 'I' principle and the principle of non-contradiction. This move exemplifies what Heidegger identifies as the twofold foundation of modern science, namely, the experience of working with the things, and the generalisation on the basis of those experiences of an all-encompassing metaphysical projection of the thingness of things. Heidegger's historical attention weighs more on the second aspect, but he does briefly consider ways in which Descartes' philosophical initiatives related to the scientific work of his period. These considerations touch on the work of Galileo and Newton, as well as on the nature of the early-modern experiment, and they will be discussed further in Chapter Six. For present purposes, the next step in Heidegger's history focusses on Kant's response to Descartes's introduction of pure reason as the basis for thinking and knowledge.

---

80 Heidegger (1967), *What Is a Thing?*, p. 107.
81 Heidegger (1967), *What Is a Thing?*, p. 108.

6. Heidegger's Phenomenological History of Logic: Kant

In the evolving historical dialectic between thinking and things, Descartes represents for Heidegger a key moment in the early-modern determination of the whatness of the thing on the basis of a mathematically construed, pure reason. The next stage in Heidegger's history, then, comes with Immanuel Kant's critique of pure reason. The counter-concept arising from this critique is Kant's notion of the thing-in-itself, to which we gave detailed attention in Chapter Three. As we saw there, Kant conceived of the thing-in-itself as a property-bearing substance. This marks a strong continuity between Descartes and Kant, despite the latter's criticism of the former. Indeed, Heidegger argues that, on the basis of this shared commitment to substance ontology, Kant likewise reproduced the Cartesian position that thinking is an attribute of substance.[82] According to Heidegger, then, Kant's critique of pure reason, while departing from Descartes in important ways, nevertheless left unquestioned Aristotle's original introduction of the proposition as a model by which to formalise accounts of thinking and things, an introduction from out of which the ontological concept of substance first grew. Furthermore, Heidegger alleges that Kant uncritically adopted from Descartes a concept of the 'I' as an isolated, individual subject.[83] Although he submitted the 'thinking' of the 'I think' to ontological critique, he left the ontology of the 'I' largely unexamined. The point where Kant radically departed from Descartes was with his move to submit the 'I' to phenomenological investigation. He begins to unpack the phenomenal content of the 'I think,' of the reasoning or judging individual, in a way which Descartes did not. The consequences of this move were, in Heidegger's view, of far-reaching philosophical importance.

Recall that, according to Heidegger, one crucial influence in the transformation of the notions of thing and thinking from the ancient to the early-modern period was the rise of Christianity. In particular, productive subjectivity was reconceptualised as the creation of things

---

82   Heidegger (1962a), *Being and Time*, pp. 366 [318–19].
83   Heidegger (1962a), *Being and Time*, pp. 367 [320].

from out of nothing rather than from out of pre-existing material. This shift is powerfully represented in Descartes's philosophy. Heidegger argues that, as a result, a distinctly hierarchical order was introduced into the way things are conceived.[84] Among all that is, the highest and most real is the creative source of everything else. This is God, the uncreated creator, the *ens increatum*. Every other being is created, an *ens creatum*. Among created beings, the individual human being is most distinctive because its eternal salvation is in question. This is Descartes's *res cogitans*. What remains of created beings constitutes the world, Descartes's *res extensa*. The hierarchical ordering, in descending order of reality and perfection, is thus God, human beings, and world.

It is important to recognise what Heidegger is not arguing. He is not arguing that Cartesian philosophy is simply a transposition of Christian doctrine into a philosophical idiom. Indeed, Heidegger emphasises that the relation of early-modern philosophy to Church dogma can be 'very loose, even broken.'[85] His point is, rather, that the profound intellectual transformations of the early-modern period, to which Descartes made a key contribution, took place within a cultural context deeply suffused with orthodox Christian belief. We should thus not find it surprising that Descartes's philosophical views were consequentially influenced by the specificities of the social and material environment in which he worked. The underlying assumption of Heidegger's analysis is that philosophers are not individual and autonomous agents, whose singular ideas spring free from the social and historical soil in which they germinated but from which they no longer need to draw nourishment. This assumption is strongly at odds with the Cartesian view, and one may view its persuasiveness as bound together, in part, with the subsequent decline in influence of Christian notions of personal salvation on the intellectual milieu in which philosophers necessarily articulate and seek credibility for their work.

As we have seen, Heidegger argues that Descartes's position was influenced not just by Christianity, but also by the growing importance of mathematics in his period. On the one hand, the predominance of the Christian doctrine of personal salvation helps to explain why Descartes relocated the guiding principle, which unifies things and thinking,

---

84   Heidegger (1967), *What Is a Thing?*, p. 109.
85   Heidegger (1967), *What Is a Thing?*, p. 109.

from the 'is' of the proposition to the 'I,' the foundational subject, of the proposition. On the other hand, the growing authority of mathematical techniques helps to explain why Descartes then formulated the epistemic security of the 'I' by analogy to the axiomatic certainty of mathematics. The doctrine of salvation suggests that the individual, by overcoming the burden of sin, may slip free from the finitude and impurity of mortal existence and find eternal existence in God. Axiomatic knowledge, by comparison, provides the individual with a means by which to overcome the threat of error and falsehood, and hence to achieve an absolute certainty uncorrupted by the contingent constraints of finite physical existence. The pure reason of the *res cogitans*, the 'soul' or 'mind,' provided Descartes with an ultimate ground for knowledge, a ground which could be internally validated, independently of the worldly, and thus imperfect, influence of the *res extensa*, the 'external world.' The 'I' principle places the ground of reason in the individual thinking substance. The principle of non-contradiction ensures the internal purity of that individual substance.

On Heidegger's reading, Kant's critique of pure reason questions the alleged internal purity of the individual knower by challenging the idea that an individual can know anything at all in isolation from the 'external world.' Implicit in this challenge is a rejection of the idea that an individual may escape the finite constraints of physical existence and secure knowledge of a potentially infinite scope. In Chapter Three, we reviewed Kant's notion of finitude, and the doctrine of humility which it grounds. We can now see how finitude and humility arose for Kant in his response to the problem introduced by Descartes's location of the conditions determining the thingness of things in the pure reason of the thinking 'I.' By placing the 'I think' at the foundation of knowledge, Descartes gives urgency to the question 'What am I?' or, more generally, 'What is the human being?' The 'I' is not just one domain among others, but precisely that domain to which knowledge of all other domains must be traced back.[86] The question 'What is the thing?' thus leads back inevitably to the question 'What is the human being?' Because Kant accepts Descartes's account of the human being as a thinking substance, the question about the human being becomes for him a question

---

86   Heidegger (1967), *What Is a Thing?*, p. 110.

about thinking. According to Heidegger, Kant's primary question is thus the question of what an individual human being must be like in order to think. Descartes had already provided one clue: namely, that the individual, as a thinking substance, is bound by the law of non-contradiction. According to Heidegger, Kant picks up on this clue and develops it into an investigation of what the 'I' must be like in order to be bound by laws. In other words, Kant's critique of pure reason seeks after the conditions of possibility for law- or rule-governed thinking as such. Heidegger calls the articulation of these conditions 'a basic problem for logic.'[87] The search for a solution to this problem is a search for a 'philosophical logic, or better, *the metaphysical foundations of logic.*'[88]

On Heidegger's reading, then, Kant's enquiry into the law-governedness of thinking was a metaphysical enquiry into the ontological conditions which make scientific knowledge as such possible. He claims that this enquiry is prior to, and more fundamental than, the investigations typical of psychology, anthropology, ethics, or sociology.[89] Whereas these latter domains of enquiry, in Heidegger's view, take for granted the specific projection of the thingness of things which determines their particular subject matter, a more general ontological enquiry must address the conditions which make such determinations possible. Furthermore, Heidegger argues that '[o]nly as phenomenology is ontology possible.'[90] Hence, an enquiry into the existential conditions determining the possibility of logic is, for Heidegger, a phenomenological enquiry into the foundations of science. For him, logic is a particular kind of science, the 'science of the rules of thought.'[91] Accordingly, he argues that one signal achievement of Kant's critique of pure reason was to open up the 'thinking' of Descartes's 'I think' to phenomenological investigation. In particular, Heidegger reads Kant as exploring the ontological basis for a subject's ability to follow the laws structuring scientific thinking, chief among them being the law of non-contradiction. With this move, the law of non-contradiction, which had been treated by Descartes as the highest principle of all knowledge, is 'removed from its position of dominance.'[92]

---

87 Heidegger (1984a), *Metaphysical Foundations of Logic*, p. 20.
88 Heidegger (1984a), *Metaphysical Foundations of Logic*, p. 21.
89 Heidegger (1984a), *Metaphysical Foundations of Logic*, p. 17.
90 Heidegger (1962a), *Being and Time*, p. 60 [35].
91 Heidegger (1984a), *Metaphysical Foundations of Logic*, p. 104.
92 Heidegger (1967), *What Is a Thing?*, p. 184–85.

What now dominates are those objective features of subjectivity which allow the 'I' to submit itself to the rules governing thinking.

By opening thinking up to phenomenological investigation, Kant reopens a question that was explicitly addressed by Plato and Aristotle but then closed off by Descartes: namely, the question of *logos* as that which functions to meaningfully combine concepts. As we saw earlier, Plato locates this phenomenon in the idea of the good, conceptualised by analogy to the demiurge, the producer pure and simple. Aristotle, in contrast, conceptualised it by analogy to the 'is' of the proposition, thereby characterising it not as a thing but rather as that according to which things are combined. Descartes, by relocating the phenomenon of ontological combination from the 'is' to the subject of the proposition, then came to treat it as a thing the nature of which resists further analysis. Kant, by investigating this phenomenon as an active, non-thinglike feature of the subjectivity of the thinking subject, began to shed phenomenological light on what Descartes had formerly cloaked in darkness.

According to Heidegger, Kant located the phenomenon of ontological combination in the individual subject's power of imagination [*Einbildungskraft*].[93] For Kant, imagination is a 'faculty of forming.'[94] This harkens back to Plato's conceptualisation of the relation between things and thinking by analogy to production. Just as a craftsperson, or demiurge, forms disparate materials into a complete and unified whole, so too does the imagination of the subject form the jumble of brute sensation into an ordered and intelligible experience. Imagination thus makes intelligible experience possible. In the *Critique of Pure Reason*, Kant argued that '[t]he conditions of the possibility of experience in general are at the same time conditions of the possibility of the objects of experience.'[95] Heidegger argues that this passage reflects a 'fundamental posture' in human history, one which it is impossible for us to avoid.[96] Both an intelligible experience and the things which meaningfully present themselves in that experience — that is, both thinking and the

---

93   Martin Heidegger (1997 [1929]), *Kant and the Problem of Metaphysics*, 5th edn, enlarged, trans. by Richard Taft (Bloomington: Indiana University Press), p. 90.

94   Heidegger (1997), *Kant and the Problem of Metaphysics*, p. 91.

95   Immanuel Kant (1998), *Critique of Pure Reason*, ed. and trans. by Paul Guyer and Allen W. Wood (Cambridge: Cambridge University Press), p. 283 (line A158/B197).

96   Heidegger (1967), *What Is a Thing?*, p. 183.

things thought — are dependent on the faculty of imagination. We saw in Chapter Three that, for Kant, an intelligible experience involves the structured unity of two distinct faculties: spontaneity and receptivity, or what we have also discussed as constructivity and affectivity. According to Heidegger, Kant's notion of imagination describes the common root from out of which the distinct stems of spontaneity and receptivity both grow.[97] The power of imagination is, he writes, the 'original unity' of the two.[98] Chapter Three focussed on receptivity, especially as it relates to the issues of human finitude and epistemic humility. In this chapter, our concern lies more with the constructive faculty which lends order and intelligibility to experience, thereby allowing us to make sense of the things around us.

Kant described the constructive aspect of imagination in terms of a 'faculty of rules.'[99] Furthermore, he argued that rules, insofar as they are objective, may also be called 'laws.'[100] The power of imagination, then, and especially its constructive aspect, provides the conditions of possibility for following, among the other fundamental rules of logic, the law of non-contradiction. Logic thus finds its phenomenological, which is to say its ontological, origins in the imagination of the individual subject. However, Heidegger emphasises that, for Kant, the subjectivity of the individual subject is not constitutive of the law. Rather, the law is that towards which the subject directs herself, and it is through such directedness that the individual first realises the possibilities for action

---

97   Heidegger (1997), *Kant and the Problem of Metaphysics*, p. 97.

98   Heidegger (1997), *Kant and the Problem of Metaphysics*, p. 107. This reading of Kant has been strongly criticised. Michael Friedman, for example, argues that Heidegger, in this passage, is 'turning Kant's original problematic entirely on its head' (Michael Friedman (2000), *A Parting of the Ways: Carnap, Cassirer, and Heidegger* (Chicago: Open Court), p. 61). In his preface to the second edition, Heidegger notes that '[r]eaders have taken constant offense at the violence of my interpretations' (Heidegger (1997), *Kant and the Problem of Metaphysics*, p. xx). On the other hand, the book's translator, Richard Taft, comments that '[o]ver the years, *Kant and the Problem of Metaphysics* has emerged as the cornerstone of an important and original (if controversial) direction in Kant interpretation that continues to assert an influence today' (in Heidegger (1997), *Kant and the Problem of Metaphysics*, p. xii). Whatever the case, I am here less concerned with the 'accuracy' of Heidegger's interpretation of Kant, than with the role played by that interpretation in the development of Heidegger's own account of logic.

99   Kant (1998), *Critique of Pure Reason*, p. 242 (line A126); cf. Heidegger (1997), *Kant and the Problem of Metaphysics*, p. 105.

100 Kant (1998), *Critique of Pure Reason*, p. 242 (line A126).

open to her as a unique person.[101] Strictly speaking, then, the law is not a purely spontaneous construction of the individual subject. But neither, strictly speaking, is it purely an object of the sensibility, or receptivity, of the individual. The law, writes Heidegger, is nothing empirical.[102] It is not a thing awaiting discovery in the world. It is, rather, a structure found in reason. In this sense, then, the individual subject discovers the law within herself as something she is ultimately free to either follow or disregard, but which she nevertheless encounters as an obligation. Heidegger thus interprets Kant as arguing that reason is a 'receptive spontaneity' wherein the rules governing it are not freely constructed by the individual subject but rather encountered by her as the necessary constraints making reason, as such, possible. To accept these constraints, then, means to submit oneself to a 'self-given necessity.'[103] Heidegger writes that, for Kant, in submitting to the law, 'I submit to myself.'[104] We thus experience the rules of reason as self-validating. We come to recognise them through the obligation we feel towards them, and we likewise experience them as the source of that obligation. According to Heidegger, it is because the subject's relation to rules is marked by both constructivity and receptivity that Kant located rules in the power of imagination, a power which provides a common root for both.

It is a matter of logical necessity that a thing cannot be, for example, both a bird and not a bird. If S is P, then S cannot at the same time be not-P. This would be a logical impossibility, a contradiction. If we want to reason logically, then we must follow the principle, or law, of non-contradiction. This law sits at the foundation of logic. It puts an *a priori* constraint on thinking as Aristotle construed it on the model of the proposition. The necessity of the law is thus conjoined with the necessity of thinking in strictly propositional terms. We are obliged to follow the law of non-contradiction only so long as we also feel obliged to express ourselves in a logically sound manner. A poet, for example, may describe a thing in a way which is ambiguous between its being and not being a bird. Poetic thinking need not be logical thinking. We are thus free to ignore the law of non-contradiction to the extent that we

101 Heidegger (1982a), *Basic Problems of Phenomenology*, pp. 135, 138; Heidegger (1997), *Kant and the Problem of Metaphysics*, p. 111.
102 Heidegger (1982a), *Basic Problems of Phenomenology*, p. 135.
103 Heidegger (1997), *Kant and the Problem of Metaphysics*, p. 109.
104 Heidegger (1997), *Kant and the Problem of Metaphysics*, p. 111.

feel free to think in ways not determined by the logic of the proposition. Logical rules may compel us to think in certain strictly determined ways, but the compulsion itself is not a determination. We may, after all, choose to act otherwise, whatever the cost may be.

Heidegger draws this insight from Kant's discussion of the role of rules in theoretical reasoning, finding there the idea that '[f]reedom already lies in the essence of pure understanding, i.e., of pure theoretical reason, insofar as this means placing oneself under a self-given necessity.'[105] Furthermore, because freedom is a question of action, theoretical reason is importantly tied to practical reason. Hence, Kant writes in the *Critique of Pure Reason*: 'Everything is practical that is possible through freedom.'[106] Heidegger thus reads Kant as undertaking to explain theory in terms of action by giving priority to practical over theoretical reason. In his view, however, this undertaking remained ambiguous, as Kant presupposed an either/or of receptivity and spontaneity, of passivity and activity, thus failing in the end to fully realise the task of tracing both back to a common source in the power of imagination.[107] Heidegger writes that Kant, in unveiling the receptive spontaneity of our relation to rules, 'saw the unknown,' and 'had to shrink back.'[108] Kant's retreat is evinced in his shift of focus between the A and B editions of the *Critique of Pure Reason*. In the first edition, rule-governed thinking springs from the unified and irreducible receptive spontaneity of the imagination. In the second edition, Kant moved away from this unified stance, and instead assimilated rule-governedness, and imagination in general, to pure spontaneity.[109] With this, Kant backed away from his original insight that logic stems from the synthetic unity of our non-empirical constructive power, on the one hand, and our existence as finite creatures necessarily affected by things and persons, on the other. Heidegger argues that, subsequent to Kant's important insight in the first edition of the *Critique of Pure Reason*, 'pure reason as reason drew him increasingly under its spell': 'Kant awoke to the problem of now searching for finitude precisely in the pure, rational

105 Heidegger (1997), *Kant and the Problem of Metaphysics*, p. 109.
106 Kant (1998), *Critique of Pure Reason*, 674 (A800/B828); cf. Heidegger (1997), *Kant and the Problem of Metaphysics*, p. 109.
107 Heidegger (1984a), *Metaphysical Foundations of Logic*, p. 184.
108 Heidegger (1997), *Kant and the Problem of Metaphysics*, p. 118.
109 Heidegger (1997), *Kant and the Problem of Metaphysics*, p. 114.

creature itself, and not first in the fact that it is determined through "sensibility."'[110]

Kant thus managed to momentarily lift the ontological principle unifying things and thinking from out of the substance-subject where Descartes had buried it. In doing so, he exposed the phenomenological character of this unifying source as the receptive spontaneity of imagination, but he then allowed this insight to collapse back into the Cartesian subject. Consequently, imagination, as the ground for logical reason, gets reentrenched as the purified property of an individual and autonomous thinking substance. This means that the world towards which thinking, as pure reason, directs itself is likewise conceptualised in terms of pure substance, as an independent property-bearing object standing opposite a thinking subject. Sceptical doubts about the existence of a world construed in this way, which we considered in Chapter One, then emerge to make no end of philosophical mischief.

Thus, although Kant failed to grasp the implications of his insight into the phenomenology of rule-following, he broke new ground and provided the hints on which Heidegger was then able to build. Heidegger located one important such hint in Kant's remarks on the phenomenology of 'respect' [*Achtung*].[111] Kant approached respect as the 'feeling of my existence' or as a 'moral feeling' which one experiences in response to the laws governing moral action.[112] In my own terms, respect is an instance of the affectivity of the subject, and thus has an elementally emotional structure. Through this existential feeling of respect for the law one comes not only to recognise the law as law, one also comes to understand oneself as a person who can be guided by such laws. To be guided by a law or rule is to be affected by it. Guidability thus entails affectivity. The feeling that one's existence is rule-governed is thus a sign of one's finitude. I do not freely create the conditions governing my judgements or actions. On the contrary,

---

110 Heidegger (1997), *Kant and the Problem of Metaphysics*, p. 118. In a lecture one year later, Heidegger would add that 'Kant himself, as the second edition of the *Critique of Pure Reason* reveals, helped to prepare the turn away from an uncomprehending finitude toward a comforting infinitude' (Martin Heidegger (1995a [1983]), *The Fundamental Concepts of Metaphysics: World, Finitude, Solitude*, trans. by William McNeill and Nicholas Walker (Bloomington: Indiana University Press), pp. 208–09).

111 Heidegger (1982a), *Basic Problems of Phenomenology*, p. 133.

112 Heidegger (1982a), *Basic Problems of Phenomenology*, p. 133.

in judging or acting I am constantly affected by existential pressures which are not solely of my own autonomous creation. On Heidegger's reading, Kant argues that the law offers one a resource by which to repel those impulses or inclinations which threaten to disrupt moral action. Hence, the law may play a negative role, for example, by disabling the immediate sensible feelings of pleasure or pain which threaten to interfere with sound judgement. However, Heidegger points out that, for Kant, the disabling force exercised by the law against immediate, sensible feelings is itself also a feeling, an emotion. Indeed, he identifies this view, which he also ascribes to Spinoza, with the claim that 'an emotion can be overcome only by an emotion.'[113] In phenomenological terms, the repulsion of sensible feelings in the pursuit of rule-governed, or principled, action is itself motivated by a feeling. For Kant, this feeling is one of respect for the law, and it differs from sensible feelings like pleasure and pain in that it has an intellectual rather than somatic ground. Heidegger interprets Kant as arguing that respect for the law is produced by reason alone, that we possess it independently of, and prior to, sensible experience: 'Reason, as free, gives this law to itself.'[114] Furthermore, Heidegger emphasises Kant's claim that respect for the law, as a non-empirical feeling, is not directed towards things but rather towards persons. He quotes Kant's statement in the *Critique of Practical Reason* that '[r]espect always goes to persons alone, never to things.'[115]

---

113 Heidegger (1982a), *Basic Problems of Phenomenology*, p. 134. Cf. Heidegger's statement in *Being and Time*: 'when we master a mood, we do so by way of a counter-mood; we are never free of moods' (Heidegger (1962a), *Being and Time*, p. 175 [136]). Note, too, this statement by scientist-philosopher Ludwik Fleck, written in the early 1930s: 'The concept of absolutely emotionless thinking is meaningless. There is no emotionless state as such nor pure rationality as such. [...] There is only agreement or difference between feelings, and the uniform agreement in the emotions of society is, in its context, called freedom from emotions' (Ludwik Fleck (1979), *Genesis and Development of a Scientific Fact*, trans. by Fred Bradley and Thaddeus J. Trenn, ed. by Thaddeus J. Trenn and Robert K. Merton (Chicago: University of Chicago Press), p. 49). Note, furthermore, this more recent statement from feminist epistemologist Alison Jagger: 'emotional attitudes are involved on a deep level [...] in the intersubjectively verified and so supposedly dispassionate observations of science' (Alison M. Jagger (1989), 'Love and Knowledge: Emotion in Feminist Epistemology,' in *Women, Knowledge, and Reality: Explorations in Feminist Epistemology*, ed. by Ann Garry & Marilyn Pearsall (Boston: Unwin Hyman), pp. 129–55 (p. 138)).

114 Heidegger (1982a), *Basic Problems of Phenomenology*, p. 135.

115 Heidegger (1982a), *Basic Problems of Phenomenology*, p. 135.

Heidegger paraphrases this statement as: 'Respect as respect for the law relates also, in its specific revelation, to the person.'[116] More specifically, he interprets Kant as arguing that the person to whom respect for the law relates is 'myself.' Respect for the law is, on this reading, a 'self-subjection': 'In subjecting myself to the law, I subject myself to myself as pure reason.'[117]

There is, however, an interpretive anomaly here. Kant writes that respect for the law is a feeling directed toward *persons*. Heidegger paraphrases this as the claim that respect for the law is a feeling directed toward the *person*, namely, oneself. Where Kant used the plural, Heidegger transposes the singular. On the basis of this misinterpretation, Heidegger then accuses Kant of failing to overcome Descartes's burial of the rules binding things and thinking in the individual 'I.' Yet, immediately following the passage Heidegger cites from the *Critique of Practical Reason*, Kant writes that the example of a virtuous person 'holds a law before me,' that this person 'provides me with a standard.'[118] On Kant's account, it is towards another person, a virtuous person, that respect is felt. The behaviour of that person, in exemplifying the law, commands my respect for the law. Respect for the law is thus really respect for a person whose actions are seen to exemplify the law. Put in a nutshell, there is a distinctly intersubjective element in Kant's phenomenological account of respect. Only by considering the way in which we are affected by the behaviour of others can we begin to understand the compulsive respect for the rules which motivate and guide our own behaviour.

A more felicitous reading of Kant's critique of Descartes's pure reason should thus attribute to him two important insights. First, as Heidegger observes, Kant opened up the Cartesian 'I' to phenomenological enquiry, revealing the receptivity at the core of thinking. For Kant, this receptivity was especially manifest in a feeling of respect towards rules or the law. Second, Kant furthermore challenged the autonomy of the Cartesian 'I' by recognising that respect for the law is constituted by respect towards

---

116 Heidegger (1982a), *Basic Problems of Phenomenology*, p. 135.
117 Heidegger (1982a), *Basic Problems of Phenomenology*, p. 135.
118 Immanuel Kant (1956), *Critique of Practical Reason*, trans. by Lewis White Beck (New York: Macmillan), pp. 79–80.

other persons. Heidegger obscures this second insight by interpreting Kant's respect for persons as a respect for oneself, thereby foisting onto Kant a Cartesian construal of the 'I' as an isolated, self-referring individual. On this basis, Heidegger then argues that the 'I' should not be understood in terms of isolation, but rather in terms of community, or being-with-others, a point which will be considered in more detail later in this chapter. In making this argument, however, Heidegger is developing a point already present, if only in germinal form, in Kant's own account. Where Heidegger more clearly departs from Kant is in his rejection of the substance ontology which the latter took over from Descartes. Whereas Kant conceived of the 'I' as a substance, or thing, Heidegger describes it in terms of existence. For Heidegger, thinking is not the attribute of a substance-subject, but rather of a form of existence, an intentional act which is necessarily directed towards things and other persons, and which can therefore never be understood independently of those things and other persons.

This brings us to the end of Heidegger's history of the thematisation of thinking and its formalisation as logic. To recap, Plato and Aristotle had conceived of intentionality as the unifying principle which binds together things and thinking. For Plato it was the mythical demiurge, while for Aristotle it was the 'is' of the *logos*, construed narrowly as the proposition. Descartes shifted attention from the 'is' to the subject position of the proposition, thereby closing intentionality up in the pure reason of an isolated 'I' and thus supressing the essential link between thinking and things. Kant opened the 'I' back up, arguing that the rules which govern thinking entail the receptivity of the thinking 'I' towards other persons. Heidegger, for his part, builds on Kant's insights. As we saw in Chapter Three, he more fully opens up the affectivity of the 'I' to include things as well as persons. Later in this chapter, we will see how Heidegger also elaborated on the affective receptivity of the individual 'I' towards other persons through the methods of existential phenomenology. We will also consider some of the limitations of that method and examine the ways in which SSK can help to overcome them. In particular, SSK can help lead Heidegger towards a more fully developed account of the social foundations of logic. For the time being, let us directly address Heidegger's phenomenological account of the existential foundations of logic.

7. 'The Argument Lives and Feeds on Something'

According to Heidegger, Kant's critique of pure reason focussed on the question 'What is the human being?' and sought to answer this question by uncovering the conditions of possibility for rule-governed thinking as such. The articulation of these conditions is, for Heidegger, a basic problem for philosophical logic, a problem which can be solved only through metaphysical enquiry. As a metaphysical task, this enquiry refuses to take for granted the ancient Aristotelian projection of thinking as structured by the proposition. Heidegger thus understands logic as grounded in metaphysics, a position which puts him at odds with the prevailing view concerning the grounds of logic, a view he describes as self-evident, as 'immediately intelligible on the level of common sense.'[119]

This prevailing, common-sense argument views logic as a 'free-floating' and 'ultimate' form of thinking, one which holds primacy over all the sciences, broadly construed.[120] By classifying metaphysics as a science, proponents of this argument claim that logic also holds primacy over metaphysics. All knowledge, including metaphysical knowledge, is necessarily grounded in logic. Metaphysical thinking presupposes logic, because without logic it could never be justified or carried out.[121] Yet, Heidegger observes that, for the sake of consistency, this argument must also be applied to logic itself, as the science of thinking, thus yielding the discomforting conclusion that logic too presupposes logic.[122] It thus becomes difficult to see how logic could be justified without falling into a vicious circle. Heidegger seeks to soften this circle with the counter-argument that logic is grounded in metaphysics, and that metaphysics holds primacy over logic. He thus attempts to turn the prevailing view that metaphysics presupposes logic on its head. Indeed, at one point he even makes the provocative suggestion that logic is only contingently related to philosophy: 'Logic originated in the ambit of the administration of the Platonic-Aristotelian schools. Logic is an invention of schoolteachers, not of philosophers.'[123]

---

119 Heidegger (1984a), *Metaphysical Foundations of Logic*, p. 103.
120 Heidegger (1984a), *Metaphysical Foundations of Logic*, p. 103.
121 Heidegger (1984a), *Metaphysical Foundations of Logic*, p. 104.
122 Heidegger (1984a), *Metaphysical Foundations of Logic*, p. 104.
123 Heidegger (2000), *Introduction to Metaphysics*, p. 128.

Heidegger's response to the common-sense argument for the ultimacy of logic is a diagnostic one. Indeed, he cautions against attempts to directly attack the argument by means of a formal refutation.[124] The problem with such attempts is that they implicitly presuppose that thinking is, at its base, necessarily governed by the formal rules of logic. Hence, in their very method, these attempts surrender to the prevailing account the very ground they had hoped to contest. Yet, although Heidegger refuses to cede this ground, he does accept as inevitable the circularity of any argument which seeks to probe the foundations of thinking as such.[125] This inescapable circle is not, however, a formal circle, not a circle of logic, but a hermeneutic circle, a circle of interpretation. It is, moreover, an inherent feature of thinking. Heidegger argues that attempts to overcome the hermeneutic circle fail 'from the ground up' to grasp the way in which thinking actually works.[126]

Heidegger's considerations are driven by the insight that all acts of thinking are acts of interpretation. This is just another way of recognising that thinking is necessarily constructive. To make sense of an experienced thing means to determine its whatness. An act of sense-making is an interpretive act which implicates certain presuppositions about the thingness of the thing experienced. As we saw above, Heidegger describes the presuppositions guiding interpretation in terms of the metaphysical projection of the thingness of things. Metaphysics concerns the range of possibilities available for intelligibly determining the whatness of things experienced. Those possibilities may be viewed as the range of presuppositions on the basis of which one can make sense of one's experience in the world. The hermeneutic circle springs from the fact that interpretation necessitates the prior existence of understanding. In order to achieve a determinate understanding of something through an act of interpretation, one must already possess some understanding relevant to that thing. For example, in order to achieve a determinate understanding of a word in a sentence, one must already possess some understanding of the overall meaning of the sentence. It thus appears that understanding presupposes understanding, that thinking presupposes thinking.

---

124 Heidegger (1984a), *Metaphysical Foundations of Logic*, p. 106.
125 Heidegger (1962a), *Being and Time*, p. 27 [7].
126 Heidegger (1962a), *Being and Time*, pp. 194 [153], 363 [315].

If we then introduce into this circle the further claim that thinking is necessarily governed by formal logic, we end up with the assertion that logic presupposes logic, that logic grounds logic, which is just to say that logic needs only itself in order to be what it is. Here, then, we have the nub of the common-sense argument that logic is 'free-floating' and 'ultimate.' Logic is both the ground for all other forms of thinking, as well as the ground for itself. The primacy of logic is, in this way, self-evident. Heidegger's diagnostic response to this argument is to deny the claim that thinking is necessarily grounded in logic. This leaves in place the interpretive circle essential to thinking, but resists its further determination as a circle of logic. Indeed, Heidegger argues that the thinking presupposed in an act of interpretation 'has nothing to do with laying down an axiom from which a sequence of propositions is deductively derived.'[127] Thinking is not, at its base, structured by axioms, and thus interpretation cannot, at its base, be an act of inference from axioms, or first principles. Interpretation is not, in other words, inferential, but constructive or projective. Heidegger argues that when thinking turns to the interpretation of itself, when it thematises itself as an object of attention, it 'put[s] itself into words for the very first time, so that it may decide of its own accord whether, as the thing which it is, it has that state of Being for which it has been disclosed in the projection with regard to its formal aspects.'[128] What Heidegger is describing here is a process of self-interpretation by which thinking comes to exploit the possibility for articulating itself in formal or axiomatic terms, that is, in terms of logic. To treat thinking as fundamentally structured by axioms is to overlook this process of interpretation, and hence to confuse the end product of the interpretation — logic, as the science of thinking — for the thing being interpreted — thinking itself. In fact, according to Heidegger, logic is but one possible way in which thinking may come to understand itself. Only when the contingency of this particular self-articulation is acknowledged can thinking 'decide of its own accord' whether this articulation is an adequate representation of itself. Concealing this contingency, by erasing the interpretive processes which underlie it, risks forcing on thinking a representation of itself which it lacks the resources to question. Under such circumstances, it

---

127 Heidegger (1962a), *Being and Time*, p. 28 [8].
128 Heidegger (1962a), *Being and Time*, p. 362 [315].

becomes a matter of common sense to view thinking in terms of logic, and logic as something free-floating and ultimate.

Heidegger's proposed diagnostic refutation to the common-sense argument for the ultimacy of logic is meant to demonstrate two points: first, why the argument is necessary under certain presuppositions; second, why the argument is at all possible.[129] On the first point, we can now see that the argument becomes necessary once one presupposes both that thinking is necessarily grounded in fundamental axioms, as well as that the articulation of the formal propositional structure of thinking is not a constructive act of interpretation, but rather one of deductive derivation from those fundamental axioms. On the second point, the argument becomes possible only once these two presuppositions have been formally articulated and rendered credible. Heidegger's history of the evolving relationship between things and thinking, recounted above, is meant to describe this process of articulation. It was Aristotle's narrow interpretation of *logos* in terms of the proposition which founded logic as the formal study of propositionally structured speech acts. Descartes uncritically adopted this foundation, and was furthermore guided by mathematics, and perhaps also by the Christian doctrine of salvation, to shift the unifying locus of the proposition from the 'is' to the subject position, which, on the basis of the presumed axiomatic truth of the statement 'I think,' he narrowly construed in terms of the *ego*, the 'I.' This history is meant to suggest that the two presuppositions underpinning the common-sense argument did not, as it were, jump fully formed from the brow of Zeus, but instead emerged through a complex, protracted and controversy-laden historical process.

Heidegger claims that his diagnostic argument offers a 'refutation' of the common-sense argument, but this is not true.[130] It is, after all, not impossible for the existence of a presupposition to be a matter of historical contingency while its justifiedness remains a matter of necessity. However, although Heidegger has not conclusively proven that the presuppositions of the common-sense argument are only contingently justified, he has, I think, succeeded in loosening up intuitions about their presumed necessity. In light of Heidegger's historical analysis, concerns about the vicious circularity of the common-sense argument

---

129 Heidegger (1984a), *Metaphysical Foundations of Logic*, p. 106.
130 Heidegger (1984a), *Metaphysical Foundations of Logic*, p. 106.

are thus less likely to be assuaged by pointing to the conviction that the argument must, in any case, be accepted because there is no other alternative. Indeed, Heidegger's analysis opens the field for a consideration of his own hermeneutic approach, which seeks to soften the circle by grounding logic in metaphysics. He writes that the common-sense argument for the ultimacy of a free-floating logic 'lives and feeds on something, something which the argument itself not only cannot produce but which it even believes it must deny.'[131] This 'something' is, in his view, the existential foundation of logic, a foundation which he seeks to disclose through phenomenological analysis. The common-sense argument cannot produce these extra-logical foundations because its mode of thinking is restricted to the domain of logic. And it furthermore denies the existence of such foundations because it contends that all other modes of thinking must ultimately reduce to logic. Heidegger's existential-phenomenological account of thinking disputes this lattermost claim.

These considerations, focussing on logic as the science of thinking, find their place within Heidegger's broader existential conception of science. This broader conception was given detailed attention in Chapter Two. Recall from there that Heidegger draws a distinction between logical and existential conceptions of science. The logical conception views science, in ideal terms, as a coherent body of true propositions, a conceptual scheme. It is important to emphasise, once again, that Heidegger does not reject the logical conception of science, but instead seeks to elucidate how it could have become at all possible, that is, to illuminate the grounds for its existence. He does this through a phenomenological enquiry into the existential conditions which enable the emergence of the propositional stance presupposed by the logical conception. Heidegger described this process in terms of a change-over from immersed involvement in a work-world, where things are taken up non-deliberatively as ready-to-hand, to a theoretical knowing which disengages from and objectifies things as present-at-hand within-the-world, thus determining them as a subject matter for propositional knowledge claims. In this chapter, we have seen how the change-over in the way one understands the thingness of things also

---

131 Heidegger (1984a), *Metaphysical Foundations of Logic*, p. 106.

profoundly relates to the way thinking itself is understood. As things come to be understood as property-bearing substances, or objects, thinking correspondingly comes to understand itself as a specific kind of object, one structured by propositions and formal rules. Logic, as the science of thinking, thus comes to understand itself logically, as being necessarily grounded in logic. In so doing, it loses sight of the historical change-over from which it emerged, thus closing itself off from an understanding of its own existential grounds. As a consequence, the science of thinking comes to misunderstand itself as a necessary and exclusive interpretation of thinking rather than as but one possible and contingent way in which thinking may be thematised.

Heidegger accepts Kant's claim that thinking must be fundamentally understood in terms of a faculty of rules, arguing that rule-following is an inevitable part of the thinking process.[132] On this basis, he defines logic, as the science of thinking, more narrowly as the 'science of the rules of thought,' and observes that one may follow rules without having a scientific understanding of them: 'the inescapability of rule usage does not in itself immediately imply the inescapability of logic.'[133] Heidegger thus draws a clear line between rules, on the one hand, and logic, on the other. The latter arises through the scientific study of the former, but a science is neither presupposed nor entailed by the existence of its object of study. Indeed, Heidegger argues that rules, in the act of thinking, 'are not grasped as something at hand "in consciousness."'[134] While one is immersed in the act of thinking, the rules guiding that act are understood unreflectively in their readiness-to-hand; they are not thematised as objects of deliberative attention. Because thinking, as rule-following, may proceed in the absence of logic, thinking does not necessarily presuppose logic. Thus the vicious circularity of the common-sense argument — that logic presupposes logic — is softened with the phenomenological argument that logic presupposes rules, but rules do not entail logic. It follows, therefore, that logic is not ultimate, and hence that metaphysics, as a form of thinking, need not presuppose logic. Indeed, Heidegger claims that his phenomenological investigation into the existential grounds of possibility for logic reveals

---

132 Heidegger (1984a), *Metaphysical Foundations of Logic*, p. 104.
133 Heidegger (1984a), *Metaphysical Foundations of Logic*, p. 104.
134 Heidegger (1997), *Kant and the Problem of Metaphysics*, p. 108.

the metaphysical foundations of logic. It is logic which depends on metaphysics, not the other way around.

The common-sense argument thus 'lives and feeds' on the prior existence of unthematised and inescapable rules for thinking. Attention to these unthematised rules, however, provides only a description of what thinking fundamentally is, but not an explanation of how it could at all be possible. Such rules, on their own, encompass neither the metaphysical foundations of logic, in particular, nor of thinking, in general. As we saw in the previous section, according to Heidegger, one of Kant's principal achievements was that he pointed the way forward for such an explanation. Kant sought the conditions of possibility for thinking as such, consequentially locating them in the affectivity of the subject, that is, in the phenomenon of respect before the law. The phenomenology of respect is meant to explain the normative force attributable to the laws governing thinking. Let us now consider Heidegger's attempts to build on Kant's insight.

8. Time and Tradition at the Existential Root of Logic

Heidegger's historical account of the relation between things and thinking began with his remarks on Plato's mythical image of the demiurge. According to Heidegger, Plato's demiurge represents the idea of the good, that is, the fundamental standard which guides thinking into proper contact with things. The good allows us to distinguish between the epistemically good and bad, between things which contribute to knowledge and those which do not. In this way, the good stands as the indispensable *a priori* element in thinking, one making scientific knowledge, as such, possible. The condition of possibility for scientific knowledge is thus the grasp of the *a priori* element which binds thinking together with things in the production of knowledge. On this basis, Heidegger declares Plato 'the discoverer of the *a priori*.'[135] However, as discussed earlier, he furthermore observes that Plato's conception of the *a priori*, expressed with the image of the demiurge, possesses a 'mythical quality' through which the philosophical significance of the

---

135 Heidegger (1982a), *Basic Problems of Phenomenology*, p. 326.

a priori fails to properly present itself. Heidegger's own account of the metaphysical foundations of logic may thus be viewed as an attempt to more fully articulate the philosophical significance of the *a priori* against the historical backdrop of its original presentation in the ancient myth of the cosmic demiurge.

As noted earlier, Plato expresses the normative contribution made by the good to acts of cognitive discernment by drawing an analogy with the light contributed by the sun to acts of visual discernment. Only in the light of day can we reliably distinguish the real from mere shadows. Likewise, only in the 'illumination' of the good can we reliably distinguish epistemically valuable phenomena from irrelevant or deceptive ones. Successful acts of discrimination require the prior existence of illumination, whether of a sensory or intellectual kind: 'What sunlight is for sensuous vision the idea *tou agathon*, the idea of the good, is for scientific thinking, and in particular for philosophical knowledge.'[136] In Plato's work, then, the *a priori* was introduced as that which provides the illumination, or enlightenment, necessary for proper acts of epistemic discernment, and hence for knowledge in general. The productive acts of the demiurge are not random acts of creation, but display an order and coherence — an inherent and compelling intelligibility — a *logos* — which directly expresses the fundamental contours of the good. As *logos*, the mythical image of the demiurge points out, and so lets us see, the relations of significance, or meaning, which bring unity and purpose to acts of production. It manifests the highest guiding principle of such action. The demiurge thus stands for the *a priori* principle without which productive action, including the production of scientific knowledge, would not be possible.

When Aristotle replaced the demiurge with the proposition, he narrowed down and rendered more conceptually precise the content of this guiding principle, and hence initiated a historical process in which the *a priori* was progressively demythologised. According to Heidegger, this was the foundational moment of logic as the science of thinking. Later in his life, Heidegger would remark that science demythologises myth in much the same way one might drain a marshland, leaving only

---

136 Heidegger (1982a), *Basic Problems of Phenomenology*, p. 283.

'dry' ground behind.[137] In the earlier works from which the present discussion draws, however, Heidegger's mood is less glum. The young(ish) Heidegger sees himself as carrying forward a torch which had passed through the hands of, among others, Aristotle, Descartes, and Kant, thus bringing the sharp light of scientific philosophy into ever remoter corners of Plato's inexhaustibly polysemous image of the *a priori* as cosmic demiurge.

Heidegger's intervention was enabled by Kant's critique of Descartes's doctrine of pure reason, including the *a priori* rules governing such reason. Kant sought to disclose the conditions making it possible for the subject to follow rules, locating those conditions in the spontaneous receptivity of the subject's imagination. We saw earlier how Heidegger criticised Kant's dependency on the Cartesian concept of the subject, arguing that the imagination is not the extant property of a thinking substance, but instead a basic structure of the subject construed as existence. In this section, we will focus on another aspect of Heidegger's critique of Kant's concept of imagination: the aspect of time. By attending to the temporality of the imagination, Heidegger sought to further explicate the normativity of rule-following as first broached by Kant in his phenomenology of respect.

According to Heidegger, Kant argued that the spontaneity (or constructivity) and receptivity (or affectivity) which are unified in thinking share a common root in the subject's imagination. Heidegger takes this argument one step further, urging that this common root is itself rooted in the existential soil of 'original time.'[138] Original time makes imagination, and hence thinking, possible. It is the existential condition of possibility for imagination, conceived as the faculty of rules, and for thinking, conceived as rule-following. On Heidegger's account, we are able to follow rules only because we exist in original time. Original time thus replaces Plato's demiurge as the normative standard providing a necessary guide for thinking.

Heidegger's conceptualisation of thinking in terms of temporality is a natural consequence of his move from treating science as a body of concepts and formalised rules to treating it as a form of existence.

---

137 Martin Heidegger (1996 [1984]), *Hölderlin's Hymn 'The Ister,'* trans. by William McNeill and Julia Davis (Bloomington: Indiana University Press), p. 111.

138 Heidegger (1997), *Kant and the Problem of Metaphysics*, p. 137.

Science is not a static framework, but a dynamic activity of thinking and doing, a way of productively comporting oneself towards nature. Once scientific thinking has been construed dynamically, as something characterised by motion or change, it is a small and natural step to view thinking as also involving a temporal aspect. Indeed, Heidegger points out that the *a priori*, by its very name, demonstrates the temporality of thinking. The *a priori* is what 'comes before' sensory experience in the thinking process. It helps us to bring coherence and meaning to that experience by allowing us to discriminate between information which is salient and information which is not salient to the task at hand. The temporality of scientific practice is evinced by the necessary relation between *a priori* rules and sensory information, both of which combine in the production of natural knowledge. Thinking thus presupposes time. Heidegger writes that '[t]ime is the way in which the mind lets itself be given anything at all.'[139] His urge to understand the conditions of possibility for scientific thinking, as such, therefore pushes him towards a phenomenological investigation of the way time is experienced in the existential act of thinking. This is an experience of what Heidegger calls 'original time.'

Before turning to an explication of Heidegger's concept of original time, it will be helpful to first examine the common understanding of time to which he contrasts it. According to Heidegger, this common understanding was expressed in the first instance by Aristotle in his remarks, first, that 'time is not movement, but only movement in so far as it admits of enumeration,' second, that 'we discriminate [...] more or less movement by time,' and, third, that time is therefore 'a kind of number' or something 'countable.' Aristotle furthermore observed that time, insofar as it involves the 'before and after,' is measured in terms of the 'now.' Conceived of as motion enumerated in terms of the 'now,' time thus comes to be understood as a 'perpetual succession' of nows.[140] During a 1927 lecture, Heidegger demonstrated the plausibility of Aristotle's definition by pulling his watch out of his vest pocket. 'I

---

139 Martin Heidegger (2010a [1976]), *Logic: The Question of Truth*, trans. by Thomas Sheehan (Bloomington: Indiana University Press), p. 280.

140 Aristotle (1941c), *Physica*, trans. by R. P. Hardie and R. K. Gaye, in *The Basic Works of Aristotle*, ed. by Richard KcKeon (New York: Random House), pp. 213–394 (p. 291 [lines 219a3–12]); cf. Heidegger (1982a), *Basic Problems of Phenomenology*, pp. 238–39.

take my watch out of my pocket and follow the change of place of the second hand, and I read off one, two, three, four seconds or minutes. This little rod, hurrying on, shows me the time, points to time for me, for which reason we call it a pointer, a hand. I read off time from the motion of the rod.'[141] Yet, Heidegger also observes that this common understanding of time does not tell us what time is, but only how we may point to it, or read it off. Aristotle's definition is, he says, an 'access definition' or 'access characterisation'; it tells us only how time may become accessible.[142] Indeed, returning his watch into his pocket, Heidegger asks: 'Where then is this time? Somewhere inside the works, perhaps, so that if I put the watch into my pocket again I have time in my vest pocket? Naturally not.' He will raise the same puzzle again in a 1935 lecture: 'This clock is set according to the German Observatory in Hamburg. If we were to travel there and ask the people where they have the time, we would be just as wise as before our journey.'[143] The use of a watch, clock, or observatory only provides us with a means by which to precisely measure time in terms of the now. An understanding of time as a perpetual succession of nows is already presupposed in these acts of measurement. Moreover, Aristotle defines the unit of measure by which time becomes accessible — the now — in terms of the 'before and after.' Responding to this, Heidegger observes that 'the experience of the before and after intrinsically presupposes, in a certain way, the experience of time, the earlier and the later.'[144] Hence, the common understanding of time presupposes time. Aristotle's definition thus appears circular: '[t]*ime is time.*'[145]

This circularity is roughly homologous with the circularity of the argument addressed in the previous section, that logic presupposes logic. There we saw how Heidegger softened the circle by differentiating between logic, as the science of rules, and the rules themselves. A science of rules entails the prior existence of rules. A thematic, formal understanding of rules presupposes a non-thematic, existential understanding of rules. The difference is between rules present-at-hand,

141 Heidegger (1982a), *Basic Problems of Phenomenology*, p. 240.
142 Heidegger (1982a), *Basic Problems of Phenomenology*, pp. 256–57.
143 Heidegger (1967), *What Is a Thing?*, p. 22.
144 Heidegger (1982a), *Basic Problems of Phenomenology*, p. 247.
145 Heidegger (1982a), *Basic Problems of Phenomenology*, p. 241.

taken up as objects of study, and rules ready-to-hand, taken up as 'equipment' in the act of thinking. In similar fashion, Heidegger distinguishes the common time expressed in clock usage from the original time presupposed by such usage. Common time is original time thematised and rendered countable in terms of the now. The common understanding of time as a perpetual succession of nows thus 'lives and feeds' on a prior existential understanding of original time. Just as Aristotle narrowed thinking down to what can be explicitly expressed in the proposition, so too did he narrow time experience down to what can be explicitly expressed as part of a succession of countable nows.

These remarks on rules and time merge in the case of an *a priori* rule. In its very name, as what 'comes before,' an *a priori* rule would seem to possess an inherently temporal aspect. Yet, when the temporality of the *a priori* is understood in terms of common time, difficulties arise. Under this common understanding, we seem forced to posit the *a priori* as the first, or earliest, now in a perpetual succession of extant nows. But the *a priori* is not properly characterised by its being the first in a series of such events, but rather by its being the condition of possibility for such events. The *a priori* is not a countable now, but something which must come before all countable nows. Indeed, the extant events of common time presuppose the original temporality of the *a priori*. Hence, when time experience is narrowed down to what can only be expressed explicitly in terms of a succession of countable nows, the *a priori* must be viewed as somehow existing beyond this perpetual string of extant temporal moments. Seen from the perspective of common time, then, the *a priori* must exist beyond time — it must be timeless. Hence, Heidegger writes that, when orienting ourselves towards the common concept of time, it becomes 'consistent to deny dogmatically that the *a priori* has anything to do with time.'[146] Furthermore, seen from the perspective of common time, thinking, insofar as it is grounded in *a priori* rules, must itself also be viewed as a consequentially timeless phenomenon. On such an account, it also becomes consistent to dogmatically deny that thinking, at its innermost, rule-governed core, has anything to do with time. This would appear to give grounds for asserting that scientific

---

146 Heidegger (1982a), *Basic Problems of Phenomenology*, p. 325.

thinking draws its ultimate justification from a realm of rules which somehow exists beyond time.

The momentum behind such dogmatism loses its strength, however, once one recognises a role for original time in addition to, and as distinct from, common time. This recognition comes with the awareness that, just as there is a non-thematic way of experiencing rules, so too is there a non-thematic way of experiencing time. Experienced non-thematically, *a priori* rules need not appear to be uprooted from the soil of time. We discover the original temporality of *a priori* rules in our use of them as 'equipment' in the task of thinking, rather than in our thematisation, or foregrounding, of them as distinct and determinate objects of thinking.

Viewed phenomenologically, original time is experienced in the course of our immersed involvement in a work-world. As we saw in Chapter Two, a work-world is structured by the equipmental relations in which the significance, or meaning, of a particular task is manifest. When we are immersed in a task, our attention is focussed not on the equipment we are manipulating or otherwise using, but on the work towards which we employ the equipment. In the midst of our undisturbed immersion in an equipmental context, we understand the things we use as ready-to-hand rather than present-at-hand within-the-world. Heidegger observes that '[a]t the basis of this undisturbed imperturbability of our commerce with things, there lies a peculiar temporality which makes it possible to take a handy equipmental contexture in such a way that we lose ourselves in it.'[147] With this, he draws our attention to the familiar experience of losing track of time while being immersed in a task. His point is that time does not disappear when we lose track of it. Our actions do not, in other words, suddenly become timeless. Rather, time is still there in the way we experience our actions, but we do not experience it numerically as a countable succession of nows, or as any kind of present-at-hand thing which is 'passing by,' but instead in a 'peculiar' way, as the temporality of what is ready-to-hand in our productive acts of thinking and doing. Heidegger thus argues that temporality is the 'condition of the possibility of all understanding.'[148] Hence, we do not experience original time as an aspect or feature of the equipment we use in the course of work.

---

147 Heidegger (1982a), *Basic Problems of Phenomenology*, p. 309.
148 Heidegger (1982a), *Basic Problems of Phenomenology*, p. 274.

Original time is, rather, that which makes our equipmental dealings, our basic understanding of things ready-to-hand, possible as a basic mode of existence. Heidegger characterises the way in which original time becomes intelligible to us in the course of such equipmental dealings in terms of two kinds of relation: the 'in-order-to' (*Um-zu*) and the 'for-the-sake-of-which' (*Worumwillen*).[149] Each relation picks out a central aspect of a ready-to-hand thing, for example, a pen. The 'in-order-to' of the pen is its purpose: the task of writing. Yet the pen is not the sole piece of equipment which contributes to this task. Heidegger writes that '[e]quipment is *in terms of* [*aus*] its belonging to other equipment: ink-stand, pen, ink, paper, blotting pad, table, lamp, furniture, windows, doors, rooms.'[150] Before an individual piece of equipment can 'show itself' in this arrangement, 'a totality of equipment has already been discovered.'[151] The work-world in which the task of writing becomes practically intelligible is thus a relational manifold of ready-to-hand things the totality of which facilitates the task of writing. Heidegger argues that the 'whatness' of a ready-to-hand thing is constituted by the multiple relations in which the thing is able to show its use, its handiness. The whatness of a piece of equipment is constituted by its 'assignedness' [*Bewandtnis*] within an overall equipmental context.[152]

The in-order-to relation, however, only describes the assignedness of equipment; it does not explain whence that assignedness comes. This is where the 'for-the-sake-of-which' relation comes into play. Heidegger writes that '[o]nly so far as the for-the-sake-of a can-be is understood can something like an in-order-to (a relation of

---

149 Heidegger (1982a), *Basic Problems of Phenomenology*, p. 271.
150 Heidegger (1962a), *Being and Time*, p. 97 [68].
151 Heidegger (1962a), *Being and Time*, p. 98 [69].
152 Heidegger (1982a), *Basic Problems of Phenomenology*, pp. 292–93. In this context, I have chosen to translate *Bewandtnis* as 'assignedness,' in preference to Hofstadter's translation, 'functionality.' 'Assignedness' better captures the connotation of 'directedness' present in *Bewandtnis*, emphasised in Heidegger's close comparison of *Bewandtnis* with *Verweisen*, 'assignment' or 'reference' (Martin Heidegger (1927), *Sein und Zeit* (Tübingen: Max Niemeyer Verlag), pp. 83–84; cf. Heidegger (1962a), *Being and Time*, p. 115 [84], where *Bewandtnis* is translated as 'involvement'). 'Assignedness' also blocks unhelpful connotations of functionalism, which may be suggested by Hofstadter's choice. I further address this delicate translation issue in Chapter Five, where I introduce and defend a context-specific translation of *Bewandtnis* as 'end-directedness.'

assignedness) be unveiled.'[153] The assignedness relation of the pen can show itself only if one has already understood the for-the-sake-of-which relation which makes possible the practical intelligibility of the pen as equipment for writing. Only after one has understood this relation can one be in a position to 'let' the pen be used as a pen. Heidegger writes that '[l]etting-be-assigned [*Bewendenlassen*], as understanding of assignedness, is that projection which first of all gives to the Dasein the light in whose luminosity things of the nature of equipment are encountered.'[154]

Heidegger's use of the metaphor of illumination in this passage was surely deliberate. The whatness, or what-being, of equipment is projected as an illumination in which things may be 'seen' in their assignedness. Seeing, in this context, does not mean observing, but using. For Heidegger, the using and manipulating of equipment 'has its own kind of sight, by which our manipulation is guided.'[155] He calls this sight 'circumspection,' and argues that circumspection 'dwells' in the multiple relations of an equipmental totality.[156] Furthermore, Heidegger identifies the for-the-sake-of-which with the source of light which lets equipment be seen in its assignedness. He thus intended the for-the-sake-of-which to be a critical reinterpretation of Plato's idea of the good, which Plato had introduced by drawing an analogy to the sun. Heidegger writes that 'the *idea tou agathon* [i.e., the idea of the good] is the for-the-sake-of-which,' and that 'the for-the-sake-of-which excels the ideas, but, in excelling them, it determines and gives them the form of wholeness, *koinonia*, communality.'[157] Like Plato's idea of the good, Heidegger's notion of the for-the-sake-of-which refers to a unifying principle, one which grants coherence to our experience by guiding us in discriminating cognitively valuable from irrelevant or deceptive phenomena. Just as the productive acts of Plato's demiurge display an order and coherence — an inherent and compelling intelligibility — so too do constructive acts of thinking, guided by the for-the-sake-of-which,

153 Heidegger (1982a), *Basic Problems of Phenomenology*, p. 295; translation modified.
154 Heidegger (1982a), *Basic Problems of Phenomenology*, p. 293; translation modified.
155 Heidegger (1962a), *Being and Time*, p. 98 [69].
156 Heidegger (1962a), *Being and Time*, pp. 98 [69], 122 [88].
157 Heidegger (1984a), *Metaphysical Foundations of Logic*, pp. 184–85.

express an order and coherence, whether of a practical or theoretical kind. Like Plato's mythical image of the demiurge, the for-the-sake-of-which stands for the *a priori* element binding together things and thinking in the production of knowledge.

Through this critical reinterpretation of the demiurge, Heidegger attempts to further spell out, in phenomenological terms, the philosophical significance of Plato's mythical image. Unfortunately, he seems not to have got very far in realising this attempt, certainly not as far as he may have thought he had. Recall from earlier that Heidegger credits Kant with taking an important step forward on this path by locating the conditions of possibility for rule-following, as such, in the feeling of respect. We are able to follow rules because we are capable of being compelled by those rules, of being affected by them. In this way, the rules of thinking help us to make sense of our experience, including our experience of ourselves. According to Heidegger, Kant argued that respect is specifically a feeling of respect for oneself, that is, for the self as Cartesian 'I.' However, as we saw, Kant actually claimed that respect is a feeling felt towards another person, a virtuous person whose actions exemplify the law or rule in question. Heidegger's own emphasis on the intersubjective nature of our faculty of rules is thus not an innovative step beyond Kant, but rather a further elaboration of Kant's own original insight. On this account, then, the apriority of rules for thinking depends on the ontological priority of social relations over relations that 'I' have with myself. One can understand oneself as an 'I,' as an individual thinking substance, only because one is already immersed in interpersonal relations which enable one to thematise and articulate one's subjectivity in this way. Heidegger writes that 'Being with Others belongs to the Being of Dasein [...] [A]s Being-with, Dasein "is" essentially for the sake of Others. This must be understood as an existential statement as to its essence.'[158]

The for-the-sake-of-which, as the *a priori* phenomenon binding together things and thinking in knowledge production, is thus a fundamentally social phenomenon. Its temporality is expressed non-thematically as the original time we experience in the course of our

---

158 Heidegger (1962a), *Being and Time*, p. 160 [123].

non-deliberative immersion in a work-world. This fundamental mode of existence is the *a priori* social and historical condition for any and all possible acts of knowing, both practical and theoretical. Heidegger holds that we experience these conditions as 'tradition' [*Überlieferung*]. Tradition, he writes, is the 'innermost character of our historicity,' and we come to know it as the manifest 'lore' [*Kunde*] of the historical world in which we find and understand ourselves.[159] The lore in which tradition is manifest 'carries and leads the historical being of the era' in which we live.[160] Thus, on Heidegger's account, it is tradition which turns out to provide the *a priori* unifying phenomenon which bind together things and thinking, thus providing the basic conditions of possibility for the production of knowledge in its broadest possible sense. The soil in which Plato's demiurge, Kant's imagination, and Heidegger's for-the-sake-of-which are finally rooted, that soil from which the existential rules governing thinking draw their sustenance, is the soil of tradition. Hence logic, through its reliance on existential rules, turns out to itself be ultimately grounded in tradition. According to Heidegger, then, tradition provides the existential foundation for logic. As a consequence, the origins and essence of scientific thinking must lie not in an abstract and timeless realm of *a priori* logical rules, but in the rich social and historical fabric of our shared and largely unthematised co-existence as finite human beings. '[O]ur spiritual history,' writes Heidegger — using an adjective (*geistig*) which might just as well be translated as 'intellectual' or 'mental' — 'is bound 2000 years back.'[161]

---

159   Heidegger (2009), *Logic as the Question Concerning the Essence of Language*, p. 98.

160   Heidegger (2009), *Logic as the Question Concerning the Essence of Language*, p. 78. For this reason, James McGuire and Barbara Tuchanska err in writing that 'Heidegger does not squarely consider history as a succession of generations, a process within which the handing down and inheriting of cultural heritage from predecessors occurs' (James E. McGuire and Barbara Tuchanska (2000), *Science Unfettered: A Philosophical Study in Sociohistorical Ontology* (Athens OH: Ohio University Press), p. 74). 'Handing down' is just what *Überlieferung* means. They also fail to appreciate Heidegger's claim that being-with-others is elemental for the subject: 'If the only ontological ground for cognition lies in the existential structures of Dasein, cognition remains an individual activity' (p. 80). Hence, they falsely conclude that, for Heidegger, 'science is a way of being of Dasein, not of communities' (p. 81), missing the fact that Heidegger's subject is an essentially social entity, 'essentially for the sake of Others.'

161   Heidegger (2009), *Logic as the Question Concerning the Essence of Language*, p. 7.

Plato's discovery of the *a priori* was a discovery of the way in which tradition guides and gives shape to thinking, hence making possible its intelligible combination with the things of experience. The *a priori* provides the light in which thinking lets things be experienced as what they are. It is, furthermore, the source of the semantic, syntactic, and practical rules which make language use possible as a form of intelligible behaviour.[162] Or, to take a non-linguistic artefact, the *a priori* is the basis in tradition through which a pen may become what it is as an instrument for writing.

Having employed the methods of phenomenological history to trace the ontological origins of science back to tradition, Heidegger seems to have slowly lost any further interest in this project. Perhaps this reveals the limitations of his existential-phenomenological method, with its focus on an archaeological investigation into the subject's interior layers of historically sedimented understanding. Such a method would seem not especially well-suited for a more detailed exploration of the social and historical dimensions of science conceived as tradition. Once he had excavated the scientific subject down to its existential base in the bedrock of tradition, Heidegger's interests shifted increasingly towards those possibilities latent in the tradition but excluded by Aristotle's narrowing of the *logos* down to what can only be expressed within the formal structure of the proposition. He thus turned his attention more and more to non-scientific modes of subjectivity. Rather than further attending to the language of science, as determined by the proposition, Heidegger would instead declare in 1934 that the 'original language is the language of poetry.'[163] Many of his subsequent reflections on science would focus more negatively on scientific thinking as a historically contingent constraint imposed on the original scope and power of poetry as an essential expression of human experience. Hence, for the remainder of this chapter, we will trade the methods of phenomenology for those of sociology in order to more fully elaborate some of the positive implications of Heidegger's conclusion that logic is rooted in the social soil of tradition.

---

162 Heidegger (1962a), *Being and Time*, p. 120 [87].
163 Heidegger (2009), *Logic as the Question Concerning the Essence of Language*, p. 141.

9. From the Phenomenology of Thinking to the Sociology of Knowledge

In a 1928 lecture, Heidegger argued that a fundamental philosophical study of subjectivity 'remains prior to every psychology, anthropology, and characterology, but also prior to all ethics and sociology.'[164] He seems to have held that the human and social sciences would remain unclear or in error until the object of their study — human being, whether taken individually or collectively — had been properly clarified in its fundamental structure. This clarification was the task Heidegger took up in *Being and Time*, which had appeared in 1927. Indeed, not only was the bulk of that book concerned with a phenomenological description of human subjectivity, the fourth chapter specifically addressed 'being-with' (*Mitsein*), that is, the essential relation between the subject's being-in-the-world with its equally fundamental existential state of being-with-others.

Section 27 of that chapter focussed on the 'they,' which badly translates the German neologism *das Man*. The concept of the 'they' is meant to pick out the way in which one uncritically blends in with one's fellows in the course of everyday life: 'In utilizing public means of transport and in making use of information services such as the newspaper, every Other is like the next.'[165] The point seems a bit overstated, but the idea is sound: those who live together in a common society are continuously, and often imperceptibly, pushed into the same settled and uniform patterns of behaviour. Furthermore, these social conditions directly affect, and thus serve to organise, our intellectual and emotional lives: 'We take pleasure and enjoy ourselves as *they* [*man*] take pleasure; we read, see, and judge about literature and art as *they* see and judge; likewise we shrink back from the "great mass" as *they* shrink back; we find "shocking" what *they* find shocking.'[166] Heidegger goes on to observe that a person, by critically reflecting on her embeddedness in the 'they,' may discover less everyday possibilities for thinking and acting in the world, possibilities which may then enable her to express her personhood in a more unique or individualised manner.

---

164 Heidegger (1984a), *Metaphysical Foundations of Logic*, p. 17.
165 Heidegger (1962a), *Being and Time*, p. 164 [126].
166 Heidegger (1962a), *Being and Time*, p. 164 [126–27].

The distinguishing features of a unique individual are thus not to be taken as a sign of her separateness from the group, but of her success at exploiting the less well-trodden possibilities made available by her participation in a shared tradition. Heidegger's discussion is sometimes read as suggesting that the achievement of such individuality is a rare feat, performed only by an elite few, but this is neither here nor there. The interesting theoretical observation is that individuality derives from sociality, that individualism is enabled and maintained by and in an irreducibly social existence.

In a 1947 essay, Heidegger reflected back on his earlier discussion of the 'they' (*das Man*), writing that '[w]hat has been said in *Sein und Zeit* [*Being and Time*], §§27 and 35, about the word "*man*" (the impersonal one) is not simply meant to furnish, in passing, a contribution to sociology [*soll keineswegs nur einen beiläufigen Beitrag zur Soziologie liefern*].'[167] Section 35 is a narrower application of the argument from §27 to the specific case of everyday versus unconventional forms of speech, and need not delay us here. Heidegger's 1947 comment affirms that §27 of *Being and Time* was meant to make a fundamental, rather than an incidental, contribution to sociology.[168] There is thus no clear evidence that Heidegger harboured an animosity toward sociology. Indeed, in a 1934 lecture, when addressing the definition of *das Volk*, or 'the people,' he suggested that '[w]e could for this purpose follow a new science, sociology, that is, the doctrine of the forms of society and community.'[169] He declines to go this route, however, alleging that sociology's search for definitions presupposes that *das Volk* is a present-at-hand thing, a substance with fixed and timeless properties, a thing without a history. In sum, Heidegger does not reject sociology, but

---

167 Martin Heidegger (1962b [1949]), 'Letter on Humanism,' trans. by Edgar Lohner, in *Philosophy in the Twentieth Century*, vol. 3, ed. by William Barrett and Henry D. Aiken (New York: Random House), pp. 271–302 (p. 273); my brackets, taken from Martin Heidegger (1949), *Über den Humanismus* (Frankfurt: Vittorio Klostermann), pp. 9–10.

168 This point gets somewhat lost in the more widely available translation, by Frank A. Capuzzi and J. Glenn Gray, of the 1947 passage: 'What is said in *Being and Time* (1927), sections 27 and 35, about the "they" in no way means to furnish an incidental contribution to sociology' (Martin Heidegger (1993b [1947]), 'Letter on Humanism,' trans. by Frank A. Capuzzi and J. Glenn Gray, in *Basic Writings*, revised and expanded edn, ed. by David Farrell Krell (New York: HarperCollins), pp. 217–65 (p. 221)).

169 Heidegger (2009), *Logic as the Question Concerning the Essence of Language*, p. 59.

rather criticises it for allegedly failing to recognise that the existential conditions governing social relations, including the relations governing knowledge production, are historically contingent conditions, that is, conditions rooted in a tradition.

Heidegger thus meant for his existential phenomenology to, among other things, provide sociology with a more defensible account of the existential conditions of possibility governing its core subject matter. We will set aside the question of whether Heidegger's criticism offers a fair portrayal of the sociology of his time. What seems clear, however, is that it does not find much traction against the more recent sociology of knowledge — SSK in particular — which recognises the historical and social contingency of the conditions determining its own scientific method. Here, then, Heidegger's scientific philosophy does resonate with the more recent scientific sociology. Both methods regard science as a finite, social and historical practice, and both methods attempt to reflexively style themselves in accordance with this model of science. The underlying assumption in both cases is that science provides the most reliable and authoritative basis for knowledge. David Bloor has been especially clear on this point: 'I am more than happy to see sociology resting on the same foundations and assumptions as other sciences. [...] For that foundation is our culture. Science is our form of knowledge.'[170] Furthermore, both Barry Barnes and Harry Collins have noted the important influence of phenomenological methods on the way sociology of knowledge has developed since Heidegger's time.[171] There is, then, a methodological thread running consequentially from Heidegger's phenomenology of thinking to the more recent sociology of knowledge.

As was mentioned at the end of the last section, Heidegger would eventually lose his enthusiasm for scientific philosophy, and begin following other fundamental threads in the tradition of thinking. Nevertheless, his phenomenological study of logic, construed as the science of thinking, constitutes his own reflexive attempt to lay hold

---

170 Bloor (1991), *Knowledge and Social Imagery*, pp. 160–61.
171 Barry Barnes (1974), *Scientific Knowledge and Sociological Theory* (London: Routledge & Kegan Paul), pp. vii, 43; Harry M. Collins (1992), *Changing Order: Replication and Induction in Scientific Practice*, with a new forward (Chicago: University of Chicago Press), pp. 15, 25 n. 12.

of the conditions of possibility governing his own particular form of scientific philosophy. This affords us a conceptual bridge over which to now cross from the phenomenology of logic to the sociology of logic.

10. The Social Foundations of Logic

The sociology of logic challenges the view that the basic principles of logical thinking are resistant to sociological explanation. The key assumption which must be cast into doubt is that such principles enjoy an absolute ground impervious to the vagaries of historical and social life. This chapter has attempted to reconstruct Heidegger's account of how this assumption came into being, and how it has subsequently fallen into doubt.

In the first instance, Descartes applied a mathematical interpretation to the proposition, which may have been motivated in part by the Christian doctrine of salvation, in order to establish thinking on a pure and absolute ground. He furthermore argued that the pure and absolute core of thinking, the individual 'I,' must act in accordance with an unrestricted law of non-contradiction. In the second instance, Kant addressed the question of what the 'I' must be like in order to be bound by rules like the law of non-contradiction. This question, observed Heidegger, is the basic problem of logic. By asking it, Kant effectively removed the law of non-contradiction from its position of dominance, instead explaining laws in terms of the faculty of imagination. Heidegger subsequently rooted the imagination in the soil of tradition, which is manifest, and hence made available for deliberate transmission, as an accumulated body of lore. It is through our participation in a historical tradition that the world becomes intelligible for us, and tradition does this by enabling us to properly follow the rules which give structure to thinking. Tradition allows us to make sense of things, to let things be what they are, by supplying us with the informal existential rules we need to bring our thinking into proper contact with things. These existential rules are historically manifest as lore, and may be subsequently formalised as logic.

There are striking similarities between Heidegger's account of logic and the one afforded by SSK practitioners. Indeed, Barnes and Bloor have characterised logic as a 'body of conventions and esoteric traditions,' as well as a 'learned body of scholarly lore.'[172] Barnes has also written, more generally, that the teacher of science is concerned primarily with 'transmitting lore.'[173] Furthermore, like Heidegger, Barnes and Bloor distinguish between the 'varied systems of logic as they are developed by logicians,' on the one hand, and the 'informal intuitions upon which they all depend for their operation,' on the other.[174] Bloor has elaborated an account of this distinction as one holding between formal and informal kinds of knowledge, and emphasised 'the priority of the informal over the formal.'[175] This appears similar to Heidegger's distinction between thematised and unthematised forms of thinking, as well the existential priority he gives to the latter over the former.

Bloor's central claims are that the formal rules of reasoning are always the instruments of informal reasoning, and hence that any particular application of a formal rule is always a potential subject for informal social negotiation.[176] In some cases, the application of a formal rule is not actually subject to negotiation because it is embedded in a highly standardised but informal context of shared practice. As Barnes and Bloor observe, in such cases everyone, or nearly everyone, just accepts the application of the rule without question. The application appears self-evidently appropriate, and thus requires no reasoned justification.[177] As one example of a rule which enjoys this kind of status, Bloor identifies the law of non-contradiction.

Bloor begins his discussion of the law of non-contradiction by turning to Edward Evan Evans-Pritchard's classic study of witchcraft among the Azande of Central Africa. According to Evans-Pritchard, the Azande treat witchcraft as an inherited physical trait, passed on from fathers to sons or from mothers to daughters.[178] He comments

---

172 Barry Barnes and David Bloor (1982), 'Relativism, Rationalism and the Sociology of Knowledge,' in *Rationality and Relativism*, ed. by Martin Hollis and Steve Lukes (Oxford: Blackwell), pp. 21–47 (p. 45).

173 Barnes (1974), *Scientific Knowledge and Sociological Theory*, p. 64.

174 Barnes and Bloor (1982), 'Relativism, Rationalism and the Sociology of Knowledge,' p. 44.

175 Bloor (1991), *Knowledge and Social Imagery*, p. 133.

176 Bloor (1991), *Knowledge and Social Imagery*, p. 133.

177 Barnes and Bloor (1982), 'Relativism, Rationalism and the Sociology of Knowledge,' p. 46.

178 Edward E. Evans-Pritchard (1937), *Witchcraft, Oracles and Magic among the Azande* (Oxford: Clarendon), p. 23.

that '[t]o our minds it appears evident that if a man is proven a witch the whole of his clan are *ipso facto* witches, since the Zande clan is a group of persons related biologically to one another through the male line.'[179] Yet, the Azande do not see things this way. They recognised the theoretical sense of Evans-Pritchard's argument, but in practice they only consider the close paternal kinsmen of a known witch to be witches. Their practice would thus seem to contradict the theoretical implications of their beliefs about the inter-generational transmission of witchcraft substance. Evans-Pritchard concludes from this that practical negotiations of belief 'free Azande from having to admit what appear to us to be the logical consequences of belief in biological transmission of witchcraft.'[180] For example, the Azande reason that even if a man is the son of a witch, and so has inherited witch-substance, that substance may remain 'cool' throughout his life. If his witchcraft is never activated, then he can hardly be considered a witch.[181] Hence, Evans-Pritchard observes that 'Azande do not perceive the contradiction as we perceive it because they have no theoretical interest in the subject, and those situations in which they express their beliefs in witchcraft do not force the problem on them.'[182]

Bloor's response to this has a positive and a negative side. On the negative side, he argues that Evans-Pritchard's analysis involves two central ideas: first, that there really is a contradiction in Zande views whether the Azande see it or not; and second, that if the Azande were to recognise the contradiction, one of their major social institutions would be untenable. The first idea has to do with the uniqueness of logic, and the second with the authority of logic.[183] Bloor rejects both ideas. In the first instance, he argues that there is no contradiction in Zande beliefs about witchcraft because the Azande may be seen to use a different logic. Hence, Evans-Pritchard's premise of uniqueness must be rejected.[184] There is a standard Zande logic and a standard Western logic. In the second instance, Bloor argues that both logics exercise little

---

179 Evans-Pritchard (1937), *Witchcraft, Oracles and Magic*, p. 24.
180 Evans-Pritchard (1937), *Witchcraft, Oracles and Magic*, p. 24.
181 Evans-Pritchard (1937), *Witchcraft, Oracles and Magic*, p. 25.
182 Evans-Pritchard (1937), *Witchcraft, Oracles and Magic*, p. 25.
183 Bloor (1991), *Knowledge and Social Imagery*, p. 139.
184 Bloor (1991), *Knowledge and Social Imagery*, p. 139.

authority over the practical negotiations of those who invoke them.[185] With the Azande, logic does not threaten their institution of witchcraft because one piece of logic can always be used against another, and even this is only necessary when logic is used by someone to pose a threat. But then it is the person, not the logic, who threatens.[186] In such cases, informal reasoning, or practical negotiation, takes priority over formal reasoning. Hence, Bloor alleges that Evans-Pritchard was also wrong to think that logical contradictions, where they exist, must necessarily threaten social stability if they are allowed explicit articulation.

As observations about the nature of logic and its relationship to informal reasoning, both of these points seem convincing. However, they do not work as a negative assessment of Evans-Pritchard's own views. As can be seen from the passages just quoted, Evans-Pritchard emphasised that the Zande doctrine of witchcraft contains 'what appears to us' to be a logical contradiction; he writes that 'to our minds' there is a contradiction in Zande beliefs. Elsewhere, he has argued that '[w]hat appear to be hopeless contradictions when translated into English may not appear so in the native language.'[187] Furthermore, in an appreciation of Lucien Lévy-Bruhl, Evans-Pritchard writes that 'primitive mystical thought is organized into a coherent system with a logic of its own.'[188] This is no assertion of uniqueness. As Mary Douglas has observed, Evans-Pritchard in fact strongly opposed the idea that those peoples who fail to employ standard Western logic are 'illogical,' or 'non-logical,' or 'pre-logical.'[189] Furthermore, Douglas argues that Evans-Pritchard also rejected the idea that formal logic should be granted authority over informal practical negotiation. She writes that '[w]ell before phenomenology's claim that sociological understanding must start from the negotiating activities of conscious, intelligent agents, Evans-Pritchard had seized the problem, developed a method and shown what progress can be made.'[190] Thus there is, in

---

185 Bloor (1991), *Knowledge and Social Imagery*, p. 140.
186 Bloor (1991), *Knowledge and Social Imagery*, p. 141.
187 Edward E. Evans-Pritchard (1965), *Theories of Primitive Religion* (Oxford: Clarendon), p. 89.
188 Edward E. Evans-Pritchard (1970), 'Lévy-Bruhl's Theory of Primitive Mentality,' *Journal of the Anthropological Society of Oxford* 1(2), 39–60 (p. 57).
189 Mary Douglas (1980), *Evans-Pritchard* (Sussex: Harvester Press), p. 34.
190 Douglas (1980), *Evans-Pritchard*, p. 86.

fact, a strong doctrinal continuity between Bloor and Evans-Pritchard. They are friends, not foes.[191] It must be noted, however, that Douglas's attempt to grant Evans-Pritchard precedence over phenomenology fails to acknowledge that Heidegger's argument in *Being and Time* for the existential priority of practice over theory, as well his description of the 'they' as the existential condition of possibility for social negotiation as such, pre-date Evans-Pritchard's earliest publications by several years. But there is no evidence that Evans-Pritchard had read Heidegger, nor is there any attempt in *Being and Time* to elaborate a sociological notion of negotiation. It would be better here to speak of synchronicity, rather than of priority.

Let us move on, then, to the positive side of Bloor's analysis. Like Evans-Pritchard, Bloor blocks the parochial and paternalistic charge that the Azande are guilty of ignoring a contradiction in their reasoning. The problem arises because we are insensitive to the specific existential conditions which give shape to Zande social life: 'The fact that we can imagine extending the witchcraft accusation to the whole of the clan is simply because we do not really feel the pressure against this conclusion.'[192] Furthermore, if we did feel the absurdity of the conclusion, we could easily give reasons to reject it, if this were deemed necessary. The Azande, for example, drew from their well-established notion of 'cool' witches in response to Evans-Pritchard's theoretical queries. And, Bloor argues, we make precisely the same kinds of moves in our own Western scientific culture: 'the Azande think very much as we do.'[193]

Bloor argues that the process of intellectual elaboration, where reasons are introduced to deflect unwanted conclusions, is a pervasive feature of our science. He gives two historical examples.[194] The first

---

191 In fact, both also share a legacy stretching back to the experimental psychologist Frederick Bartlett. On the one hand, Bloor recognises Bartlett as 'a prototype sociologist of scientific knowledge' (David Bloor (2000), 'Whatever Happened to "Social Constructiveness"?,' in *Bartlett, Culture & Cognition*, ed. by Akiko Saito (London: Psychology Press), pp. 194–215 (p. 196); see also David Bloor (1997a), 'Remember the Strong Program?,' *Science, Technology, & Human Values* 22(3), 373–85). On the other hand, as Douglas observes, Evans-Pritchard adopted heavily from Bartlett, his own work thus displaying a 'continuous relationship' with Bartlett's earlier studies of conventionalisation in perception (Douglas (1980), *Evans-Pritchard*, p. 26).

192 Bloor (1991), *Knowledge and Social Imagery*, p. 142.

193 Bloor (1991), *Knowledge and Social Imagery*, p. 145.

194 Bloor (1991), *Knowledge and Social Imagery*, pp. 143–45.

concerns the phlogiston theory of combustion, a chemical theory which dominated in the eighteenth century and which has since been rejected. Proponents of this theory held that a metal is comprised of phlogiston and calx, or what we now call oxide. It was thought that when a metal is burned, the phlogiston is removed from it, leaving only calx behind. It was, however, discovered that the calx was heavier than the original metal, which raised the question of how the weight of a substance could be increased by taking something away from it. Some historians have argued that these circumstances logically imply that phlogiston must have a negative weight, an absurd conclusion which therefore doomed the theory. Yet, Bloor writes, most of those who adhered to phlogiston theory did not draw this conclusion. They instead reasoned that when phlogiston leaves the metal another, heavier, substance must take its place. The best candidate for this replacement was thought to be water. Hence, the conclusion that phlogiston must have a negative weight was deflected. It now only became necessary to hold that phlogiston was much lighter than water. The logic of subtraction — the removal of phlogiston from the metal through combustion — was thus traded for a logic of replacement — water substitutes for phlogiston during the combustion of the metal. Advocates of phlogiston theory were not contradicting a tacitly held logic of subtraction; they were using another sort of logic altogether.

The second of Bloor's examples comes from the atomic theory of chemistry. At the turn of the nineteenth century, Joseph Louis Gay-Lussac discovered a regularity in the way gases combine. One volume of gas always combines with some small whole number of volumes of another gas, where volume measurements are controlled for temperature and pressure. Thus, two volumes of hydrogen combine with one volume of oxygen to produce one volume of gaseous water. Because John Dalton's atomic theory explained chemical combinations by the direct combination of atoms, Gay-Lussac's results suggested that two hydrogen atoms must combine with one atom of oxygen to produce one compound atom of water. Each volume of gas therefore contains the same number of atoms. The trouble with this hypothesis, however, was that one volume of nitrogen could combine with one volume of oxygen to produce two volumes of nitric oxide. This suggested that one volume of nitric oxide contained only half the number of atoms of the original volumes of nitrogen and oxygen. The idea that volumes contained the same number of atoms could now only be maintained if one assumed that atoms could be split in half. But Dalton resisted this conclusion, as he was not prepared to give up the doctrine of the indivisibility of the atom. There was a contradiction between this doctrine and the alleged regularity discovered by Gay-Lussac. Perhaps Gay-Lussac was wrong?

As Bloor observes, the conclusion that atoms are divisible is easily avoided without rejecting Gay-Lussac's results. One need only assume that each atom of gas is really a particle composed of two atoms. When nitrogen and oxygen combine, they swap atoms. The combination is not the result of addition, as when hydrogen and oxygen combine to form water, but rather of substitution. We now know this solution as Avagadro's hypothesis. This was a simple elaboration which supplanted the earlier logic of addition with a logic of substitution, thus allowing for a negotiation around a potential contradiction without giving up either Gay-Lussac's useful empirical results or the doctrine of atomic indivisibility.

Bloor argues that this negotiation is of a piece with both eighteenth-century negotiations around alleged contradictions in phlogiston theory and twentieth-century negotiations around alleged contradictions in Zande beliefs about witchcraft. He emphasises that the Azande and Western scientists think in the same general way, writing that 'they have the same psychology [...] but radically different institutions.'[195] An institution is, on Bloor's definition, 'a collective pattern of self-referring activity.'[196] A rule, or a group of rules, a logic, is an institution in this sense. To say that a logic is self-referring is to recognise that its validity is internal to the collective patterns of activity in which the logic becomes manifest. So, when we accuse the Azande of contradiction, our statement in fact refers, not to Zande logic, but to our own. We treat the Azande as if they were a part of our own community. Because the accusation carries force only within our own community, it is not surprising that it left the Azande unmoved. By the same token, the accusation that Gay-Lussac's discovery contradicts the doctrine of atomic indivisibility presupposes a kind of chemical thinking in which gas particles are composed of only one atom. But this way of thinking was renegotiated

---

195 Bloor (1991), *Knowledge and Social Imagery*, p. 145.
196 Bloor (1997b), *Wittgenstein, Rules and Institutions* (London: Routledge), p. 33.

with the introduction of Avagadro's hypothesis. Hence the accusation carries force only in reference to a kind of chemical thinking which, through a process of negotiation, has now become obsolete.

The circular validity of self-referring logical domains is captured in Heidegger's observation that logic, when treated as the ultimate ground for all thinking, must also ground itself. As discussed earlier, Heidegger characterised this as the argument that logic is 'free-floating' and 'ultimate,' that it grounds all other forms of thinking without requiring a further ground of its own. Heidegger challenged this argument by differentiating logic, as a system of formal rules, from informal, existential rules. The former is grounded in the latter. He furthermore conceptualises existential rules in terms of original time, which is manifest in two types of relation: in-order-to relations and for-the-sake-of-which relations. This distinction between relations plays a role similar to the distinction Bloor makes between a self-referring practice and what 'primes' that practice. It is, he writes, a characteristic of self-referring practices that a separate account must be given of their origin, of how they get going, and this account must be presented in terms lying outside the domain of the practice itself.[197] In other words, the priming element must exist prior to the practice; it is the *a priori* element which makes the practice possible.

Bloor's self-referring practices, his institutions, recall Heidegger's description of the multiple relations of the in-order-to. The assignedness of a work-world is manifest in the in-order-to relations making up that world. These relations represent the significance, or meaning, of a work-world. The for-the-sake-of-which relation, on the other hand, provides the origins of that assignedness. In-order-to relations thus depend for their own existence on the prior existence of the for-the-sake-of-which. The latter provides the conditions of possibility for being assigned, for significance, as such. This looks a lot like Bloor's priming element, which works to get the self-referring practice going. As discussed earlier, Heidegger's notion of the for-the-sake-of-which is a phenomenological reinterpretation of Plato's idea of the good. It serves as the fundamental feature of thinking enabling us to discriminate between epistemically

---

197 Bloor (1997b), *Wittgenstein, Rules and Institutions*, p. 32. Bloor adopts this notion of priming from Barry Barnes (1983), 'Social Life as Bootstrapped Induction,' *Sociology* 17(4), 524–45 (p. 529).

relevant and irrelevant phenomena, and hence to connect thinking with things in a coherent and meaningful way. The for-the-sake-of-which serves, in other words, as the normative principle guiding thinking.

Bloor writes that 'the mystery of logical compulsion is [...] the mystery of normativity.'[198] Normativity emerges as individuals fall into regularised patterns of self-referential activity. Only as participants embedded in such patterns of regularised activity will individuals feel compelled to follow those patterns. The Azande interviewed by Evans-Pritchard in the 1930s, for example, were not embedded in the patterned activities constitutive of Western logic, and so they did not feel the force of the contradiction alleged by Evans-Pritchard to exist within their doctrine of witchcraft. They could see the sense of it, but it did not move them, did not affect them. Normativity thus provides a necessary basis for logic as an institution, but the specific norms governing a particular logic will carry force only within the finite bounds of the specific social and historical setting in which that logic is sustained as a form of thinking. According to Heidegger, Kant cracked open the problem of normativity by asking what the human being must be like in order to be compelled by a rule. He suggested that human beings must be such that they are affected by the law, that they feel respect before the law as it is manifest in the actions of others. This claim resonates with Bloor's own statement that '[w]e are compelled by rules in so far as we, collectively, compel one another.'[199] Bloor furthermore specifies that 'the compelling character of rules resides merely in the habit or tradition that some models be used rather than others.'[200] This invites comparison with Heidegger's claim that the for-the-sake-of-which, as the normative element in thinking, is itself rooted in the soil of tradition. Both Bloor and Heidegger seek to dispel the mystery of normativity by tracing its origin back to tradition. For both, what primes an informal domain of logic, what compels its users into regularised patterns of common activity, is the tradition into which those users have been socialised.

As mentioned earlier, Heidegger's analysis of logic comes to an end at this point. He identified the source of normativity in logic, but did not ply deeper into its structure. Bloor has, on the other hand, done

---

198 Bloor (1997b), *Wittgenstein, Rules and Institutions*, p. 2.
199 Bloor (1997b), *Wittgenstein, Rules and Institutions*, p. 22.
200 Bloor (1991), *Knowledge and Social Imagery*, p. 138.

just that. He emphasises that the socialisation process, because it is the source of normativity, cannot itself be described in normative terms.[201] The logic student is not guided by her own dispositions when learning to draw the correct inference, she is guided by her teacher. Indeed, only through such training does the student acquire the specific dispositions necessary for carrying on alone in proper logical fashion. By becoming habituated to a particular tradition of thinking, the student learns to act instinctively in response to certain kinds of stimuli. She will feel repelled by actions violating the rules to which she has become habituated, and attracted by actions which accord with them. She will feel herself compelled to act in accordance with those rules, and will have thus now also acquired the competence to socialise others into her tradition by transmitting to them the practical and theoretical lore she has made her own.[202]

So much agrees with, and also serves to extend, Heidegger's account. However, Bloor's sociological explanation of logical compulsion also includes a move which appears to conflict with Heidegger's account. Bloor argues that we need to explain how 'non-intentional, non-normative responses' can give rise to the regularised pattern of an informal self-referring practice.[203] The conflict arises because, for Heidegger, there can be no non-intentional responses. At the root of the conflict lie differing conceptions of intentionality. Bloor seems to understand intentionality as being inextricably tied up with semantic content. Hence, because 'we always encounter the dependence of semantic content on a more basic, non-semantic level,' Bloor concludes that the more basic non-semantic level must be free from 'intentional elements.'[204] We must, he cautions, resist the temptation to assume that our automatic responses are already endowed with 'propositional and logical content.'[205] Bloor thus seems to view intentionality as propositional content belonging to a substance-subject. To be sure, Bloor has noted that groups, rather than individuals, may be the bearers of intentions.

---

201 David Bloor (2004b), 'Institutions and Rule-Scepticism: A Reply to Martin Kusch,' *Social Studies of Science* 34(4), 593–601 (p. 595).
202 The emotional dynamics which help keep epistemic groups together will receive slightly more systematic attention in Chapter Seven.
203 Bloor (1997b), *Wittgenstein, Rules and Institutions*, p. 135.
204 Bloor (2004b), 'Institutions and Rule-Scepticism,' p. 597.
205 Bloor (2004b), 'Institutions and Rule-Scepticism,' p. 598.

He has written that '[o]n the institutional theory intentionality, plural and singular, is not in the head.'[206] Combined with the conviction that intentions must include propositional content, this would seem to push Bloor into viewing the group likewise as a substance-subject bearing propositional content, content which is not 'in the head' but, strange as it may seem, somewhere else. All the difficulties faced at the individual level thus threaten to reappear at the group level. Bloor is, of course, not insensitive to these difficulties, and thus he attempts to resolve them by furthermore arguing that intentions must be explained reductively in non-intentional terms.

Heidegger, in contrast, viewed intentionality in terms of our existential relation to things and persons in the world. Because one basis of human existence is being-in-the-world, we necessarily find ourselves always already among the other entities also inhabiting the world. As noted earlier, Heidegger viewed intentionality as the 'directedness' of our relation to these other entities. It is thus not a propositional content belonging to a substance-subject, whether individual or group, but a directional relation belonging to our existence as being-in-the-world. On this view, a basic feature of our existence is that 'there is always an entity and an interconnection with an entity already somehow unveiled, without its being expressly made into an object.'[207] The key points here are that, for Heidegger, our being in the world entails intentionality, and intentionality need not have an object. This departs from more standard theories of intentionality, but we have already discussed the conceptual background which drives this move. In Heidegger's view, perceiving a thing as an object is a derivative, theoretical way of understanding the whatness of that thing. A more fundamental mode of understanding is one in which the thing is encountered non-thematically as a piece of equipment taken up in the course of a practical task performed in a work-world. In this latter case, we are directed towards the equipment, because we use it, but the intentionality of the relation includes no propositional content. Only when the intentional relation also takes on the character of a thematising do we then come to perceive that thing as an object, that is, as a property-bearing substance. This is not a transition

---

206 David Bloor (1996), 'Idealism and the Sociology of Knowledge,' *Social Studies of Science* 26(4), 839–56 (p. 850).

207 Heidegger (1982a), *Basic Problems of Phenomenology*, p. 157; translation modified.

from a non-intentional to an intentional state, but from one mode of intentionality to another.

It is difficult to see that Bloor would lose anything crucial by adopting Heidegger's account of intentionality.[208] On the other hand, the benefits are clear. Bloor writes: 'I certainly want to see the links between the intentional level and its non-intentional basis made stronger.'[209] An explanation of this link may be strengthened in two ways. First, Bloor could shift his definition of intentionality in the way just mentioned. Second, he could adopt Heidegger's phenomenological analysis of the change-over in the way we experience and understand things. This is a change-over in intentionality. Recall that the transformation is from the way things are practically experienced during immersed involvement in a work-world to the way things are theoretically experienced as objects, and as the subject matter for propositional statements. A detailed account of the change-over was given in Chapter Two. As a rough recap, Heidegger analyses the change-over in terms of four stages. First, after an interruption in our immersed involvement with things, we step back and just look at them. This pure looking then becomes a thematising in which things now become experienced as the distinct objects of a disengaged form of perception. Third, we then begin to deliberate over these distinct objects so as to determine their properties. Finally, these determinate objects then become the subject matter for propositional assertions, that is, for the statements of theory and logic.

This stepwise description provides a more nuanced analytical tool than does Bloor's bipartite distinction between informal and formal knowledge. It thus allows us to better understand how non-deliberative, non-propositional thinking shifts into deliberative, propositionally structured thinking. This shift can take place quickly and on an individual level, as was discussed in Chapter Two. However, as the present chapter has argued, a shift on the individual level is itself enabled by a collective history of the way the relation between things and thinking has come

---

208 Bloor might worry that Heidegger's account of intentionality threatens his own commitment to causal explanation. But Barnes has nicely demonstrated how the sociologist can incorporate intentional phenomena into a counterfactual account of causation (Barnes (1974), *Scientific Knowledge and Sociological Theory*, pp. 71–78). Of course, such accounts are far from uncontroversial, but I make no claim that the proposed negotiation will be easy to achieve in pragmatic terms.

209 Bloor (2004b), 'Institutions and Rule-Scepticism,' p. 600.

to be thematised, and so theoretically understood, over the course of centuries. The way an individual makes sense of her relation to things can be explained by reference to the tradition which structures the way she thinks about those things, and, more concretely, to the processes by which she becomes socialised into that tradition. The explanation thus has both historical and social components.

Bloor argues that explaining the links between intentional phenomena and their non-intentional base is a difficult and complex task, and he compares it with attempts to explain the links between the phenomenal qualities of water and their molecular base in hydrogen and oxygen.[210] His point is that chemists are not normally criticised for attempting such an explanation. He thus suggests that sociologists, in turn, 'may reasonably ask that what is granted to the chemists be granted to them as well.'[211]

Maintaining the analogy to chemistry, it will be useful to compare Bloor's hypothesis to the hypothesis of Gay-Lussac, discussed earlier in this section. On the basis of experimental findings, Gay-Lussac proposed that every fixed volume of gas included the same number of atoms. As we saw, this hypothesis directly contradicted the established doctrine of the indivisibility of the atom. Was Gay-Lussac's hypothesis thus rejected in order to protect this doctrine? No. As Bloor argued, a negotiation occurred which swapped a logic of addition for a logic of substitution.

Move now to Bloor's hypothesis that intentional phenomena may be explained in terms of non-intentional phenomena, the latter of which may also include 'ordinary human interactions.'[212] This hypothesis directly contradicts, and so threatens, the doctrine that human action must include an irreducibly intentional element. Should this doctrine be rejected in order to preserve Bloor's hypothesis? Bloor evidently thinks so. But Heidegger's account of intentionality allows for a simple negotiation which would preserve both this doctrine and Bloor's hypothesis. Consider that Bloor's hypothesis employs a logic of addition. Non-intentional phenomena combine to produce intentional phenomena much as hydrogen and oxygen combine to produce water.

---

210 Bloor (2004b), 'Institutions and Rule-Scepticism,' p. 600.
211 Bloor (2004b), 'Institutions and Rule-Scepticism,' p. 600 n. 3.
212 Bloor (1997b), *Wittgenstein, Rules and Institutions*, p. 133.

One assumption of this logic is the identification of intentionality with propositional content. As just argued, we can replace this assumption with Heidegger's claim that intentionality may either include or not include propositional content. By doing so, we trade Bloor's logic of addition for a logic of substitution — whereby one mode of intentionality is substituted for another — thus preserving both his hypothesis as well as the doctrine that human actions are irreducibly intentional.[213] The key move in this negotiation is the redefinition of intentionality from being a property of a mental substance to its being a structure of existence. Seen against the backdrop of the history sketched out in this chapter, this frees the concept of intentionality from the glass-bulb model of an internally organised mental substance, where it had been shielded from philosophical scrutiny by Descartes in the seventeenth century. Intentional acts are not necessarily mental acts, sealed up in a purified propositional space, and trying desperately to somehow break out and connect with the elusive things of an external world. They are existential acts which necessarily take place in a world shared with other persons and populated by all manner of things ceaselessly stimulating us into ever recurrent and reconfigurable patterns of thought.

11. Conclusion

The cumulative effect of this and the preceding chapters should have been to convince readers of the benefits of comparing and combining the tools and insights of SSK with those of Heidegger's existential phenomenology. As we have seen, there are many striking similarities between Heidegger's earlier concerns and the more recent concerns of SSK practitioners, as well as many ways in which their respective methods may be fruitfully combined. We have now explored a few of these. Absent from our considerations thus far, however, has been an explicit focus on issues relevant to the history of the natural sciences.

---

213 The proposed negotiation is consistent with Martin Kusch's observation that 'Bloor only succeeds in reducing one intentional phenomenon to another' (Martin Kusch (2004a), 'Reply to My Critics,' *Social Studies of Science* 34(4), 615–20 (p. 618)). However, Kusch appears to also identify intentionality exclusively with propositional content: 'an intentional fact involves concepts like beliefs and desire' (Martin Kusch (2004b), 'Rule-Scepticism and the Sociology of Scientific Knowledge: The Bloor-Lynch Debate Revisited,' *Social Studies of Science* 34(4), 571–91 (pp. 579–80)).

This wrinkle will be smoothed out in Chapters Five and Six. As will have already become clear in the present chapter, Heidegger was himself deeply engaged in historical research. The phenomenological history of logical thinking presented in this chapter represents just one thread running through the tapestry of Heidegger's life-long engagement with the history of philosophy. Furthermore, as is evident from the preceding discussion, this engagement was no mere antiquarian interest in the past, but was instead driven by Heidegger's desire to resolve perceived confusions in the present by excavating and developing the unrealised possibilities of the past. His longing after a new method by which to build a more stable foundation for the sciences, bounded as it was by a commitment to the irremediable finitude of human being, pushed Heidegger into ever deeper historical reflection. For him, the reservoir from which to draw insight lay not in a timeless and immaterial realm of distinct ideas or formalised propositional structures, but rather in the untapped possibilities latent in his own rich and variegated historical tradition.[214]

A key moment in Heidegger's phenomenological history of logic was the early-modern mathematical interpretation of Aristotle's construal of thinking in terms of the proposition. Heidegger broadly located this moment in the early seventeenth century, taking Descartes as his exemplar. In his view, the mathematicisation of the science of thinking, that is, of logic, coincided with the emergence of modern natural science. Indeed, as mentioned earlier, Heidegger discussed Descartes's innovations together with the equally decisive work of Galileo and Newton. Heidegger's remarks about Galileo and Newton will be considered in Chapter Six. In that chapter, as well as the one immediately ahead of us, we will pick up the second part of Heidegger's two-part description of the existential foundations of modern science, a part left largely unaddressed in this chapter. Recall that Heidegger distinguished modern science from its predecessors by pointing to what 'rules and determines' the basic activities of the former in contrast to the latter. The first part of this determination was the distinctive

---

214 As Reinhard May has demonstrated, Heidegger's European tradition also included a long-standing, transcultural engagement with East Asian thought (Reinhard May (1996), *Heidegger's Hidden Sources: East Asian Influence on His Work*, trans. by Graham Parkes (London: Routledge)).

way in which early-modern scientists came to experience their work with things. The second part was what Heidegger described as early-modern scientists' metaphysical projection of the thingness of things. This description concerned the ways in which scientists' productive comportment towards things served to determine, *a priori*, the whatness of those things. As we will see, these two parts feed into one another: the way a scientist experiences her work with things is shaped by her *a priori* projection of the thingness of those things, and her *a priori* projection of the thingness of things is likewise influenced by her past experience of working with those things. This interplay is possible because, as was argued in this chapter, the *a priori* does not derive its authority from a rarefied realm lying outside the everyday work-world, but is instead firmly rooted in the tradition which both enables and is sustained by that work-world.

In this chapter, the focus has been on the second part of Heidegger's two-part description of the foundations of modern science. Indeed, this is where Heidegger placed almost all of his own attention. He thus never managed to give a full account of the interplay which he himself had identified and placed at the centre of his phenomenological account of science. Missing in Heidegger's work is a more thoroughgoing exploration of the way in which the everyday work-world becomes thoroughly implicated in the metaphysical deliberations of scientists. This is another place where the empirical work definitive of SSK can help to fill out and strengthen Heidegger's abstract reflections with detailed historical studies of concrete scientific work. We will see, in turn, that attention to Heidegger's broader theoretical reflections can help protect historians of science from the methodologically induced parochialism which may threaten any specialised intellectual practice occupying itself so much with the trees that it comes to neglect certain important features of the forest.

Chapter Five

Mathēsis and the Emergence of Early-Modern Science

1. Introduction

Place a grain of wheat on the ground in front of you. Does this amount to a heap of grain? No, of course not. So, add a second grain. Is it a heap now? No. A third grain? No. A fourth grain? No. A fifth grain? No. And so it goes also after a sixth grain, a seventh grain, an eight grain, and a ninth grain. When a tenth grain is added, we still do not have a heap.

But we do have a pattern. The pattern is this: if some grains do not make a heap, then adding one more grain will not turn them into a heap. On the basis of this pattern, we can now make more specific predictions. We can predict that adding an eleventh grain will not make a heap, nor will adding a twelfth grain, a thirteenth grain, or a fourteenth grain. The pattern, then, has the character of a general prediction: when you add one grain to a non-heap, the result will be a non-heap. A single grain cannot determine the difference between a non-heap and a heap.

The second-century Greek physician, Galen of Pergamon, exploited the predictive power of this pattern when he wrote:

> I know of nothing worse and more absurd than that the being and not-being of a heap is determined by a grain of corn. And to prevent this absurdity from adhering to you, you will not cease from denying, and

  http://dx.doi.org/10.11647/OBP.0129.05

will never admit at any time that the sum of this is a heap, even if the number of grains of wheat reaches infinity by the constant and gradual addition of more. And by reason of this denial the heap is proved to be non-existent, because of this pretty sophism. And so it follows from this sophism that the mountain also does not exist.[1]

The 'you' to whom Galen addressed his comment was the 'dogmatic' physician. Galen argues that the dogmatist cannot reject the implication of the pattern without appearing silly. However, the dogmatist will then be forced to reject the existence of heaps, which is also silly. The dogmatist is thus faced with a paradox for which no solution is obvious.

This is a case of turnabout being fair play, because the dogmatist had already laid out the same kind of argument against Galen and other empirical physicians. In the above passage, Galen is demonstrating that the paradox cuts both ways, causing potential problems for the dogmatist as well.

Galen tells us that the dogmatist has challenged the empiricist physician's claim that a belief will be credible if it is supported by evidence which has been seen 'very many times.' To this, the dogmatist asks: How many times? Ten? The empiricist says no, ten times is not enough. The dogmatist then asks: Eleven times? No, says the empiricist, eleven times is also not enough. The pattern has now appeared, and the dogmatist leads the empiricist into the paradox. If the empiricist insists on the method of 'seeing very many times,' she will be forced to admit the impossibility of empirical knowledge.

> For if something that was seen forty-nine times and yet in all these times was not accepted nor considered to be true, now by the addition of this one single time comes to be considered acceptable and true, it is obvious that only by being seen a single time has it become acceptable and true. The inevitable conclusion is that seeing a thing once — although at the outset this was not accepted and considered true — has on this occasion such force that when added to something which was not acceptable and not considered true as to make it acceptable, and vice versa.[2]

The dogmatist argues that one observation is enough to produce knowledge. The 'force' of this single observation derives from the

---

1 Galen of Pergamon (1944), *On Medical Experience*, 1st edn of the Arabic version, with English trans. by Richard Walzer (Oxford: Oxford University Press), p. 116.

2 Galen (1944), *On Medical Experience*, p. 97.

physician's 'know[ing] from the very beginning what things have to be eliminated and disregarded as being superfluous and unnecessary, and what things have to be examined and to be judged carefully as to their usefulness and their necessity.'[3] What the dogmatic physician is insisting on here is the need for an *a priori* standard by which to differentiate between epistemically desirable and epistemically undesirable phenomena. Moreover, the force of this standard in use will be immediate, manifesting itself on the basis of a single act of observation. At best, the empiricist's little-by-little method fails to reflect the immediacy and constructivity of reason. Today, this kind of dogmatism is more commonly called 'rationalism.'

The paradoxes generated by this pattern of argument pose a threat to empiricists and rationalists alike. More generally, they threaten confidence in both inductive and deductive forms of inferential reasoning. Scepticism about induction is well known, and has often been discussed in terms of the underdetermination of theory by data, a topic which we earlier addressed in Chapter Three. No matter how many times the empiricist observes the phenomena, no knowledge will arise without the mediation of some additional, unobserved element. For the rationalist, this additional element is the *a priori*. For the sociologist of knowledge, it is grounded in social convention. As we saw in Chapter Four, Heidegger grounds the *a priori* in a historical tradition, a stance which is compatible with that of SSK. By posing the socio-historical contingency of the *a priori*, the latter two positions reject the absolutism of the rationalist. For them, scientific knowledge is objective, but not absolutely so.[4]

---

3   Galen (1944), *On Medical Experience*, p. 93.
4   David Bloor uses the paradox of the heap to challenge a rationalistic belief in the absolute (i.e., exceptionless) validity of deduction (David Bloor (1991 [1976]), *Knowledge and Social Imagery*, 2nd edn (Chicago: University of Chicago Press), pp. 182–83). The paradox seems to offer an exception, because the repeated application of deduction to apparently true premises ('adding one grain to a non-heap results in a non-heap' and 'one grain is added to this non-heap') seems to result in a false conclusion ('this is a non-heap'). Defensive reactions to the paradox can make explicit the social labour required to maintain belief in the absolute validity of deduction. An example can be found in a paper by Colin Howson, wherein he seeks to suppress the paradox as I present it in my own work (Colin Howson (2009), 'Sorites Is No Threat to *Modus Ponens*: A Reply to Kochan,' *International Studies in the Philosophy of Science* 23(2), 209–12; Jeff Kochan (2008), 'Realism, Reliabilism, and the "Strong Programme" in the Sociology of Scientific Knowledge,' *International*

Galen's second-century response to the dogmatic physician is similar to these responses. He rebuffs the dogmatist's charge that the empiricist cannot produce 'technical' — that is, objective — knowledge: 'you would be acting both unjustly and wrongly in pestering us to specify, of a thing which when seen once only is not "technical" according to your argument, how many times it must be seen in order to become "technical."'[5] The injustice lies in ridiculing the empiricist 'because we cannot state with exactitude the precise number [...], but are only able to give a general notion.'[6] Based on this argument, the dogmatist will insist that a definition of 'heap,' in order to be objective, must be precise rather than general; it must exactly specify how many grains make a heap. Yet, as Galen has already shown, the dogmatist cannot do this without appearing silly. Indeed, by specifying the precise number of grains with which a non-heap becomes a heap, the dogmatist reveals his purportedly 'technical' knowledge of heaps is, in fact, lamentably subjective.

Galen then expands this critique to encompass the dogmatist's medical knowledge in general. He points out that the dogmatist's *a priori* standard 'is not uniform, universal, comprising all of you, because you have different views and each one of you holds an opinion completely contradictory to the opinion of the others.'[7] The implication is that objective knowledge is marked, to some substantial degree, by social agreement. The paradox of the heap relies on this idea, since, because most people agree that heaps exist, the dogmatist will appear silly if he claims that they do not.

> If you wish, speak, it will not cause me to be angry with you; if, however, you should say of something which people continually see under the same conditions throughout their lives, that it is non-existent, it will not help you at all. [...] I for my part adhere to and follow that which is known to men, and accept what is obvious without inquiring into the cause of each individual thing. Therefore I say of what has been seen

---

      Studies in the Philosophy of Science 22(1), 21–38). I respond to Howson in Jeff Kochan
      (2009a), 'The Exception Makes the Rule: Reply to Howson,' *International Studies in
      the Philosophy of Science* 23(2), 213–16.
5   Galen (1944), *On Medical Experience*, p. 118.
6   Galen (1944), *On Medical Experience*, p. 119.
7   Galen (1944), *On Medical Experience*, p. 104.

but once, that it is not 'technical,' just as a single grain of wheat is not a perfect heap; if, however, it is a thing that is seen many times in the same way, then I call it 'technical.'[8]

For Galen, objective knowledge has a social aspect — it is 'that which is known to men' — and a historical aspect — it is based on that which is 'seen many times in the same way.' However, he also notes that such knowledge is 'obvious without inquiring into the cause of each individual thing.' When it comes to heaps, this may be true. But what about disease? At least in the case of unfamiliar diseases, it would seem that close attention to causes may help in the development of an effective treatment. This was a point made by Galen's dogmatic opponent.

> But just because the number of concomitants of diseases is so great and there is such variety in what causes evacuation and what is vomited up and what is introduced into the organism, while those things that affect it from the outside are still more numerous, the Empiricist is still less able to judge which of them are beneficial and which harmful.[9]

Here the dogmatist reinforces the claim that the physician requires a standard by which to reliably distinguish between causes which are, and those which are not, necessary for the understanding and effective treatment of disease. Galen accepts the need for such a standard, but dismisses the idea that it be subjected to systematic enquiry.

> [I]t has been found that what has been seen many times becomes 'technical.' With regard to the cause, however, which makes it completely 'technical' and when it begins to be completely 'technical,' I am of opinion that it is idle to demand this. For I find that not a particle of harm befalls arts and men in their modes of life and activities for being ignorant of such things.[10]

Objective medical knowledge is produced through serial, disciplined observation, but the question of what norms guide those observations is of little interest to Galen. For him, it is enough to know those norms in a vague and general way, much as we do the norms used to distinguish

---

8   Galen (1944), *On Medical Experience*, p. 119.
9   Galen (1944), *On Medical Experience*, p. 93.
10   Galen (1944), *On Medical Experience*, p. 121.

between a heap and a non-heap. There is, he thinks, no great need to specify them in a systematic or precise way. In Heidegger's term, Galen has no interest in thematising the norms which guide successful medical thinking. In SSK's terms, he is happy to rely on those norms as a resource, but resists turning them into a topic. Galen's attitude may thus strike one as distinctly unscientific. Moreover, it seems to be an attitude hostile to epistemology, in general, and, insofar as Galen attributes a social aspect to medical knowledge, to SSK, in particular.

In fact, as we will see in this chapter, later physicians became increasingly interested in understanding the methods by which reliable medical knowledge is produced. More than a millennium after Galen, Renaissance empirical physicians began to topicalise and systematically investigate the informal logic which they thought must underpin their techniques of medical discovery. As historians have long recognised, these physicians were innovators in the rationalisation of scientific method, and hence key contributors to the rise of early-modern empirical science.

The Renaissance empiricists examined in this chapter accepted the dogmatist's view that physicians must know something in advance, from the very beginning, in order to successfully diagnose and treat disease. Moreover, they agreed with the dogmatists that this *a priori* knowledge was a knowledge of causes. However, they resisted the dogmatist's claim that, on the basis of this knowledge, only one observation is needed to properly diagnose a disease. Indeed, the empiricists continued to defend their doctrine of serial observation, of medical knowledge gained little by little through a practice of 'seeing many times in the same way.'

The epistemology of these empirical physicians thus appears to have been circular: they claimed both that they already possessed knowledge of health and disease at the very beginning of their enquiry, and that their knowledge of health and disease was the consequence of their method of serial observation. They argued, however, that this epistemic circle was not a vicious one. Like Galen over one thousand years earlier, these Renaissance physicians argued that their *a priori* knowledge was of only a general, imprecise sort. Their empirical method, then, was meant to transform this general and imprecise knowledge into specific and precise knowledge. Hence, their account of knowledge was not viciously circular, because it involved two different sorts of knowledge, one linked to the other by an empirical method of investigation.

As already noted, this method of investigation bears striking similarities to what Heidegger called a thematising articulation, a topic to which we gave detailed attention in Chapter Four. One key aspect of this process of articulation is the transformation of informal, often tacitly held knowledge into explicit, formalised knowledge. The efforts of Renaissance physicians to understand this process, as a central feature of their own medical practice, provides further historical support for Heidegger's existential conception of science. Moreover, this historical case also allows us to more fully explore one particular facet of Heidegger's existential conception, namely, the mathematicisation of natural knowledge. With this, we expand on a topic first addressed in Chapter Two under the label of 'the mathematical projection of nature.' The role of mathematics in the emergence of early-modern science has been a key point of debate among contemporary historians of science, a debate which, in many ways, reiterates the long-standing feud between empiricists and rationalists alluded to above. Heidegger's concept of *mathēsis*, or 'the mathematical,' as well as its relation to what is commonly referred to as the 'Scientific Revolution,' adds a further perspective to this debate, one which seeks to combine the best insights from both camps. In addition, as we will see near the end of this chapter, Heidegger's account of modern science as *mathēsis* also challenges the historiographic commonplace that the Scientific Revolution coincided with the expungement of Aristotelian 'final causes' from scientific practice in the seventeenth century. On Heidegger's account, final causes were not abandoned, but instead radically transformed — to wit, *mathematicised*. This claim will lay the ground for a more detailed discussion in Chapter Six about the rise of seventeenth-century experimental philosophy. For now, let us start with a review of Heidegger's account of *mathēsis*, and then move on to consider the deliberations of empirically minded physicians during the three centuries prior to the emergence of early-modern science. These topics will lead us into the heart of key historiographic debates over the existence and nature of the Scientific Revolution.

2. Modern Science as *Mathēsis*

As discussed in previous chapters, the thing has been typically defined as a property-bearing substance. Heidegger writes that this definition seems 'natural,' in the sense of 'what is understood without further ado and is "self-evident" in the realm of everyday understanding.'[11] Yet he challenges this construal of the 'natural' as being self-evident, and thus not amenable to further analysis, arguing instead that '[t]he "natural" is always historical.'[12] The prevailing definition of the thing did not 'just fall absolutely from heaven, but would have itself been based on very definite presuppositions.'[13] Indeed, as was argued in Chapter Four, one key presupposition determining the prevailing definition of the thing was the Aristotelian claim that the structure of the thing may be usefully modelled on the structure of the proposition. According to Heidegger, this proposition-based account of the thing has played a central role in the development of the modern scientific understanding of the thing as an object of investigation.

One consequence of this definition is the treatment of a thing in abstraction from its concrete circumstances and other unique features. A definition of the thingness, or whatness, of the thing, *as such*, is a generalised definition which deliberately overlooks all the peculiarities distinctive of any one particular thing. Thus Heidegger writes:

> [A] botanist, when he examines the labiate flower, will never be concerned about the single flower as a single one: it always remains an exemplar only. That is also true of the animals, for example, the countless frogs and salamanders which are killed in a laboratory. The 'this one' (*je dieses*) which distinguishes every thing, will be skipped over [übersprungen] by science.[14]

In this section, we will consider in detail Heidegger's account of modern science as a form of understanding which 'skips over' the individual specificities of the things it investigates. Heidegger elaborates this account by characterising modern science in terms of the ancient

---

11   Martin Heidegger (1967 [1962]), *What Is a Thing?*, trans. by William B. Barton, Jr., and Vera Deutsch (Chicago: Henry Regnery), p. 39.

12   Heidegger (1967), *What Is a Thing?*, p. 39.

13   Heidegger (1967), *What Is a Thing?*, p. 40.

14   Heidegger (1967), *What Is a Thing?*, p. 15; translation modified.

Greek concept of *mathēsis*, or what he also calls 'the mathematical' (*das Mathematische*). These considerations, arising out of a 1935–1936 lecture course, present a further development of his earlier account of the mathematical projection of nature. What he had earlier treated as a central element of science, in general, now becomes a defining feature of modern science, in particular.

For Heidegger, *mathēsis* refers to that fundamental characteristic of modern science which distinguishes it from both ancient and medieval science. He addresses, and rejects in turn, attempts to distinguish modern science on the grounds that it begins with facts about things rather than with speculative propositions and concepts, that it uses experiments to get information about the behaviour of things, and that it relies on calculation and measurement in its investigations of things. On all three counts, Heidegger argues, there is no substantive difference between ancient and medieval science, on the one hand, and modern science, on the other. First, ancient and medieval science also observed the facts, and modern scientists also rely on speculative propositions and concepts. Second, the use of experiments, in the broad sense of controlled tests to gain information about things, was already familiar in the ancient and medieval periods. Heidegger comments that '[t]his kind of experience lies at the basis of all technological contact with things in the crafts and the use of tools.'[15] Third, ancient science also made use of measurement and number.

Heidegger argues that it is not reliance on facts, as such, which is decisive for modern science, but 'the way the facts are conceived.' Likewise, it is not the experiment, as such, that matters, but 'the manner of setting up the test and the intent with which it is undertaken and in which it is grounded.' And so too with calculation and measurement: 'it is a question of how and in what sense calculating and measuring were applied and carried out, and what importance they have for the determination of the objects themselves.'[16] Hence, it is not *that* facts, experiments, calculation and measurement are deployed, but *how* and *to what end* they are deployed, which distinguishes modern science. This points us towards the historically specific existential conditions of possibility governing what 'essentially and decisively rules the basic

---

15 Heidegger (1967), *What Is a Thing?*, p. 67.
16 Heidegger (1967), *What Is a Thing?*, pp. 66–68.

movement' of modern science.[17] It points us, in other words, towards the phenomenon of *mathēsis*. Heidegger suggests that *mathēsis* is something which governs modern science at a basic level. It thus includes a strongly normative component.

Heidegger defines *mathēsis* as 'the teaching' (*die Lehre*), in the sense of 'the doctrine' or 'the apprenticeship,' which in turn has a double sense: 'teaching' as entering into an apprenticeship (*in die Lehre gehen*) and then learning or studying; and 'teaching' as what is taught. Heidegger means here teaching and learning 'in a broad and at the same time essential sense, not in the more recent, narrow, hackneyed sense of "schools" and "scholars."'[18] He furthermore distinguishes two fundamental features of modern science as *mathēsis*: (1) 'work experiences,' or 'the direction and way of controlling and using [or manipulating] what is'; and (2) 'metaphysics,' or 'the projection of the fundamental knowledge of being, upon which what is establishes itself as knowable.'[19] These two fundamental features are 'reciprocally related,' and always occur together in the activities of scientifically engaged human beings.[20]

We have already encountered this twofold structure in Chapter Four, where our attention was primarily on the metaphysical projection of the thingness of things, as a constructive aspect of thinking, rather than on scientists' work experiences in respect of those things. Nevertheless, already there we encountered the historical interplay between things, experienced as property-bearing substances, and thinking, construed as propositionally structured rule-following. In this chapter, and in Chapter Six, we will treat the experience of things more directly as *work* experience, that is, as experiences which arise from the direct engagement with and manipulation of material things. Our focus in

---

17   '[...] was die Grundbewegung der Wissenschaft [...] gleichursprünglich maßgebend durchherrscht' (Martin Heidegger (1984b [1962]), *Die Frage nach dem Ding: zu Kants Lehre von den transzendentalen Grundsätzen* (Gesamtausgabe, vol. 41) (Frankfurt: Vittorio Klostermann), p. 68). Cf. Heidegger (1967), *What Is a Thing?*, p. 68.

18   Heidegger (1984b [1962]), *Die Frage nach dem Ding*, pp. 69–70. Cf. Heidegger (1967), *What Is a Thing?*, p. 69.

19   Heidegger (1967), *What Is a Thing?*, p. 66; translation modified, my brackets. Cf. Heidegger (1984b), *Die Frage nach dem Ding*, p. 66.

20   Heidegger (1967), *What Is a Thing?*, p. 66. Cf. Heidegger (1984b), *Die Frage nach dem Ding*, p. 66. Thus, Heidegger's usage of *mathēsis* differs from that of Michel Foucault, who defines it as 'the science of calculable order' (Michel Foucault (1970), *The Order of Things: An Archaeology of the Human Sciences* (New York: Pantheon Books), p. 73).

this chapter and the next will, in other words, be directed more towards the existential conditions enabling physical experimentation in modern science. As we will see, the experimental manipulation of things is, indeed, reciprocally related to the metaphysical projection of the thingness of things. The experimental set-up is a material instantiation of, as well as the material basis for, the metaphysical projection. Hence, Heidegger's account of modern science in term of *mathēsis* recognises a central place for the material practices of experimental science without rejecting a role for metaphysics. His account can thus be assimilated to neither of the conventional empiricist and rationalist interpretations of the Scientific Revolution. It must be immediately emphasised, however, that Heidegger did not develop this strand of his existential conception of science in any great detail. Additional material from the history of science will allow us to develop and refine Heidegger's account of modern science in a way which he did not, and which is, I think, consistent with his intentions.

According to Heidegger, the word *mathēsis* stems from the Greek word *mathēmata*, the name for a specific kind of thing. These are things insofar as they are learnable, or amenable to study.[21] However, as he also points out, we do not, strictly speaking, learn a thing. We learn instead how to relate to a thing, for example, how to observe it, how to use it, or how to produce it. Hence, when we learn a thing, we are really learning something about our relation to it. Heidegger illustrates this point with the example of a rifle. We do not learn a rifle, he writes, but its usage. The acquisition of the usage happens through the usage itself, that is, through 'exercise' or 'practice' (Übung).[22] In our practice with the rifle, we learn to load it, to control its trigger, and to aim it.

However, Heidegger argues that 'practising' (Üben) is only one kind of learning. There is another, more fundamental, kind of learning which actually makes it possible for us to learn through practice. This more basic learning allows us to perceive what a weapon is, what a use-item is, and, most generally, what a thing is. Heidegger argues that we do not learn the 'what' of a weapon only once we have learned the 'how'

---

21 Heidegger (1984b), *Die Frage nach dem Ding*, p. 71; cf. Heidegger (1967), *What Is a Thing?*, p. 71.

22 Heidegger (1984b), *Die Frage nach dem Ding*, p. 71; cf. Heidegger (1967), *What Is a Thing?*, p. 71.

of its usage. On the contrary, '[w]e already know [what a weapon is] beforehand, and must know it; otherwise we could not even recognise the rifle as such,' we could not, that is, tell the difference between a rifle and a non-rifle.[23]

Yet, although we must first know what a rifle is if we want to learn its usage, this prior knowledge is something we need only have in a 'general' and 'indefinite' way. In other words, when it comes to learning the usage of a thing, prior knowledge of the whatness of the thing is certainly necessary, but we need only possess it in a tacit way, as a kind of general and indeterminate background knowledge. When, in contrast, it comes to learning in the sense of *mathēsis*, we must deliberately 'take note' of (*zur Kenntnis nehmen*) what the thing is, doing so '*specifically and in a determinate way.*' Such deliberate 'taking note' is the very ground of learning as *mathēsis.*[24] This is a kind of learning in which — by taking note of what we already know — we begin to transform our pre-existing knowledge of what a thing is from a general and indefinite state into a specific and determinate state. Such is the case, Heidegger claims, when a specific rifle model is brought into existence: '[w]hen it becomes essential, in a general sense, to make available a thing like the one whose usage we are practising, that is, when it becomes necessary to produce it.' This requires 'a becoming-acquainted [*Kennenlernen*] with what fundamentally belongs to a firearm and with what a weapon is.' Compared to the knowledge gained of a rifle's usage, this is 'a more basic acquaintance, one which must be learned beforehand, so that such a rifle type and its corresponding tokens may come to exist at all.'[25]

So what is it about the rifle which both the sharpshooter and the gunsmith must know in advance, but of which only the gunsmith need take note in order to fulfill her task? More generally, of what does one need to take note in order to learn in the sense of *mathēsis*? The answer, Heidegger tells us, is that 'the producer must know beforehand what

---

23 Heidegger (1984b), *Die Frage nach dem Ding*, p. 73; cf. Heidegger (1967), *What Is a Thing?*, p. 72.

24 Heidegger (1984b), *Die Frage nach dem Ding*, p. 73; cf. Heidegger (1967), *What Is a Thing?*, pp. 72–73.

25 Heidegger (1984b), *Die Frage nach dem Ding*, p. 72; cf. Heidegger (1967), *What Is a Thing?*, p. 72.

Bewandtnis fundamentally accompanies the thing.'[26] The German word *Bewandtnis* is a tricky one to translate. In this passage, Heidegger uses it without pausing to explain its meaning. However, if we turn to an earlier discussion, in §18 of *Being and Time*, we can get a better sense of what the term means in the context of Heidegger's existential account of science. Crucially, Heidegger treats the noun *Bewandtnis* together with the closely related verb *bewenden*. Below is my translation of the key passage in *Being and Time* in which the term *Bewandtnis* first appears.[27] As will be immediately evident, Heidegger is introducing the term in the context of his discussion of readiness-to-hand, a concept we have already encountered numerous times in previous chapters.

> That the being of the ready-to-hand has the structure of a reference or assignment [*Verweisung*] — means: it has in itself the character of directedness [*Verwiesenheit*]. What-is is thereby discovered as that which is directed towards something. With what-is, at this something, there is an end [*es hat... sein Bewenden*]. The being-character of the ready-to-hand is end-directedness [*Bewandtnis*]. In end-directedness, there is a letting-be [*bewenden lassen*] with and at something. The relation of 'with... at...' [»*mit... bei...*«] will be denoted by the term 'assignment.'[28]

In respect of Heidegger's phenomenological account of scientific practice, the meaning of *Bewandtnis* is best understood as 'end-directedness.' I do not recommend this translation for general application throughout Heidegger's work, but only as the best translation when it comes to his reflections on science as *mathēsis*. This is a philosophical translation with a particular, narrow aim.[29]

---

26 '[...] welche Bewandtnis es überhaupt mit dem Ding hat' (Heidegger (1984b), *Die Frage nach dem Ding*, p. 72); cf. Heidegger (1967), *What Is a Thing?*, p. 72.

27 Ernst Tugendhat comments that there is probably no word in any other language which includes the same constellation of meanings as *Bewandtnis*. Thus, he concludes, Heidegger's argument in §18 of *Being and Time* must resist intelligible translation (Ernst Tugendhat (1967), *Der Wahrheitsbegriff bei Husserl und Heidegger* (Berlin: de Gruyter), p. 290 n. 6). *Caveat emptor!*

28 Martin Heideyger (1927), *Sein und Zeit* (Tübingen: Max Niemeyer Verlag), pp. 83–84; cf. Martin Heidegger (1962a [1927]), *Being and Time*, trans. by John Macquarrie and Edward Robinson (Oxford: Blackwell), p. 115 [83–84].

29 Other philosophical translations for *Bewandtnis*, as used by Heidegger, include: 'involvement' by John Macquarrie and Edward Robinson (in Heidegger (1962a), *Being and Time*), and by Hubert Dreyfus (in Hubert L. Dreyfus (1991b), *Being-in-the-World: A Commentary on Heidegger's* Being and Time, *Division I* (Cambridge,

Being and Time provides the following concrete example for what Heidegger means by end-directedness, as well as the 'with... at...' relation of assignment: '*with* the ready-to-hand thing we call "hammer," there is the end-directedness of being *at* work, hammering; *with* hammering, the end-directedness of being *at* fortifying; *with* fortifying, the end-directedness of being *at* sheltering against bad weather.'[30] Transposed to our example of the rifle, we can say that *with* the rifle there is the end-directedness of being *at* work, shooting; *with* shooting, the end-directedness of being *at* impacting at a distance; and *with* impacting at a distance, the end-directedness of being *at* subjugating others. Put otherwise, the task *at which* the sharpshooter works *with* the rifle is shooting. *With* the rifle, *at* this task, there is an end: proximally, impacting at a distance; more distally, subjugating others. The being of the rifle, its readiness-to-hand, is a directedness towards this end, it is that *towards which* the rifle is, essentially, directed. Heidegger writes that the 'at-which' [*Wobei*] of an end-directedness is also a 'towards-which' [*Wozu*], and that with this towards-which there can be yet another end-directedness.[31] Hence, shooting is the end towards which the rifle is directed. With the towards-which of the rifle, proximally directed towards shooting, there is a further, more distal, end: impacting at a distance. The end-directedness of the rifle points towards shooting, and shooting, in turn, points towards impacting at a distance, and impacting at a distance points towards subjugating others.

However, this iteration of assignment finally comes to an end in a 'primary' towards-which. Heidegger writes that the 'primary

---

MA: The MIT Press)); 'relevance' by Joan Stambaugh (in Martin Heidegger (2010b), *Being and Time*, trans. by Joan Stambaugh, revised edn (Albany: SUNY Press)); 'relevance/involvement' by Daniel Dahlstrom (in Daniel O. Dahlstrom (2013), *The Heidegger Dictionary* (London: Bloomsbury 2013)); 'how it works' by William Barton and Vera Deutsch (in Heidegger (1967), *What Is a Thing?*); 'functionality' by Alfred Hofstadter (in Martin Heidegger (1982a [1975]), *Basic Problems of Phenomenology*, trans. by Albert Hofstadter (Bloomington: Indiana University Press)); and 'role' by John Haugeland (in John Haugland (2013), *Dasein Disclosed: John Haugeland's Heidegger*, ed. by Joseph Rouse (Cambridge, MA: Harvard University Press)). At the time of writing, the poplar on-line *LEO* German-English dictionary translated *Bewandtnis* as 'matter' or 'reason' (*dict.leo.org*). The 2001 print edition of the well-respected *PONS* English-German dictionary offers 'reason' and 'explanation.'

30   Heidegger (1927), *Sein und Zeit*, p. 84; my emphases. Cf. Heidegger (1962a), *Being and Time*, p. 116 [84].

31   Heidegger (1927), *Sein und Zeit*, p. 84. Cf. Heidegger (1962a), *Being and Time*, p. 116 [84].

"towards-which" is a "for-the-sake-of-which" [*Worum-willen*].'[32] For example, in the case of the hammer, the iteration bottoms out in 'protection.' The hammer points towards hammering, which points towards fortifying, which points towards sheltering, which points, finally, towards protection. That *for the sake of which* one hammers, fortifies, shelters, is the subject's protection, which Heidegger describes as a 'possibility' of the subject's existence.[33] Returning, once more, to our rifle example, we may take the primary towards-which of the rifle — that for the sake of which the rifle, ultimately, exists — to be sovereignty. With the rifle, one shoots, impacts, subjugates, for the sake of the subject's sovereignty, an ontological possibility of its existence. Ultimately, *at* work, *with* the rifle, there is a directedness towards sovereignty as end. According to Heidegger, in order to work with the rifle in a way which lets it be what it is, one must already have a knowledge of its end-directedness, and, hence, the final end towards which it points. This is an antecedent knowledge which the sharpshooter and the gunsmith both possess in a general and indefinite way, and of which only the latter need deliberately take note, developing it, through *mathēsis*, into a more specific and determinate knowledge.

Now consider the example of a plant collector. A plant collector is more like a sharpshooter. In order to pick a plant and put it in her collection, she must already know what a plant is. If she did not know this, then she would not even be able to identify a plant, to tell the difference, for example, between a plant and a platypus. But this prior knowledge need only be of a general and indeterminate sort. On its basis, the plant collector can take a *particular* plant for her collection, without deliberately taking note of that plant's general plantness, or whatness. In contrast, the botanist will deliberately take note of the general plantness of the plant. In the course of her work, she 'skips over' the individual specimens, taking them only as tokens of a general type. Nevertheless, it is *with* these individual specimens, by being concretely *at* work with them, that the botanist learns — develops a specific and determinate knowledge of — the type. This learning is *mathēsis*, and, according to Heidegger, it marks the difference between ancient and medieval science, on the one hand, and modern science, on the other.

---

32   Heidegger (1962a), *Being and Time*, p. 116 [84].
33   Heidegger (1962a), *Being and Time*, p. 116 [84].

It should be emphasised that, in the examples of both the plant collector and the botanist, the particular, concrete plant is being treated as something ready-to-hand in a work-world. In other words, we are considering it as it is experienced by the one who is at work with it. From this perspective, then, there is no basic ontological difference between the plant and the rifle. Both are things ready-to-hand within a work-world. On Heidegger's account, then, in being at work with either the particular plant or the particular rifle, there is an end-directedness. Furthermore, in both cases this end-directedness ultimately bottoms out in that for the sake of which the ready-to-hand thing is what it is. This for-the-sake-of-which is an existential possibility of the subject.

In the case of the hammer, this existential possibility is the subject's protection. In the case of the rifle, it is the subject's sovereignty. What about the plant? Here the answer is less obvious. On first blush, there does not seem to be anything for the sake of which the plant is what it is. This doubt is tied to our intuition that the plant is, in the first instance, not something ready-to-hand, but something present-at-hand. Yet this intuition was challenged in Chapters One and Two. According to Heidegger, a thing within the world is experienced, most immediately, as ready-to-hand. Only on this basis can it be subsequently experienced as present-at-hand within in the world. Hence, the what-being of a scientific thing should be understood fundamentally in terms of its readiness-to-hand, which is to say, in terms of its end-directedness. From this it follows that the scientific thing — that which in scientific work is let be what it is — *is* for the sake of an existential possibility of the subject. Heidegger argues that we already know this possibility, and must know it, when we intelligibly experience the scientific thing, as such, not to say when we start working with it. This possibility is the ultimate end towards which the what-it-is of the scientific thing is directed. The hammerness of the hammer points, finally, towards protection, and the rifleness of the rifle, finally, towards sovereignty. To what, then, does the plantness of the plant, finally, point?

Heidegger does not answer this question. Instead, he concentrates on the more general question of to what existential possibility the thingness of the scientific thing itself finally points. We will consider the specific content of Heidegger's answer to this question in Chapter Six. In the meantime, in order to properly appreciate the grounds for that answer, we must first gain a firmer grip on the more formal characteristics of Heidegger's answer: namely, that the scientific thing — as something with which scientists are at work — possesses an end-directedness pointing towards a final end. We must know this end-directedness beforehand if we are to successfully identify and work with the thing. In scientific work, we deliberately take note of this end-directedness, bring it into the foreground, and thereby seek to articulate our prior general and indeterminate knowledge of it into a more specific and determinate knowledge. This is how we study *mathēmata*, things insofar as they can be learned. This is a process of *mathēsis*, whereby we learn what we already know.

Formally, then, Heidegger views the scientific thing in terms of its end-directedness, a feature which we deliberatively experience only insofar as we relate to the thing through *mathēsis*, that is, only insofar as we study or learn it. A further formal feature of the scientific thing is, therefore, epistemic circularity: we can only know it because we already know it. This may raise the worry that Heidegger's account of *mathēsis* attributes to science a fallacious form of reasoning. On this account, so the worry goes, the conclusion of a scientific inference is already among its initial premises; hence, science gives us no reason to accept its conclusions as valid.

But we are already familiar, from Chapter Four, with Heidegger's response to this worry in respect of logic, construed as the science of the rules of reason. The worry there was that, since science is grounded in rules of reasoning, and since logic is identical with those rules, then logic grounds logic.

In response, Heidegger rejected the premise which equates logic and rules. Logic may presuppose rules, but rules do not entail logic. Indeed, only when the rules of reason have been rendered in a specific and determinate way do they count as rules of logic. The job of logic, as the science of thinking, is to deliberatively take note of the general and indeterminate rules which implicitly govern informal reasoning, and then to study them in a way which specifies and determines them, in other words, clearly explicates them in a formal system. Hence, it is not the case that the conclusions of the science of logic are already among its premises, because the relation between premises and conclusion is interpretative rather than inferential. The science of logic is a practice in which informal thinking becomes formalised, in which general and indeterminate rules of reason are rendered specific and determinate. If there is a circularity here, then it is not vicious.

We can now see that logic is a particular case of the more general scientific practice of *mathēsis*. Our past discussion of this case, in turn, shows why we need not worry about the circularity of *mathēsis*: this circularity is interpretative rather than inferential, hermeneutical rather than logical, and so virtuous rather than vicious. It is a fundamental feature of science conceptualised as an ultimately informal existential practice, rather than as a logically determinate system of precisely defined concepts. Furthermore, science, as a form of reasoning, is guided by rules. Logic may seek to thematise and define those rules, but the natural sciences will be less concerned with explicating and determining the dynamics of their own thinking, and more interested in reliably explicating and determining the whatness of the material things they take to be the object of that thinking. According to Heidegger, these sciences ultimately attend to the end-directedness of natural things. The rules of reason governing this attention help scientists to reliably distinguish relevant from irrelevant phenomena vis-à-vis their understanding of what the thing is. In other words, these rules play a normative role in scientific research.

As a scientific practice, then, *mathēsis* is ruled by norms. As we also saw in Chapter Four, Heidegger traces philosophical descriptions of the norm governing the relationship between thinking and things back to Plato's mythic and polysemous image of the cosmic demiurge. According to Heidegger, this image — as well as its cognates, the sun and the idea of the good — mark Plato's discovery of the *a priori* element in our understanding of things. The image of the demiurge, in particular, reflects the impulse of some early Greek thinkers to construe thinking and things in terms of craft production. As Heidegger observes in this context, '[a]ll forming of shaped products is effected by using an image, in the sense of a model, as guide and standard.'[34] Hence, the normative aspect of *mathēsis*, as a productive practice, stands as an *a priori* image in the experience of working with things, as a projective image guiding the pursuit of a specific and determinate knowledge of those things. Heidegger describes this image as a *Grundriss*, a 'ground

---

34   Heidegger (1982a), *Basic Problems of Phenomenology*, p. 106.

rendering' or 'basic blueprint.' For present purposes, we should note the dual nature of the blueprint representing the normative element in scientific practice. On the one hand, a blueprint helps guide the act of production. On the other hand, it represents, schematically, the end result of production. A blueprint is thus both a set of directions, and an image of that towards which those directions point. As a model of scientific practice, it captures, consequentially, both the how and the what of that practice. To experience the scientific thing in terms of its end-directedness thus means to experience it in light of an image of both its direction and its end. This image enables, is the condition of possibility for, that experience.

That *mathēsis* is guided by a basic blueprint lies at the core of Heidegger's account of modern science. Heidegger views this blueprint in terms of a 'measure.' He writes that '[t]his basic plan (*Grundriss*) [...] provides the measure [*Maßstab*] for laying out of the realm, which, in the future, will encompass all things of that sort.'[35] Hence, the ground plan regulates scientific practice by imposing a general measure, a measure meant to apply to all things falling within the plan. Note that, although Heidegger introduces the concept of 'measure' in the context of his discussion of the mathematical, we are not meant to view this measure in quantitative terms. One can, for example, take the measure of a thing without judging it according to some quantitative unit. For example, the common phrase 'taking the measure of a man' need not imply quantification. On the other hand, one cannot judge a thing quantitatively without also taking its measure. In Heidegger's account, the mathematical and the quantitative are connected, but the former is a broader category than the latter: '[i]n no way [...] is the essence of the mathematical defined by numberness.'[36] Hence, while the practice

---

35   Heidegger (1967), *What Is a Thing?*, p. 92.

36   Martin Heidegger (1977a [1952]), 'The Age of the World Picture,' in *The Question Concerning Technology*, by Martin Heidegger, trans. by William Lovitt (New York: Harper & Row), pp. 115–54 (p. 119); cf. Heidegger (1967), *What Is a Thing?*, p. 70. Cf. also: 'When we hear of measure, we immediately think of number and imagine the two, measure and number, as quantitative. But the *nature* of measure is no more a quantum than is the *nature* of number' (Martin Heidegger (1971a), '"... Poetically Man Dwells...,"' in *Poetry, Language, Thought,* by Martin Heidegger, trans. by Albert Hofstadter (New York: Harper & Row), pp. 213–29 (p. 225)). The *OED* defines 'measure,' in one substantive sense, as a 'standard or rule of judgement; a criterion, test; also, a standard by which something is determined or regulated.'

of *mathēsis* may include numerical calculation, it cannot be reduced to such calculation. To render a general and indeterminate knowledge more specific and determinate does not necessarily mean to render it more numerically precise.

In addition to circularity and normativity, there is a third parallel between the present discussion and our earlier discussion of Heidegger's phenomenological history of logical practice. This has to do with the source of normativity, of the measure regulating modern scientific practice. Recall from Chapter Four that Heidegger describes the temporal aspect of scientific work experience in terms of 'original time,' that is, time as experienced in the course of our immersion in a work-world. Original time becomes intelligible to us in terms of two types of relation: in-order-to relations; and for-the-sake-of-which relations. The manifold of in-order-to relations is an existential space within which a particular piece of equipment is let be what it is in its readiness-to-hand. This manifold, in turn, is revealed only in light of the for-the-sake-of-which relation. Only on the basis of the for-the-sake-of-which do we experience the intelligibility of things within a work-world. As we saw in Chapter Four, Heidegger also viewed the for-the-sake-of-which as a social phenomenon, rooted in, and continually nourished by, tradition. Hence, according to him, the source of normativity, of the measure regulating scientific practice, is tradition.

The multiplicity of in-order-to relations serves the same role as the iteration of assignment, or what Heidegger also calls the 'totality of end-directness' (*Bewandtnisganzheit*), in which a particular thing is let be what it is in its end-directedness.[37] And just as the multiplicity is revealed only in light of the for-the-sake-of-which, so too does this totality of end-directedness bottom out in the for-the-sake-of-which. In being at work with the thing in a way which lets it be what it is, we must already possess an at least general and indefinite knowledge of the thing's final end. Heidegger describes this ultimate end as an existential possibility of the subject. We may now more decisively identify it as a possibility afforded by the historical tradition in which the subject finds itself. According to Heidegger, the final end of modern scientific practice, that towards which it is ultimately directed, is rooted in an existential

---

37 Heidegger (1927), *Sein und Zeit*, p. 84; cf. Heidegger (1962a), *Being and Time*, p. 116 [84].

possibility of the subject's own socio-cultural history. Furthermore, this end has the formal structure of a ground plan, a basic blueprint laying out the measure against which all things falling within its domain — all potentially scientific things — will be judged.

In one last link to the discussion in Chapter Four, the socio-historical provenance of the ground plan of modern scientific practice suggests a solution to the sociological problem of 'priming.' Recall Bloor's distinction between a self-referring practice, in which things acquire their meaning, and the priming element which gets that practice going. I argued that the for-the-sake-of-which, as the *a priori* element which grounds the multiplicity of in-order-to relations, serves as such a priming element. It is just a small step to now apply that argument to Heidegger's description of the for-the-sake-of-which as an *a priori* ground plan which both directs, and serves as the final end for, the modern scientific practice of *mathēsis*.

As we will see in what follows, this move has consequential implications for the historiography of early-modern science. Specifically, it offers an at least partial answer to the question of how the Scientific Revolution got going. Here, the Scientific Revolution is understood as the emergence of a new cognitive and material domain in which to make sense of the things of nature. On Heidegger's account of *mathēsis*, this new domain traces its origin back to a socio-historically conditioned possibility within the everyday work-world of early-modern subjectivity. Furthermore, because Heidegger argues that this socially contingent possibility is manifest in experience, whether informally or formally, as an image — a ground rendering or basic blueprint — we may treat it as a social image. As we will see later in Chapter Six, this allows us to connect Heidegger's existential conception of science in yet another way to SSK, specifically, to David Bloor's linkage between knowledge and social imagery.

Let us now return to Heidegger's argument that the decisive difference between ancient and medieval science, on the one hand, and modern science, on the other, cannot be explained by the claim that modern science uniquely emphasises facts, measurement, and experiment as the grounds for natural knowledge. For Heidegger, it is rather the difference in the way the facts are conceived, the way the measurement or experiment is done, which is decisive. The question is not whether facts, measurement, and experiment are employed — they are employed in all three periods — but to what end they are employed. According to Heidegger, it is the kind of end-directedness possessed by modern scientific things which distinguishes them from their ancient and medieval predecessors, namely, their directedness towards a basic blueprint, a single regulative ground plan, in respect of which they are let be what they are. In establishing facts about things, in measuring and experimenting on those things, *mathēsis* skips over their token specificities, instead taking note of the regulative ground plan according to which the scientific thing, in general, becomes intelligible as what it is.

The similarities and differences between ancient science, at least as exemplified in the earlier example of Galen, and *mathēsis* seem straightforward enough. Despite their specific disagreements, Galen's empiricist and rationalist physicians agree, in general, that medical knowledge should be 'technical' or objective, that is, grounded in fact. They also agree on the necessity of an *a priori* standard according to which physicians take the measure of, and are thus able to discriminate between, epistemically relevant and irrelevant phenomena. In a broad sense, then, ancient science is concerned with both facts and measures. However, Galen's empiricist and rationalist appear to disagree on the need for experiment. Recall the rationalist physician's argument that reliable judgements can be made on the basis of a single observation, thereby dismissing the necessity of working with things over time. For the empiricist, in contrast, one must proceed 'little by little,' slowly acquiring knowledge over time through a method of disciplined serial observation of the things.

This disagreement between the ancient rationalist and empiricist over the need for a method of enquiry seems conjoined with their further disagreement over the necessary degree of determinateness of the physician's *a priori* knowledge of the measure of medical knowledge. The rationalist requires such knowledge to be fully determinate, a condition which then enables immediate judgement on the basis of one observation. Galen's empiricist, on the other hand, is content to leave such knowledge indeterminate, or obscure, arguing that 'not a particle of harm befalls arts and men […] for being ignorant of such things.' In contrast to *mathēsis*, then, Galen's empirical method does not seek to turn a general and indeterminate knowledge of the thingness of things into a particular and determinate knowledge. To his credit, Galen may thus avoid charges of circularity. However, as already mentioned, it is hard to see the scientific merit of his position. Indeed, how should reliable discrimination be achieved in the absence of an increasingly determinate knowledge of the standard which guides such discrimination?

As we will see in the next section, this question also troubled Renaissance physicians. They answered with the argument that a determinate knowledge of the norms guiding their medical practice was both necessary and to be gained little by little through an incremental method for working with the things. As a consequence, these Renaissance physicians faced the same worry about circularity which arises in Heidegger's account of modern scientific practice as *mathēsis*. And their practical response to this worry turns out to have been not so very different from Heidegger's own. This suggests that the decisive difference between *mathēsis* and medieval scientific practice was, perhaps, not quite so straightforward as Heidegger had imagined. Yet, without this difference, *mathēsis* can no longer provide an explanation for the rise of a definitively modern science, in short, for the Scientific Revolution. So, let us now consider this Renaissance method in some detail, before then comparing it with Heidegger's account of early-modern *mathēsis*.

3. Renaissance *Regressus* and the Logic of Discovery

According to John Randall, Renaissance scholars at the University of Padua, in Northern Italy, developed an account of scientific enquiry which they described as a 'double process.'[38] According to this account, a proper scientific method will begin with some observed effect, seek the cause of that effect, and then use that cause to explain the effect. This amounts to a recommendation that the effect be explained in terms of a cause which can itself be known only through that effect. The explanation of the effect thus seems to presuppose knowledge of

---

38 John Herman Randall, Jr (1940), 'The Development of the Scientific Method in the School of Padua,' *Journal of the History of Ideas* 1(2), 177–206 (p. 190).

that very same effect. Hence, scientists appear to explain only what they already know. As a consequence, there would seem to be no need for a method by which scientists acquire knowledge because they already possess the knowledge in question.

According to Randall, this double process had already been described in 1334 by Urban the Averroist.[39] A more concretely developed account can be found later in the writings of Jacopo da Forlì (ca. 1364–1413/14), who taught medicine and natural philosophy at Padua. Forlì wrote:

> [I]f when you have a fever you first grasp the concept of fever, you understand the fever in general and confusedly. You then *resolve* the fever into its causes, since any fever comes either from the heating of the humor or of the spirits or of the members; and again the heating of the humor is either of the blood or of the phlegm, etc.; until you arrive at the specific and distinct cause and knowledge of that fever.[40]

From this description, it is clear that Forlì did not consider the circularity of the espoused method to be irremediably vicious. Indeed, as he describes, the knowledge of the effect — the fever — undergoes a transformation in the course of the method. It is not, strictly speaking, the same knowledge in the end that it was at the beginning. Forlì emphasised the need for a procedure by which a general and confused knowledge of the effect is 'resolved' into a specific and distinct knowledge of that same effect. Randall describes this as 'a clear case of the method of medical diagnosis.'[41]

---

39   Randall (1940), 'The Development of the Scientific Method,' p. 190. Alistair Crombie suggests that the Paduan 'double process' was imported from Oxford around 1400 (Alistair C. Crombie (1962), *Robert Grosseteste and the Origins of Experimental Science (1100–1700)* (Oxford: Clarendon Press), p. 297). However, Randall's dating indicates that its earliest known appearance in Padua was prior to 1400.

40   Cited in Randall (1940), 'The Development of the Scientific Method,' p. 189; originally from *Jacobi de Forlivio super Tegni Galeni*, Padua, 1475, comm. Text I.

41   Randall (1940), 'The Development of the Scientific Method,' p. 189. As I have shown elsewhere, something like the Paduan double procedure — and, indeed, Heideggerian *mathēsis* — can also been found in Ludwik Fleck's historical account of the medical diagnosis of syphilis (Jeff Kochan (2015c), 'Circles of Scientific Practice: *Regressus, Mathēsis, Denkstil*,' in *Critical Science Studies after Ludwik Fleck*, ed. by Dimitri Ginev (Sophia: St. Kliment Ohridski University Press), pp. 83–99; reprinted as Jeff Kochan (2016), 'Circles of Scientific Practice: *Regressus, Mathēsis, Denkstil*,' in *Fleck and the Hermeneutics of Science* (Collegium Helveticum Heft 14), ed. by Erich Otto Graf, Martin Schmid and Johannes Fehr (Zürich, 2016), pp. 85–93). See Ludwik Fleck (1979 [1935]), *Genesis and Development of a Scientific Fact*, trans. by Fred Bradley and Thaddeus J. Trenn, ed. by Thaddeus J. Trenn and Robert K. Merton (Chicago: University of Chicago Press).

During the fifteenth century, there was an increasing focus on this double process, which came to be called *regressus* in order to emphasise its exclusion from the charge of being a *circulus vitiosus*, a vicious circle. Paul of Venice (ca. 1369–1429) launched an early defence of the *regressus* against this charge, arguing that:

> Scientific knowledge of the cause depends on a knowledge of the effect, just as scientific knowledge of the effect depends on a knowledge of the cause, since we know the cause through the effect before we know the effect through the cause. This is the principal rule in all investigation, that a scientific knowledge of natural effects demands a prior knowledge of their causes and principles. — This is not a circle, however. [...] [T]he knowledge of why (*propter quid*) the effect is, is not the knowledge that (*quia*) it is an effect. Therefore the knowledge of the effect does not depend on itself, but upon something else.[42]

This 'something else,' which distinguishes the *regressus* from a vicious circle, was addressed by Agostino Nifo (ca. 1473–1538/45), a student and later a teacher of medicine and philosophy at Padua. He called it *negotiatio*, and included it as the third of the four kinds of knowledge which comprise the scientific method:

> The first kind is of the effect through the senses, or observation; the second is the discovery (*inventio*) of the cause through the effect [...]; the third is knowledge of the same cause through an examination (*negotiatio*) by the intellect, from which there first comes such an increased knowledge of the cause [...]; the fourth is a knowledge of the same effect propter quid, through that cause known so certainly [...]."[43]

Nifo further specified *negotiatio* as an intellectual act of 'composition and division': '*negotiatio* is directed toward the cause as a [...] definition. But since a definition is discovered only through composition and division, it is through them that the cause is discovered in the form [...] from which we can then proceed to the effect.' What Forlì earlier identified as the process by which a general and confused knowledge of a thing is 'resolved' into a specific and distinct knowledge of that same

---

42 Cited in Randal (1940), 'The Development of the Scientific Method,' p. 191; originally from *Summa philosophiae naturalis magistri Pauli Venti*, Venice, 1503, I, cap. ix.

43 Cited in Randall (1940), 'The Development of the Scientific Method,' p. 192; originally from *Augustino Niphi philosophi suessani exposition... de Physico auditu*, Venice, 1552, I, com. Text 4.

thing is now identified by Nifo as an intellectual process of *negotiatio*, or definition. Through *negotiatio*, an originally confused and general knowledge of a cause is rendered more definite.

Nifo's concept of *negotiatio* was further developed by Jacopo Zabarella (1533–1589), who also taught philosophy at Padua, but, unlike those of his predecessors mentioned above, did not have a degree in medicine.[44] About the role of *negotiatio* in the *regressus*, Zabarella wrote:

> When the first stage of the procedure has been completed, which is from effect to cause, before we return from the latter to the effect, there must intervene a third intermediate process (*labor*) by which we may be led to a distinct knowledge of that cause which so far has been known only confusedly. Some men knowing this to be necessary have called it a *negotiatio* of the intellect. We can call it a 'mental examination.' [...] [S]till they have not shown how it leads us to a distinct knowledge of the cause, and what is the precise force of this *negotiatio*.... There are, I judge, two things that help us to know the cause distinctly. One is the knowledge that it is, which prepares us to discover what it is. [...] The other help, without which this first would not suffice, is the comparison of the cause discovered with the effect through which it was discovered, not indeed with the full knowledge that this is the cause and that the effect, but just comparing this with that. Thus it comes about that we are led gradually to the knowledge of the conditions of that thing; and when one of the conditions has been discovered we are helped to the discovery of another, until we finally know this to be the cause of that effect.[45]

According to Zabarella, in order to gain knowledge of *what* a cause is, we must first know *that* it is. This knowledge-that enables a knowledge-what of the cause, a knowledge which is not immediately grasped, but which only becomes clear and distinct through a fragile method of mental comparison. By this method, we are 'led' to a distinct knowledge of the 'conditions' of the thing under investigation. For Zabarella, these conditions can also be understood as the 'principles' from which the relation between cause and effect may be securely determined. As our knowledge of these conditions or principles improves, we acquire an

44   Eckhard Kessler (1988), 'The Intellective Soul,' in *The Cambridge History of Renaissance Philosophy*, ed. by Charles B. Schmitt, Quentin Skinner, Eckhard Kessler and Jill Kraye (Cambridge: Cambridge University Press), pp. 485–534 (p. 836).

45   Cited in Randall (1940), 'The Development of the Scientific Method,' p. 200–01; originally from Jacopo Zabarella, *De Regressu*, chpt. 5.

increasingly determinate knowledge of the whatness of the cause, and hence also of its necessary relation to the effect.

Randall suggests that Zabarella represents the climax of a long historical development in which scientific knowledge was increasingly recognised to rely on a form of experience which distinctly differs from 'ordinary observation,' in the sense of the 'accidental or planless collection of particular cases.'[46] Scientific experience is disciplined by method. Forlì called this method 'resolution.' Nifo called it *negotiatio.* Zabarella called it 'mental examination.' The necessity of this method undermines the claim that reasoning from effect to cause and then from cause back to effect must be viciously circular. Indeed, the method appears to be necessary just because the circle is not vicious. The knowledge of the cause at the beginning is not identical with the knowledge of the cause at the conclusion. The required method mediates between these two distinct ways of knowing the cause, joining them in a manner which then also transforms our original knowledge of the effect.

This method, then, is a method of discovery. Randall argues that the Paduan philosophers worked out 'a logic of investigation and inquiry' to accompany the existing Aristotelian theory of proof.[47] Zabarella, in particular, paid an 'ever closer attention to the way of discovery, to the careful and painstaking analysis of experience, to the method of resolution.'[48] Strikingly, Zabarella justified the need for a method of discovery by pointing to the finitude of human cognitive abilities:

> Since because of the weakness of our mind and powers the principles from which demonstration [i.e., proof] is to be made are unknown to us, and since we cannot set out from the unknown, we are of necessity forced to resort to a kind of secondary procedure, which is the resolutive method that leads to the discovery of principles, so that once they are found we can demonstrate the natural effects from them. […] It is certain that if in coming to any science we were already in possession of a knowledge of all its principles, resolution would there be superfluous.[49]

---

46   Randall (1940), 'The Development of the Scientific Method,' p. 199.
47   Randall (1940), 'The Development of the Scientific Method,' p. 201.
48   Randall (1940), 'The Development of the Scientific Method,' p. 204.
49   Cited in Randall (1940), 'The Development of the Scientific Method,' p. 197–98; originally from Jacopo Zabarella, *De Methodis*, III, xviii.

This may be read as a critical response to Galen's dogmatic physician, whom we met at the start of this chapter. Zabarella argues that because we lack the innate ability to immediately grasp, with clarity and confidence, the principles governing natural phenomena, we must rely on a method which enables us to articulate those principles on the basis of our finite sensory experience. Furthermore, Randall attributes to Zabarella the additional claim that, because this method serves to discipline and direct our experience of phenomena, it must be distinguished from the comparatively free and unstructured sensory experience indicative of more familiar, everyday modes of perception.

Yet, Randall seems to have been too optimistic in his assertion that Zabarella successfully worked out a logic or method of scientific discovery, as opposed to having just demonstrated the need for such a method. Indeed, when Zabarella turns to a discussion of different modes of discovery, his account is remarkably thin on detail. As we saw above, after first characterising 'mental examination' as the method by which a confused and indeterminate knowledge of the cause gets transformed into a clear and determinate knowledge, Zabarella then describes this transformation in terms of 'just comparing' the cause with the effect such that 'it comes about' that we are 'led gradually' to the conditions of the cause. But what rules structure this comparing, and so guide us to the correct causal conditions? Zabarella does not say. However, it appears that, if he had entertained the existence of such rules, then he would likely have considered them to be rules of reason. Consider his discussion of induction as a method of discovery. In his view, a universal stands to a particular as cause stands to effect. The relevant notion of cause is, in this context, an Aristotelian notion of formal cause. Furthermore, knowledge of the universal is gained inductively through experience of the relevant particulars. In other words, the formal cause is known through its effects. Thus, Zabarella writes: 'One says that "human" is something truly sensible, not because the senses recognise humans as something universal, but because particular individual humans are sensible.'[50] Zabarella's thus roots knowledge of universals in sensory experience:

50　Jacobi Zabarella (1985), *De Methodis; De Regressus*, ed. by Cesare Vasoli (Bologna: Cooperativa Libraria Universitaria Editrice), p. 99. Original text: *'hominem enim rem sensilem effe dicimus, non quòd hominé universalem sensus cognoscat, sed qiua singuli*

[A]ll our knowledge takes its origin from sense, nor can we know anything with our minds unless we have known it first by sense. Hence all principles of this kind are made known to us by induction [...]. [I]nduction does not prove a thing through something else; in a certain sense it reveals that thing through itself. For the universal is not distinguished from the particular in the thing itself, but only by reason [*ratio*].[51]

It turns out, then, that a crucial ingredient in the *regressus*, which serves to enable scientific knowledge, is our capacity for reason. But what role does reason play in transforming a confused knowledge that the cause is into a determinate knowledge of what it is? Zabarella writes that induction 'does not take all the particulars into account, since after certain of them have been examined our mind straightaway notices the essential connection, and then disregarding the remaining particulars proceeds at once to bring together the universal.'[52] Reason thus allows us to apply a measure of salience to the data of our sense experience, to distinguish epistemically relevant particulars from epistemically irrelevant ones. It equips us with a standard by which to discriminate the essential from the accidental, the good data from the bad data, in our search for the clear and determinate cause or causes of a particular sensible effect. Reason, in other words, provides the norms enabling proper scientific judgement. A successful method is one which, among other things, embodies these rational norms. Zabarella seems to have never addressed the discriminative power of reason explicitly in such terms. He did not, in other words, thematise this intellectual power in such formal terms as 'measure,' 'rule,' or 'norm.' Nevertheless, his inquiry does explicitly uncover a normative element in thinking which, as we saw in Chapter Four, will later be formally articulated by Kant as a 'faculty of rules.'

The question now arises of where Zabarella thought this discriminative power, these rational norms, to come from. It appears

---

individui homines sensiles sunt' (Zabarella, *De Methodis*, bk. 4, chpt. 19). I have closely followed Rudolf Schicker's German translation (see Jacopo Zabarella (1995), *Über die Methoden (De Methodis); Über den Rückgang (De Regressu)*, trans. by Rudolf Schicker (Munich: Wilhelm Fink Verlag), p. 243).

51   Cited in Randall (1940), 'The Development of the Scientific Method,' pp. 198-99; originally from Zabarella, *De Methodis*, III, xix.

52   Cited in Randall (1940), 'The Development of the Scientific Method,' p. 200; originally from Zabarella, *De Regressu*, chpt. 4.

that he took them to be, not innate and self-evident structures in reason, but the endowment of an external and mysterious intelligence. As Harold Skulsky has argued, '[t]he inductive process to which a writer like Zabarella pays tribute [...] depends ultimately, not on reason, but on the generosity, or more properly the grace, of an alien will.'[53] Eckhard Kessler similarly observes that, for Zabarella, because our irremediably finite epistemic power is capable of 'containing the universal structure only in a confused and unintelligible way, it had to be illuminated by the agent intellect [*intellectus agens*], so that the universal in the individual was rendered distinct and intelligible.' Zabarella furthermore held that the *intellectus agens* could be identified 'with God himself as the principle of intelligibility.'[54] Nicholas Jardine views this as evidence for the reliance of Zabarella's scientific method on 'divine revelation,' and he equates Zabarella's notion of *intellectus agens* with the Holy Spirit.[55] However, Jardine appears to carry this argument too far, concluding also that Zabarella took clear and determinate knowledge of the cause to be 'formed in the imagination through a merely passive observation of the world.'[56] It would seem that the point is, instead, that finite human beings may well actively observe natural phenomena, but, in the absence of normative guidelines for structuring that observation, they will not achieve proper scientific knowledge of those phenomena.[57] This has been argued by Charles Schmitt, who provides evidence that 'Zabarella did take it upon himself to go out and look at nature; and, what is more important, he observed carefully what he saw and applied it to the crucial philosophical questions in which he was interested.'[58] What

---

53 Harold Skulsky (1968), 'Paduan Epistemology and the Doctrine of the One Mind,' *Journal of the History of Philosophy* 6(4), 341–61 (p. 341).

54 Eckhard Kessler (1988), 'The Intellective Soul,' in *The Cambridge History of Renaissance Philosophy*, ed. by Charles B. Schmitt, Quentin Skinner, Eckhard Kessler and Jill Kraye (Cambridge: Cambridge University Press), pp. 485–534 (p. 531).

55 Nicholas Jardine (1976), 'Galileo's Road to Truth and the Demonstrative Regress,' *Studies in History and Philosophy of Science* 7(4), 277–318 (p. 301).

56 Jardine (1976), 'Galileo's Road to Truth,' p. 301

57 Jardine's mistaken assumption that, for Zabarella, no effort is required to move from confused to distinct knowledge of the cause, seems to underpin his questionable conclusion that the *regressus*, in general, is 'blatantly circular' (Jardine (1976), 'Galileo's Road to Truth,' p. 308). As argued above, such effort is called for just because the *regressus* is *not* viciously circular.

58 Charles B. Schmitt (1969), 'Experience and Experiment: A Comparison of Zabarella's View with Galileo's in *De Motu*,' *Studies in the Renaissance* 16, 80–138 (p. 99).

Zabarella does not do, however, 'is consciously, and with forethought, attempt to test a particular theory or hypothesis by devising a specific experiment or observational situation by which to resolve the question.'[59] Schmitt thus concludes that 'Zabarella can be called an empiricist with some justification, but he is clearly not an experimentalist,' at least not in the early-modern sense of that term.[60]

We saw earlier that Randall believed Zabarella to have drawn a clear distinction between scientific experience, on the one hand, and ordinary observation, in the sense of 'accidental or planless collection of particular cases,' on the other. On this basis, Randall attempts to turn Zabarella into a proto-experimentalist, thereby establishing a clear continuity between the method of *regressus* and the Galilean experimental method which would emerge in the years shortly afterwards. We may now conclude that the distinction was more subtle than that, too subtle, in fact, to justify naming Zabarella a precursor to the Galilean experimental method. While his approach to observation was careful and goal-oriented rather than accidental and planless, he did not seek to discipline or control, much less to create, the act of observation in the manner distinctive of subsequent early-modern experimentalists. There was something missing in Zabarella's conception of scientific method, something which prevented him from making the final step towards a modern scientific way of working with and thinking about nature. In fact, Randall recognised that absence, but he apparently minimises its significance:

> There was but one element lacking in Zabarella's formulation of method: he did not insist that the principles of natural science be mathematical. [...] With this mathematical emphasis added to the logical methodology of Zabarella, there stands completed the 'new method' for which men had been so eagerly seeking.[61]

From the perspective of Heidegger's existential conception of science, this missing mathematical element appears to be the key feature separating Renaissance *regressus* from early-modern *mathēsis*. Furthermore, Heidegger argued that the emergence of early-modern

---

59   Schmitt (1969), 'Experience and Experiment,' p. 105.
60   Schmitt (1969), 'Experience and Experiment,' p. 106.
61   Randall (1940), 'The Development of the Scientific Method,' pp. 204–05.

experimental science was possible only because natural science had itself become mathematical, in the sense of adopting *mathēsis* as its core method of discovery. On this account, it seems to follow that Zabarella could not be a precursor to early-modern experimentalism because he did not experience the things of nature mathematically, as *mathēmata*. But before we can properly delineate the dependency of early-modern experimental science on the mathematical projection of nature, we must first more carefully consider the similarities and differences between Renaissance *regressus* and early-modern *mathēsis*.

4. From Renaissance *Regressus* to Early-Modern *Mathēsis*

There are three broad features in respect of which *regressus* and *mathēsis* may be usefully compared. These are: circularity, finitude, and method. Not only do these features figure prominently in both *regressus* and *mathēsis*, the relation of each to the others is also similar in both cases. To begin with, *regressus* and *mathēsis* are both manifestly circular accounts of scientific reasoning. In both cases, too, the account can be defended against charges of vicious circularity. As we saw in the previous section, Zabarella argued that the 'weakness of our mind and powers' renders us incapable of immediately possessing, with clarity and confidence, the scientific principles determining the causes of observed natural phenomena. Indeed, according to Zabarella, if we falsely believe ourselves capable of spontaneously grasping such principles, then we might claim to possess scientific knowledge of a phenomenon simply because we have observed it. Reasoning in a tight circle, we would be attempting to justify our knowledge of the phenomenon by citing our knowledge of it. This circle of reasoning is vicious. Zabarella argues that we cannot justify such claims to clear and immediate scientific knowledge, because our cognitive powers are weak rather than powerful, because they are finite rather than infinite. His argument that the circle of scientific reasoning is virtuous rather than vicious depends on his acknowledgement that our cognitive powers are ineluctably finite in scope.

　　Although cognitive finitude is, as we saw in Chapter Three, a central element in Heidegger's existential conception of science, he did not explicitly draw a connection between this and his account of modern science as *mathēsis*. The necessary connection between *regressus* and finitude made by Zabarella may now help us to explicate a similar connection between *mathēsis* and finitude in the case of Heidegger.

The circularity of *mathēsis* lies in its being a kind of learning, or discovering, in which we learn or discover something which we already know. We do not get this knowledge out of the things themselves, by simply observing or otherwise dealing with them, but, as Heidegger writes, 'in a certain sense' we bring it already with us. When we deal with a thing, we bring with us a 'fore-conception' of the thingness of the thing.[62] More specifically, in dealing with a plant, we bring with us prior knowledge of the plant-like of the plant. In other words, the whatness of a thing is not something we get out of the things themselves, but is instead a projection which enables us to make sense of those things in terms of what they are. Heidegger argued that this projection, or fore-conception in our understanding, plays a central role in all acts of understanding. He thus defines understanding as an act of interpretation which depends on a perhaps only vaguely specified fore-conception, or prior understanding, of its subject matter. Understanding is thus a circular phenomenon: 'Any interpretation which is to contribute understanding, must already have understood what is to be interpreted.'[63] Because this circle in understanding is ineliminably present in all cognitive acts, in general, it must also be ineliminably present in all acts of scientific cognition, in particular. Heidegger recognised that this renders scientific demonstration circular, but rather than viewing this as a catastrophe, he took it to be an inevitable aspect of finite human existence. This circle of understanding, he writes, 'is the expression of the existential *fore-structure* of Dasein itself.'[64] It is, in other words, a basic structural feature of the subject's projective understanding. This existential structure is expressed in modern scientific cognition, in *mathēsis*, as a metaphysical projection of the thingness of things in terms of a basic ground plan or blueprint. As we saw in Chapter Three, Heidegger radically reinterprets the meaning of metaphysics, arguing that the basis for metaphysical

---

62 Heidegger (1962a), *Being and Time*, p. 191 [150].
63 Heidegger (1962a), *Being and Time*, p. 194 [152].
64 Heidegger (1962a), *Being and Time*, p. 195 [153].

knowledge, as such, is 'the humanness of reason, i.e., its finitude.'[65] The hierarchy of explanation in Heidegger's account of modern science thus breaks down like this: the circularity of scientific practice is to be explained in terms of the ineliminably projective element in *mathēsis*; and this ineliminably projective element is to be explained in terms of human finitude. In a nutshell, scientific demonstration is circular because scientific cognition is finite.

This is not precisely the same as the connection made by Zabarella between circularity and finitude. For him, finitude explains why the circle is virtuous rather than vicious. For Heidegger, it explains why there is any circle at all. The reason for this difference can be uncovered by addressing the third broad feature shared by both *regressus* and *mathēsis*: method. For both Zabarella and Heidegger, because we cannot immediately grasp, with clarity, the principles governing observed natural phenomena, we need to find a method which will help us to get clear on those principles. Method is thus meant, by both, to steer us towards scientific knowledge in spite of our cognitive finitude. But how it does this differs profoundly between the two. For Zabarella, method is meant to overcome the limitations of finitude. He seems to have believed that, through careful, goal-oriented acts of observation, we can prepare our minds to receive epistemic inspiration from God, the *intellectus agens*. Thus human cognitive finitude must, on Zabarella's account, be understood in contradistinction to the infinite cognitive power of God, for whom the issue of circularity never arises because omniscience makes inferential or interpretative reasoning unnecessary.

As we saw in Chapter Three, Heidegger had a profoundly different account of finitude, an account he developed specifically in contrast to the one proffered by Kant. Both Heidegger and Kant viewed cognitive experience as being comprised of two distinct faculties: the first, receptivity; the second, constructivity (or projectivity). However, whereas Kant took finitude to be a constraint on receptivity, Heidegger took it to be a constraint on projectivity. For Kant, as Heidegger reads him, finitude is a state of deprivation which prevents us from gaining cognitive access to the intrinsic, independently existing properties — the essence, or whatness — of a thing. For Heidegger, on the other hand,

---

65 Martin Heidegger (1997 [1929]), *Kant and the Problem of Metaphysics*, 5th edn, enlarged, trans. by Richard Taft (Bloomington: Indiana University Press), p. 15.

finitude is more directly connected to projectivity. The essence of a thing is not something we receive from it, but something it possesses as a result of the socio-historically conditioned metaphysical projection within which it is let be what it is. On Heidegger's account, not even an infinitely powerful intellect could grasp the intrinsic, independently existing essence of a thing, because no such essence exists. Hence, the finitude of our receptivity is not the issue; the issue is, instead, the finitude of our projectivity. The range of possible conceptualisations of a thing is conditioned by the historical tradition of the subject attempting to make sense of that thing. Only within the finite scope of possibilities enabled by the subject's tradition can it experience a thing as intelligible, not to mention develop a clearly defined understanding of what it is.

This process of articulation is advanced, for both Heidegger and Zabarella, by method. Both view method as a means of sharpening up the intelligibility of observed phenomena by clearly defining the causal conditions of those phenomena. Indeed, both even understand this process against the background of the distinction, addressed in Chapter Two, between that-being and what-being, between knowing of a thing that it is and knowing of it what it is. Recall Zabarella's argument that knowledge that a cause is enables us to discover what it is. This discovery process involves a comparison of cause and effect, through which we are gradually led to scientific knowledge of the causal principles underlying the observed effect. The method of *regressus*, by which our understanding of the observed phenomenon is rendered increasingly determinate, thus presupposes a distinction between the that-being and the what-being of the phenomenon. It presupposes, in other words, a version of the minimal realist doctrine introduced in Chapter Two.

To conclude this section, it remains only to emphasise that Zabarella's reflections on method seem to have been motivated by an account of finitude similar to the one which Heidegger attributed to Kant, namely, one developed in contrast to the notion of an infinitely powerful *intellectus agens*. For Zabarella, method helps us to painstakingly transcend our finite human condition and achieve scientific knowledge through communion with a divine intelligence. For Heidegger, method helps us, not to transcend the finitude of our existence, but to articulate the historically engendered epistemic possibilities within that existence. In this case, the self-evidence of ordinary understanding is transcended in pursuit of a more robust, conceptually clear, and critically well-grounded knowledge of nature. Hence, for both Zabarella and Heidegger, method is tied to the metaphysics of transcendence. But, unlike Zabarella, Heidegger does not treat transcendence as a solution to finitude, construed as a problem for knowledge. He views it instead as the critical exploration of the finite range of the sometimes only vaguely understood epistemic possibilities which a scientist inherits through her participation in a shared historical tradition. It was for this reason that Heidegger reinterpreted metaphysics as being grounded in finitude, rather than as being, as Zabarella ostensibly thought, a means by which to overcome such finitude.

These respective conceptions of method, including the relation between method and finitude, imply dramatically different accounts of the norms which govern that method. In the case of Zabarella, because method transcends finitude, the norms which govern it must be similarly transcendent. In the case of Heidegger, because method discloses the latent possibilities within a historical tradition, the norms governing it must be embedded within that tradition. This points to two different understandings of the source of the *a priori* norms which govern scientific practice. In one case, the ultimate source of normativity is timeless. In the other case, it is historical.[66] As we will see in the next section, these two perspectives motivate two different historiographic strategies for explaining the transition from late Renaissance to early-modern science. Insofar as that transition is construed as a process of mathematicisation, these two strategies also enroll divergent conceptions of the mathematical impulse giving rise to early-modern science. In the first case, the mathematicisation of science is an act which allows practitioners to slip free from the historical constraints of their epistemic tradition. In the other case, it is an act of critical interpretation, in which practitioners discover and exploit a possibility latent in their historical tradition, employing it as a new measure in the production of reliable natural knowledge.[67]

---

66 Recall from Chapter Four, that these two different construals of the *a priori* are not incommensurable. Indeed, Heidegger argues that the timeless construal rests on an experience of time in terms of a present-at-hand succession of 'nows.' This he then explains in terms of a more basic experience of 'original time,' which is enabled and sustained by a historical tradition.

67 These contrasting accounts of how the basic measure was set for modern science bring to mind contrasting accounts of poetic creation. On the one hand, poetry is

5. Mathematics and Metaphysics at the Cusp of the Early-Modern Period

Randall's argument for a strong continuity between the Renaissance method of *regressus* and the method distinctive of early-modern natural philosophy was quickly and influentially challenged by the historian of science Alexandre Koyré. As we have seen, Randall argued that '[t]here was but one element lacking in Zabarella's formulation of method: he did not insist that the principles of natural science be mathematical.'[68] Koyré seizes on this statement, insisting that the missing mathematical element was not as trivial as Randall implies, but instead 'forms [...] the content of the scientific revolution of the seventeenth century.'[69] For Koyré, the mathematicisation of *regressus* marks a radical historical discontinuity in knowledge-making practices, and, more specifically, a sudden and profound usurpation of Aristotelian natural philosophy by a resurgent Platonism. This was, in Koyré's view, the usurpation of empirical experience by rational theory: 'Experience is useless because before any experience we are already in possession of the knowledge we are seeking for.'[70] Koyré thus demonstrates his allegiance to a Platonic doctrine of innate ideas, that is, ideas which we somehow possess independently of, and prior to, empirical experience. He also demonstrates his allegiance to the same doctrine as Galen's rationalist critic of empiricism. This move has its merits if one believes that the circularity of the Paduan method is vicious, and that it must be broken

---

divine inspiration; on the other, it is an evocation of possibilities latent in natural language. The idea of poetry as existential measure-setting became crucial for Heidegger in the 1950s. According to Charles Bambach, for the later Heidegger, '[p]oetry measures the limits of what is appropriate for human beings, shaping the contours of our mortal fate' (Charles Bambach (2013), *Thinking the Poetic Measure of Justice: Hölderlin-Heidegger-Celan* (Albany: SUNY Press), p. 174). Indeed, Heidegger would write that poetry sets '[a] strange measure for [...] scientific ideas.' In so doing, poetry 'speaks in "images" [*Bildern*]' (Heidegger (1971a), '"... Poetically Man Dwells...,"' pp. 223, 226). This recalls, from Chapter Four, Heidegger's rooting of the origins of logic— the science of thinking — in Plato's mythic image of the cosmic demiurge. In Chapter Six, I argue that Heidegger placed the origins of the early-modern experiment in a cognate image of the scientific thing.

68   Randall (1940), 'The Development of the Scientific Method,' p. 204.

69   Alexandre Koyré (1943a), 'Galileo and Plato,' *Journal of the History of Ideas* 4(4), 400–28 (p. 406 n. 17).

70   Alexandre Koyré (1943b), 'Galileo and the Scientific Revolution of the Seventeenth Century,' *Philosophical Review* 52(4), 333–48 (p. 347).

through escape into a non-experiential realm of pure thought, an orthodoxly understood metaphysical realm lying beyond the worldly realm of physical sensation.

There is some modest reason to think that Koyré's interpretation of the history of science may have been influenced by his reading of Heidegger, that his emphasis on the mathematicisation of method tracked Heidegger's own description of the historical shift to *mathēsis*. Indeed, a 1931 French translation of Heidegger's 1929 inaugural lecture 'What is Metaphysics?' included an introduction by Koyré, in which he describes Heidegger as 'one of those great metaphysical geniuses whose influence marks an entire period.'[71] The period in question was, of course, Koyré's own. Yet, it seems that Koyré failed to properly understand both Heidegger's stance towards metaphysics, and, more specifically, towards Platonism, as well as his view of the role played by mathematics in early-modern natural philosophy. With respect to Heidegger's stance towards Platonism, recall that Heidegger reinterprets Plato's fundamental and unifying idea of the good in terms of what Heidegger dubbed the 'for-the-sake-of-which' (*Worumwillen*). The for-the-sake-of-which lends a compelling intelligibility and coherence to experience by guiding us in our discrimination between cognitively valuable and cognitively irrelevant or deceptive phenomena. Like Plato's idea of the good, captured also in the mythical image of the cosmic demiurge, the for-the-sake-of-which denotes the *a priori* rules of reason which bring together things and thinking in the production of natural knowledge. The phenomenological importance of Plato's theory of ideas, then, lies in the emphasis it puts on our compulsive feeling — our affectivity— towards the basic rules of reasoning, rules which help us to distinguish between epistemically good and bad phenomena. However, whereas Plato sought to explain this feeling of compulsion in terms of receptivity towards a supernatural realm of ideas, Heidegger attempted to instead explain it in naturalistic terms, as receptivity towards the manifold intersubjective history of a prevailing cultural tradition. In both cases, the phenomenology of this feeling of compulsion — this experience of objective necessity — is recognised

---

71 '[...] un de ces grands génies métaphysiques qui marquent de leur influence une période tout entière' (Alexandre Koyré (1931), 'L'introduction du "Qu'est-ce que la métaphysique?"' *Bifur* 8, 5–8 (p. 5)).

and described, but the respective causal explanations given for that experience are dramatically different. In the first case, the cause is supernatural; in the second case, it is natural. Indeed, for Heidegger, our affective experience of the *a priori* signals our deep epistemic dependency on a shared historical tradition, rather than our ability to transcend tradition in an act of pure reason. This manifold tradition provides the range of possibilities available to us for making sense of nature, for rendering it intelligible, and metaphysics, on Heidegger's account, is a study of the conditions of historical possibility generated by and sustained within this tradition.

Despite his enthusiasm for Heidegger's alleged metaphysical 'genius,' Koyré seems to have missed the fact that Heidegger's ambition was to deconstruct, rather than to champion, Platonic rationalism. For Koyré, the early-modern mathematisation of natural philosophy was initiated by 'some of the greatest geniuses of mankind, a Galileo, a Descartes.'[72] He argues that it was an act of 'pure unadulterated thought, and not experience or sense-perception [...] that gives the basis for the "new science" of Galileo.'[73] Koyré thus identifies mathematics with metaphysics, construed as Platonism, and hence describes Galileo unequivocally as a Platonist.[74]

During the late Renaissance and early-modern periods, however, there also existed a distinctly Aristotelian culture of mathematics. Indeed, when Randall writes that mathematics was the missing element in Zabarella's method, he had Aristotelian mathematics in mind.

> [W]ith rare exceptions the Italian mathematicians down through Galileo, when they possessed a philosophical interest at all, were not Platonists but Aristotelians in their view of mathematics, of its relations to physics, and of the proper method of natural knowledge. [...] What they constructed as 'new sciences' it remained for Descartes to interpret in the light of the tradition of Augustinian Platonism.[75]

---

72 Koyré (1943a), 'Galileo and Plato,' p. 405.
73 Koyré (1943b), 'Galileo and the Scientific Revolution,' p. 346.
74 Koyré (1943a), 'Galileo and Plato,' p. 425. In a comparison of Heidegger and Ernst Cassirer, Michael Friedman notes that Cassirer influenced Koyré's Platonic account of the Scientific Revolution (Michael Friedman (2000), *A Parting of the Ways: Carnap, Cassirer, and Heidegger* (Chicago: Open Court), p. 88).
75 Randall (1940), 'The Development of the Scientific Method,' p. 205.

In Chapter Four, I recounted Heidegger's interpretation of Descartes as first modelling his conception of the subject on Aristotle's narrow construal of *logos* as proposition, but then decisively mathematicising that construal in order to establish the indubitable certainty of pure reason. We may now more precisely specify Descartes's mathematicisation of the rules of reason as having been predominantly Platonist in its motivation. By understanding those rules as being valid independently of experience, Descartes hoped to secure them as the absolute basis for incontrovertible, universal knowledge. A Platonic interpretation of mathematical experience thus underwrote Descartes's rationalistic claim to epistemic absolutism.

That Galileo was, in fact, more motivated by an Aristotelian interpretation of mathematical experience is a point which has been recently pressed by Peter Dear: 'Galileo aimed at developing scientific knowledge [...] according to the Aristotelian (Archimedean) deductive formal structure of the mixed mathematical sciences.'[76] These mathematical sciences were 'mixed,' rather than 'pure,' because they were principally concerned with questions about the physical world rather than with abstract mathematical objects. The chief sixteenth-century examples of mixed mathematical science were astronomy and optics, the former calculating the positions and movements of celestial objects and the later studying the behaviour of light rays construed in geometrical terms. As Dear notes, sixteenth-century astronomical and optical practice also differed from the pure mathematical sciences of geometry and arithmetic in that they made wide use of specialised instruments, such as quadrants and astrolabes, in order to produce precise empirical observations.[77] Thus, according to Dear, they represent the emergence of 'something resembling "experimental science."'[78]

Mixed mathematics was, furthermore, different from natural philosophy in that the former sought to determine the quantitative properties of things through acts of uniform measurement, while the latter sought to determine what *kinds* of things they were, and hence

---

76   Peter Dear (1995), *Discipline & Experience: The Mathematical Way in the Scientific Revolution* (Chicago: University of Chicago Press), pp. 125–26.

77   Peter Dear (2006a), 'The Meanings of Experience,' in *The Cambridge History of Science*, vol. 3: *Early Modern Science*, ed. by Katharine Park and Lorraine Daston (Cambridge: Cambridge University Press), pp. 106–31 (p. 119).

78   Dear (2006a), 'The Meanings of Experience,' p. 119.

what their natural place was in a hierarchically ordered cosmos. To determine the natural kind of a thing is to determine the universal form which it instantiates. In Aristotelian terms, a thing's natural kind is thus also its 'formal cause,' its 'what-it-is.' So, for example, the formal cause of a particular oak tree is the universal oak tree, what an oak tree is *by nature,* and *by definition.* Moreover, by subsisting in its nature, by being what it is, or the kind of thing it is, an oak tree assumes its proper place within the cosmos. There is, then, a tight association between the formal cause of a thing, on the one hand, and its natural place in a heterogeneous and hierarchically ordered cosmos, on the other. Natural philosophers, by attending to formal causes, viewed themselves as the rightful surveyors of natural phenomena within this qualitatively and hence differentially ordered cosmos.

Dear observes that the main charge laid by Aristotelian natural philosophers against mathematicians was that the latter did not provide *causal* explanations of natural phenomena.[79] Indeed, he describes the explanations of mixed mathematicians as 'operational,' and he contrasts these with the explanations proffered by Aristotelian natural philosophers.[80] Yet, as we can know see, this contrast should be more strictly specified as one between operational explanations, on the one hand, and explanations in terms of *formal* causes, on the other. Indeed operational explanations are also, in a broad sense, causal explanations, and hence they were neither unknown nor unappreciated by Aristotelian natural philosophers. Recall, for example, the fourteenth-century Paduan natural philosopher and physician Jacopo da Forlì's description of *regressus* in terms of the resolution of a general and confused knowledge of fever into a specific and distinct knowledge of its causes, which will in turn lead back to a specific and distinct knowledge of the fever itself. Forlì's goal was to explain fever as being caused by such physical operations as the heating of the humor, spirits, or members of the patient's body. The type of cause at play here is not a formal cause but rather what Aristotle called an efficient cause, which

---

79   Dear (2006a), 'The Meanings of Experience,' p. 120; Peter Dear (1995), *Discipline & Experience: The Mathematical Way in the Scientific Revolution* (Chicago: University of Chicago Press), p. 36.

80   Peter Dear (2006b), *The Intelligibility of Nature: How Science Makes Sense of the World* (Chicago: University of Chicago Press), p. 2.

is 'the primary source of change or rest.'[81] Because Aristotle viewed change as movement, the efficient cause is also sometimes called the 'moving cause.' Resolving the fever into its precise causes involved discriminating between probable and improbable efficient causes, for example, between the heating of the spirits and the heating of the humors as the most likely reason for the fever. As we saw, Agostino Nifo later more precisely articulated Forlì's concept of resolution in his concept of *negotiatio*, which Zabarella then called mental examination. The *regressus* could thus involve *negotiatio* in the empirical determination of the efficient causes which produce change in natural bodies.

Dear seems to overlook this when he describes *negotiatio* as a 'mysterious process,' a 'form of contemplation' which presumes 'the mind's innate ability to grasp universals.'[82] These universals were apparently the essences, or formal causes, of the phenomena under study. According to Dear, *regressus* was a 'logical technique [...] designed to generate true scientific knowledge, which for an Aristotelian had to be certain knowledge.'[83] The Paduan *regressus* theorists allegedly believed that *negotiatio* was a strictly logical practice which allowed the contemplative mind to 'intuitively' grasp, as necessary, abstract and universal things, formal causes, which in turn were meant to correspond to something 'metaphysically *real*.'[84] In short, according to Dear, the Paduan theorists were committed to a form of rationalism.

Although this description may stick, in some degree, to Zabarella (who was not a physician), it seems mistaken as a general characterisation of the Paduan physicians. For them, *regressus* involved an empirical search for the physical causes of illness. The heating of the humors of the body was not understood to be an abstract, metaphysical phenomenon: it was understood to be a physical operation which could be studied and, hopefully, physically manipulated so as to restore the patient to health. In fact, the study of the efficient causes of disease appears to have been

---

81   Aristotle (1941c), *Physica*, trans. by R. P. Hardie and R. K. Gaye, in *The Basic Works of Aristotle*, ed. by Richard KcKeon (New York: Random House), pp. 213–394 (p. 241 [line 194b29]).

82   Dear (1995), *Discipline & Experience*, p. 28.

83   Dear (1995), *Discipline & Experience*, p. 27.

84   Peter Dear (1998), 'Method and the Study of Nature,' in *The Cambridge Companion of Seventeenth-Century Philosophy*, vol. 1, ed. by Daniel Garber and Michael Ayers (Cambridge: Cambridge University Press), pp. 147–77 (p. 152).

a common endeavour in late Renaissance medical practice. At the very least, it managed to cross north over the Alps. Indeed, as Dear himself notes, an operational emphasis on efficient causes also characterised the work of the sixteenth-century Swiss physician Paracelsus and his many followers.[85]

Moreover, as Randall makes clear, the Paduan method was not generally meant to produce rational certainty, least of all through acts of pure intellectual contemplation: 'at no time do the Paduan medical Aristotelians attribute any such perceptive power to intellect.'[86] Zabarella was not a *medical* Aristotelian, and so his case may have been different. But consider Nifo's identification of *regressus* with 'physical demonstrations,' and his claim that the empirical 'science of nature is not a science *simpliciter*, like [pure] mathematics. Yet it is a science *propter quid* [i.e., demonstrative].'[87] Nifo furthermore notes that Aristotle, in the *Meteors*, 'grants that he is not setting forth the true causes of natural effects, but only in so far as was possible for him, and in conjectural or hypothetical fashion.'[88] Since even Aristotle himself admitted that the empirical study of nature may not yield certain knowledge, it is not surprising that the medical Aristotelians of Padua demanded no more of their own method.

Dear seems to have exaggerated the epistemic difference between Paduan medical Aristotelians, on the one hand, and sixteenth-century Aristotelian mixed mathematicians, on the other. In fact, both groups appear to have been involved in empirical studies of natural phenomena, and neither side made strong claims to the sort of epistemic certainty typically attributed to the rationalistic demonstrations of logicians and pure mathematicians. Koyré, in turn, appears to have more clearly recognised the empirical and conjectural nature of the Paduan *regressus* method, but he viewed this as an ailment in need of remedy through the metaphysical salve of Platonism. Hence, where Dear

---

85 Peter Dear (2001), *Revolutionizing the Sciences: European Knowledge and Its Ambitions, 1500–1700* (Basingstoke: Palgrave), p. 51.

86 Randall (1940), 'The Development of the Scientific Method,' p. 194.

87 Cited in Randall (1940), 'The Development of the Scientific Method,' p. 194; originally from *Augustino Niphi philosophi suessani exposition...de Physico auditu*, Venice, 1552, I, comm. Text 4, *Recognitio*.

88 Cited in Randall (1940), 'The Development of the Scientific Method,' p. 194; originally from *Augustino Niphi philosophi suessani exposition...de Physico auditu*, Venice, 1552, I, comm. Text 4, *Recognitio*.

sees too much orthodox metaphysics in the Paduan method, Koyré sees too little. Their interpretations thus move in opposite directions: Koyré's away from empirical experience and deeper into metaphysics; Dear's away from metaphysics and deeper into empirical experience. In contrast to both of these interpretations, I want to suggest that the movement in question was never a movement to or from experience. The key point of contention here is the definition of metaphysics. Both Koyré and Dear understand metaphysics in the orthodox sense, as being opposed to empirical experience. On a Heideggerian reading, in contrast, metaphysics is bound together in a reciprocal relationship with experience. The mathematical projection of nature — within which things are experienced in terms of a single, basic measure — operates in continuous concert with the particular, concrete ways in which scientists work with — and, above all, seek to take the measure of — those things.

The greater the number of natural phenomena which have been successfully drawn into the realm of intelligibility circumscribed by this basic measure, the more compelling the mathematical projection becomes as the existential basis for knowing nature. In this way, by expanding the effective reach of a particular way of working with nature, scientists progressively entrench in social practice the metaphysical measure which guides and gives meaning to that work. Through this work, the projected measure increasingly becomes that for the sake of which scientific work is done. In other words, through this process, the things with which one works are progressively experienced in terms of their directedness towards this final measure. The things thus lend themselves more and more easily to an explanation in terms underpinned by that measure, the final end in light of which scientific practice lets things be what we already know them to be, if only in a general and indeterminate way.

This self-reinforcing reciprocal relation between metaphysical projection and work experience would seem to confound the more common historiographic claim that the emergence of early-modern science was the consequence of a historical swing either towards or away from either rationalism or empiricism. On a Heideggerian account, early-modern science received its impulse not from one or the other, but instead from a transformation in the existential relationship between metaphysics and experience, between abstract understanding and concrete action, between theory and experimental practice. As we will see below, this transformation was, above all, a transformation in the role played by the notion of 'final cause' in the early-modern pursuit of natural knowledge. The remainder of this chapter will thus consider the relationship between the concrete practices characteristic of, but not limited to, the mixed mathematical arts, on the one hand, and the metaphysical notion of 'final cause,' on the other. A better understanding of this relationship will put solid ground under our feet when, in Chapter Six, we go on to develop a more detailed and concrete historical analysis of the transition from late Renaissance *regressus* to early-modern *mathēsis*.

6. Nature, Art, and Final Causes in Early-Modern Natural Philosophy

It has become a historiographic commonplace that the emergence of early-modern science was accompanied by the collapse of the so-called art-nature distinction. This distinction was allegedly a barrier to the free application of the experiment in the investigation of nature. According to this widely received view, the key pillar upholding the art-nature distinction was the Aristotelian concept of final cause. Hence, the breakdown of this distinction entailed a rejection of final causes.

In this section, I will challenge this historiographic commonplace. In doing so, I will begin to apply Heidegger's notion of *mathēsis* more directly to contemporary debates in the history of early-modern science, an application which will extend into Chapter Six. My main claim here will be that no consequential breakdown in the art-nature distinction was necessary for the emergence of early-modern science, because the distinction, while important, was never as strict or inflexible as has often been suggested. As a consequence, there was no corresponding need to eliminate final causes from explanations of experimentally produced natural phenomena. Indeed, despite early-modern rhetoric to the contrary, the coherence and intelligibility of experimental manipulations of nature required that a central role be given to final causes. Without allowing room for final causes in explanations of the artful manipulation of nature, we will achieve an only partial, and perhaps not entirely coherent, understanding of early-modern scientific practice. As we will see, the claim that early-modern experimental operations cannot be properly explained without reference to final causes is a specific example of Heidegger's more general claim that early-modern *mathēsis* was a matter, not just of working with the things, but also of the metaphysical projection of the thingness of those things. According to Heidegger, the ultimate end towards which the end-directedness of scientific things points, their final end, is a basic blueprint. This blueprint is, in Aristotelian terms, the final cause of those things.

In the last section, I addressed Dear's distinction between operational and causal explanations, arguing that operational explanations, by making reference to how things happen, appeal to efficient causes, and thus are also causal explanations. But they are not causal explanations of the type most valued by sixteenth-century Aristotelian natural philosophers: they are not explanations in terms of *formal* cause. These explanations address a thing in terms of *what* it is, rather than of *how* it happens. In other words, they explain it in terms of the *kind* of thing it is, in terms of the specific *thingness* manifested in the thing. For example, a sixteenth-century explanation of fever in terms of the heating of the humors specifies that the heating causes the fever, that this is how fever happens. Yet, notice that it also says something about *what* fever is: namely, that it is the *kind* of thing which occurs when the humors are heated. There is, then, an important connection between efficient and formal causes, because operational explanations in terms of the former implicate a role for the latter.

Once fever has been defined in terms of the processes by which it occurs, it becomes possible, at least in principle, to treat it by intervening in those processes. On the medieval definition, one could mitigate a fever by artificially cooling the patient's humors: for example, by immersing her in a basin of cool water. Such an intervention may help to return the patient to a natural state of health. This is an important point. Recall that, for Aristotle, change is a kind of movement. Medical interventions, as the efficient causes of health, may also be called the moving causes of health, because to restore a patient's health means to move her back into a natural state. Aristotle wrote that 'the movement of each body to its own place is motion towards its form.'[89] We saw in

---

89 Aristotle (1941d), *On the Heavens*, trans. by J. L. Stocks, in *The Basic Works of Aristotle*, ed. by Richard KcKeon (New York: Random House), pp. 398–466 (p. 459 [lines 310a34–35]).

the last section that a thing, by being what it is, instantiates its form. In so doing, it takes its proper place in the cosmos. We can now add that a thing, by *becoming* what it is, moves towards its form, and, in so doing, also moves to its proper place — or what we may be more inclined to call its proper *state* — in the cosmos. The physician directly intervenes in the operations of the patient's body — operates on her — in order to bring her back into proper form, good shape, a natural state of health. Health is thus that *for the sake of which* the physician performs the operations, and those operations are likewise performed *in order to* restore the patient's health.

In their respective accounts of the Scientific Revolution, both Dear and SSK practitioner Steven Shapin observe that 'that for-the-sake-of-which' an operation occurs was known to Aristotelians as the 'final cause.'[90] Dear furthermore recognises a distinction between the formal cause of a thing, the *kind* of thing it is, on the one hand, and its final cause, on the other. The motivation for this distinction, he writes, 'was to understand in the most fundamental way what things *were* and why they *behaved* as they did.'[91] Yet, while Aristotle did distinguish in this way between formal and final causes, he also argued that the two are often identical, and, furthermore, that they coincide with the efficient, or moving, cause: 'the "what" and "that for the sake of which" are one, while the primary source of motion is the same in species as these.'[92]

For example, the process of growth initiated in an acorn can be explained only in light of the end, or final cause, of that process: namely, a mature oak tree. The final cause explains why an acorn grows into an oak tree rather than into an artichoke. Similarly, according to Aristotle, it explains why 'the healable, when moved and changed *qua* healable, attains health and not whiteness.'[93] A movement towards form has a directedness, a regularity, which distinguishes it from chance occurrence, and this regularity of movement is what a reference to final causes is meant to explain. Final causes explain *why something becomes* what it is: why an acorn becomes an oak tree. They thus serve

---

90 Steven Shapin (1996), *The Scientific Revolution* (Chicago: University of Chicago Press), p. 139; Dear (2001), *Revolutionizing the Sciences*, p. 13.

91 Dear (2001), *Revolutionizing the Sciences*, p. 14.

92 Aristotle (1941c), *Physica*, trans. by R. P. Hardie and R. K. Gaye, in *The Basic Works of Aristotle*, ed. by Richard KcKeon (New York: Random House), pp. 213–394 (p. 248 [lines 198b26–27]).

93 Aristotle (1941d), *On the Heavens*, p. 459 (lines 310b17–18).

to explain generative movement *qua* generative, that is, *qua* movement which we experience as organised and directed towards form. Formal causes, in contrast, explain *what something is*. Their focus is on being, rather than on becoming. What an oak tree is can be explained without reference to the generative movements which brought the oak tree into being. Final and formal causes can thus be viewed as different modes of explanation with respect to the same thing. This is a difference between a directed process and its natural outcome, between a thing's regulated actualisation and its resultant actuality. In the case of the oak tree, the operation is internal: oak trees reproduce themselves through acorns. Here, the formal and final causes are similarly located in the oak tree. They are equally situated in the very thing about which they are meant to provide an explanation.

Accordingly, Andrea Falcon has argued that, for the Aristotelian student of nature, formal and final causes were often not distinguishable in practice.[94] Indeed, Aristotle reiterates this point in *On the Generation of Animals*: 'first, the final cause, that for the sake of which a thing exists; secondly, the formal cause, the definition of its essence (and these two we may regard pretty much as one and the same).'[95] Aristotelian natural philosophers could thus be justified in speaking not of two distinct causes but of only one single 'formal/final' cause.

Crucially, the same cannot be said of Aristotelian students of art. For them, formal and final causes are separated because, in art, the source of movement lies outside the moving body, the emergent work of art. As Aristotle wrote: '[A]rt is a principle of movement in something other than the thing moved, nature is a principle in the thing itself.'[96] Hence, when the physician immerses a feverish patient in a basin of cool water, the source of the movement meant to restore the patient to health lies, at least in significant part, outside the patient, in the physician, or, more accurately, in the medical art of the physician. Furthermore, as noted

---

94 Andrea Falcon (2015), 'Aristotle on Causality,' in *The Stanford Encyclopedia of Philosophy*, ed. by Edward N. Zalta (Spring 2015 Edition), §3.

95 Aristotle (1941e), *On the Generation of Animals*, trans. by Arthur Platt, in *The Basic Works of Aristotle*, ed. by Richard KcKeon (New York: Random House), pp. 665–80 (p. 665 [lines 715a4–7]).

96 Aristotle (1941f), *Metaphysics*, trans. by W. D. Ross, in *The Basic Works of Aristotle*, ed. by Richard KcKeon (New York: Random House), pp. 689–926 (p. 874 [lines 1070a7–10]).

above, medical art is not a random activity, but an activity directed toward a specific end, namely, health. The physician is the site of both the activity (efficient cause) and the principle (final cause) which organises and gives an overall meaning to that activity. The patient, on the other hand, is both the object of treatment (material cause) and the site of health (formal cause) to which the physician endeavours to return that body. In contrast to generation in nature, the formal cause — the 'what' — and the final cause — 'that for the sake of which' — are not united in art, because they each exist in a different location. Yet, in the medical arts, there are exceptions to this general rule, and such cases serve to weaken the distinction between nature and art. Hence, Aristotle writes that 'a doctor doctoring himself: nature is like that.'[97] Or, to take another example, one walks about 'in order to be healthy.'[98] Just as with the oak tree, in this case the respective locations of the final and formal causes are now the same, and so the two causes become indistinguishable in practical terms. In addition, the efficient cause is now located, as with the oak tree, in the body being moved: the patient is also the agent, a self-mover.

One may object that, even when there is no necessary difference between nature and art with respect to location of causes, an important distinction may still be made by pointing to the deliberative character of artful movement in contrast to natural movement. According to this argument, the doctor deliberately sets out to doctor herself, while the oak tree reproduces itself automatically, that is, without deliberation. But Aristotle challenged this distinction as well. He observes that 'art does not deliberate.' Hence, '[i]f the ship-building art were in the wood, it would produce the same results *by nature*.'[99] Self-awareness thus seems to play no necessary role in Aristotle's conception of art. For him, the regulative movements of art can be just as non-deliberative as those of nature. This recalls Heidegger's own observations, discussed in Chapter Four, about the non-deliberative character of our actions when we are immersed in a work-world. The master fiddler does not

---

97   Aristotle (1941c), *Physica*, p. 874 (line 199b30).
98   Aristotle, *Physica*, lines 194b34; trans. by R. Hope and cited in Bas C. van Fraassen (1980), 'A Re-Examination of Aristotle's Philosophy of Science,' *Dialogue* 19(1), 20–45 (p. 24).
99   Aristotle (1941c), *Physica*, p. 251 (lines 199b327–29).

deliberate over the placement of her fingers while she is fiddling. She just fiddles. Hence, if the art of fiddling were instead located in the fiddle, the fiddle would play itself, in just the same way an oak tree reproduces itself through an acorn: namely, non-deliberatively, without self-awareness, but nevertheless with a directedness which serves to organise the corresponding operations. In fact, Heidegger recognises in Aristotle a distinction between the 'end' (*telos*) towards which a thing is directed, on the one hand, and the 'goal' or 'purpose' of that thing, on the other. On this basis, Heidegger concludes that for Aristotle: '*telos* is not "goal" or "purpose," but "end."' One may attribute to a thing the cause of its own activity without also attributing to it self-awareness or consciousness.[100] For both Heidegger and Aristotle, the end-directedness of both natural and artful movement may be explained without recourse to the intellectualist concepts of 'goal' and 'purpose.'

In addition to their shared interest in non-deliberative practice, there is another important parallel between Heidegger's work and Aristotelian natural philosophy. Readers will have already noticed that Heidegger, like Aristotle, uses a concept of the 'for-the-sake-of-which.' Indeed, Aristotle's observation that health is that *for the sake of which* one walks about, and that one walks about *in order to* maintain one's health, recalls Heidegger's own close association between the concepts 'for-the-sake-of-which' and 'in-order-to.' As we saw above, and more fully in Chapter Four, Heidegger links these two concepts in his discussion of equipment, or ready-to-hand things, that is, things as we experience them in use. We use a pen, for example, in order to make marks on a page, so as to communicate. The use of the pen is an in-order-to of graphic communication. Yet such communication is more than just mark-making. The marks on the page must be shaped and organised such that they convey a meaning. Only then will they count as communication. Graphic communication is thus that for the sake of which one uses the pen. Only once we have acquired the skills for

---

100 Martin Heidegger (1976 [1967]), 'On the Being and Conception of φύσις in Aristotle's Physics B, 1,' *Man and World* 9(3), 219–70 (pp. 231). See also: Martin Heidegger, Martin (1977b [1954]), 'The Question Concerning Technology,' in *The Question Concerning Technology and Other Essays*, by Martin Heidegger, trans. by William Lovitt (New York: Harper & Row), pp. 3–35 (p. 8); Martin Heidegger (2000 [1953]), *Introduction to Metaphysics*, trans. by Gregory Fried and Richard Polt (New Haven: Yale University Press), p. 63.

such communication can we use the pen in order to convey a meaning. Hence, Heidegger writes that '[o]nly so far as the for-the-sake-of a can-be is understood can something like an in-order-to (a relation of assignedness) be unveiled.'[101] This understanding of the for-the-sake-of-which — of the final cause — allows for a 'projection […] in whose luminosity things of the nature of equipment are encountered.'[102] In partly more Aristotelian terms, an understanding of the final cause of a thing opens up ('projects') a space of intelligibility in which the formal cause, or whatness, of the thing may be realised through the efficient cause, or operations, by which that thing comes to be experienced as what it is. So, the understanding, or the art, of graphic communication informs one about what kinds of marks will contribute to meaning, and hence also about how the pen is to be used in order to produce those kinds of marks. As a consequence, an understanding of the final cause, whether deliberative or non-deliberative, marks the difference between random, meaningless behaviour and results, on the one hand, and organised, meaningful practices and products, on the other. Heidegger's concept of understanding thus plays much the same role as Aristotle's concept of art with respect to production: both are directed towards the for-the-sake-of-which, towards the final cause or end. Such understanding brings with it rules for regulating our behaviour in a sensible way, for successfully choosing between cognitively good and cognitively bad courses of action in order to produce meaningful results. It explains, for example, why a carpenter *qua* carpenter produces cabinets rather than crockery.

It is worth reiterating that the rules or instructions originating in the for-the-sake-of-which need not be articulated, much less formalised, in order to perform their regulative function: they can be non-deliberative, unreflective, or tacit. They do not entail the self-awareness of the moving, or efficient, cause. On this point, Aristotle and Heidegger agree, but Aristotle goes further by applying the idea, not just to art, but also to self-generation in nature. This returns us to the issue of intentionality and naturalised epistemology, addressed at the end of Chapter Four. We saw there that Bloor's call for a naturalised account of scientific reasoning need not entail the naturalistic reduction of intentional states to

---

101 Heidegger (1982a), *Basic Problems of Phenomenology*, p. 295; translation modified.
102 Heidegger (1982a), *Basic Problems of Phenomenology*, p. 293.

non-intentional states. For such a reduction presupposes that intentional states necessarily include propositional content. Yet Heidegger rejects this presupposition, arguing that intentional actions — actions exhibiting a directedness — can also be non-propositional in character. This underpins the present claim that intentional states may be non-deliberative. Intentional acts are not necessarily conscious phenomena, and so intentional activity may be ascribed to something without supposing that thing to possess a consciousness. When Aristotle argues that nature displays intention through the directedness of its activities, but that it does not deliberate, he would appear to espouse an account of natural intention which does not presuppose a sentient nature. In Aristotle, natural things move themselves to their proper place in an ordered cosmos, but, like the master fiddler, they do not have to think about doing so — they just do it. Aristotle appears to have imagined nature as a master artist, unreflectively performing itself.

It is this 'performance' which Aristotelian natural philosophers took as their object of study. By cataloguing the final/formal causes which govern the self-movement of nature, they hoped to achieve a dense and precise understanding of the ordered cosmos. If each thing moves naturally to form, to its own proper place in the cosmos, then a catalogue of these various forms would also provide a kind of conceptual map of the heterogeneous and hierarchically ordered places which constitute the qualitatively complex structure of the world. In the most abstract of terms, this conceptual map was meant to articulate in clear and explicit terms the implicit role played by final causation in regulating natural operations in the world. It may thus be viewed as a kind of cosmic operations manual, with the proviso that the original operator, just like the master artist, has no need for such a manual. Indeed, the master artist is likely to find a manual purporting to explicate her practice as, at best, an over-simplification of what she actually does. On the other hand, she may also find it difficult to articulate, in clear and determinate propositional terms, what she finds evident in her art but lacking in the manual.

These considerations should now make us wary about the oft-made claim that early-modern practitioners can be distinguished by their rejection of the Aristotelian distinction between art and nature. Historiographically, this distinction has often been viewed as an obstacle to the rise of the mechanical philosophy on which the new experimental practice is thought to have been based. Shapin writes that 'the precondition for the intelligibility and the practical possibility of a mechanical philosophy of nature was setting aside that Aristotelian distinction.'[103] Dear, who, as we have seen, emphasises the importance of mixed mathematics in the development of experimental practice, argues that '[t]he art-nature distinction impinged on the use of artificial contrivance in the making of natural knowledge — that is, it compromised the legitimacy of using in natural philosophy the sorts of procedures used by mathematicians.'[104] Like Shapin, Dear suggests that '[t]he widespread adoption of various forms of "mechanical philosophy" went along with a drastic weakening of the art-nature distinction in philosophical thought,' and he then adds that this 'provided ontological vindication of the primacy of the mathematical sciences.'[105] It is a historiographic commonplace that early-modern experimental philosophy was enabled by the rising fortunes of the mechanical philosophy. However, Shapin and Dear also maintain that the Aristotelian distinction between art and nature was a central barrier to the rise of the new experimentalism. I think that we should not accept this claim. Let me explain my disagreement by addressing Dear's more detailed argument.

On closer inspection, it appears that the true barrier to experimental practice was, according to Dear, Aristotelian final causes: '[t]o the extent that Aristotle's natural philosophy sought the final causes of things, and thereby to determine their natures, experimental science was therefore disallowed.'[106] At the root of the rejection of the art-nature distinction was, therefore, the rejection of final causes. This would seem to make sense since, as we have seen, Aristotle distinguished art and nature by the location of their respective final and efficient causes. For the things of nature, the principle of their movement is internal, while for the things of art, the principle of their movement is external. However,

---

103 Shapin (1996), *The Scientific Revolution*, p. 31.

104 Dear (1995), *Discipline & Experience*, p. 153.

105 Dear (1995), *Discipline & Experience*, p. 151.

106 Dear (2006a), 'The Meanings of Experience,' in *The Cambridge History of Science*, vol. 3: *Early Modern Science*, ed. by Katharine Park and Lorraine Daston (Cambridge: Cambridge University Press), pp. 106–31 (p. 110).

for Aristotle this distinction was not, as Dear suggests, an 'absolute separation.'[107] Indeed, recall Aristotle's argument that nature is like a doctor doctoring herself. Hence, in some special cases, art moves its object internally, and these special cases provided Aristotle with an analogical model for his account of natural generation. There was, for him, no absolute distinction to be drawn between art and nature, because he understood natural processes, in general, by analogy to special cases of artistic process.

Yet, Dear argues that natural and artificial causes were considered distinct because the regularity of natural processes could be 'subverted' by artificial causes.[108] For example, '[a]n aqueduct [...] is not a natural watercourse; it reveals the intention of its human producer, which thwarts that of nature.'[109] But this example can also speak for an affinity between art and nature. From an Aristotelian perspective, the natural tendency of liquid water is to move as close as possible to the centre of the earth. The aqueduct does not thwart or subvert this tendency, but is entirely dependent on it for the successful delivery of water. The final cause which gives sense to the operations of the aqueduct is thus compatible with the final cause which gives sense to the operations of the natural watercourse: both facilitate the movement of liquid water to its proper place in the cosmos. This is not to say that artifice cannot be used to subvert the natural tendencies of things, but only that this possibility does not warrant an absolute separation of art and nature. Indeed, Aristotle wrote that 'if things made by nature were made also by art, they would come to be in the same way as by nature. [...] [G]enerally art partly completes what nature cannot bring to a finish, and partly imitates her.'[110] In general, then, Aristotle viewed the respective operations of art and nature as complementary rather than as contradictory.

This point is emphasised in Dear's description of the early-modern experiment as 'mimetic, not semiotic.'[111] The artifice of experiment

---

107 Dear (1995), *Discipline & Experience*, p. 155.
108 Dear (1995), *Discipline & Experience*, p. 155.
109 Dear (1995), *Discipline & Experience*, p. 155. Elsewhere, Dear assimilates 'contrived situations' to 'interference' with natural processes in the natural philosophical context (Peter Dear (1990), 'Miracles, Experiments, and the Ordinary Course of Nature,' *Isis* 81(4), 663–83 (p. 681)).
110 Aristotle (1941c), *Physica*, p. 250 (lines 199a14–17).
111 Dear (1995), *Discipline & Experience*, p. 159.

imitates nature, rather than signifying it. According to Dear, mimesis is a primary characteristic of experimental manipulation.[112] The experimental philosopher reproduces, or mimics, natural processes using artificial means, with the goal of generating new knowledge of nature, 'especially knowledge of an operational kind.'[113] Historically, this move involved what Dear describes as '[a] subtle redefinition of natural knowledge — and thus of nature itself.'[114] He refers to the 1647 book, *Discours du vuide*, by the French physician and college teacher Pierre Guiffart, as evidence for this subtle redefinition. Dear cites Guiffart as stating:

> There is a very notable difference between art and nature: art cannot produce anything without nature; it not only needs nature to furnish the material, but it also needs nature's natural inclinations to go along with it, so that thereby it supplements nature's rules and produces its own work.[115]

The subtle redefinition alleged to appear in this passage is that '[k]nowledge of nature, rather than being about identifying purposes [i.e., final causes], is now, insensibly, becoming about characterising "rules" [...] of nature.'[116] On the basis of this perceived distinction between final causes, on the one hand, and rules, on the other, Dear draws a further distinction between the 'teleological' explanations of Aristotelian natural philosophers and the 'operational' explanations of mixed mathematicians and the experimental philosophers inspired by them.[117]

This recalls Dear's distinction between the operational explanations of mixed mathematicians, on the one hand, and explanations based on formal causes, as proffered by Aristotelian natural philosophers, on the other. In response to this, I argued, first, that operational explanations are also causal explanations because they refer both to efficient causes and, at least tacitly, to final causes, and, second, that final causes and formal causes are, in the realm of Aristotelian natural philosophy, effectively identical. It follows from this that Dear's distinction between

---

112 Dear (1995), *Discipline & Experience*, p. 159.
113 Dear (1995), *Discipline & Experience*, p. 159.
114 Dear (1995), *Discipline & Experience*, p. 157.
115 Dear (1995), *Discipline & Experience*, p. 157.
116 Dear (1995), *Discipline & Experience*, p. 157.
117 Dear (1995), *Discipline & Experience*, p. 158.

operational and teleological explanations should be regarded with some scepticism. As we have seen, explanations in terms of final causes can be viewed as explanations in terms of the rules which give meaning and direction to natural processes: they explain, for example, why an acorn grows into an oak tree rather than into an artichoke. The same point is implied in the passage from Guiffart. He argues that art relies on the support of natural inclinations so that it may act to supplement nature's rule-governed activity. Yet natural inclinations seem just to be the natural tendencies definitive of final causes. Following its natural tendency, an acorn develops into an oak tree. Hence, as Guiffart suggests, because it relies on the natural tendencies of natural materials, art ends up supplementing, rather than violating, nature's rules. Natural inclinations are here conceptualised as inclinations to follow the rules governing natural movement. The crucial point to be emphasised, then, is that a focus on the rules governing natural processes is simultaneously a focus not just on material causes, but also on final causes, and so it cannot be read as a rejection of the art-nature distinction. Hence Guiffart's ability to faultlessly refer to nature's rules while still asserting a notable difference between art and nature. For Dear, Guiffart's affirmation of the art-nature distinction must be dismissed as 'lip service,' because Dear has put himself in an interpretative position where rule-based operational explanations of nature are incompatible with the Aristotelian doctrine of final cause allegedly underpinning a distinction between artificial and natural processes.[118]

Dear emphasises 'manipulation' as a key element in his explanation for the transition from late Renaissance natural philosophy to early-modern experimental philosophy, or what I have called the transition from *regressus* to *mathēsis*. The notion of artful manipulation was, he argues, a 'Trojan horse' by which the art-nature distinction could be circumvented, thereby allowing the applied techniques of mixed mathematicians to flood into early-modern natural philosophy.[119] In response to Dear's hypothesis, I have sought, in this section, to demonstrate two things. First, I have argued that the claim that early-modern experimental philosophy subverted the art-nature distinction should be treated with some scepticism. I have suggested that this

---

118 Dear (1995), *Discipline & Experience*, p. 161.
119 Dear (1995), *Discipline & Experience*, p. 161.

distinction was not, *per se*, an obstacle for experimental practice, because artful manipulation had always been viewed as potentially compatible with natural processes. Second, I have argued that manipulation alone does not sufficiently explain the early-modern mathematisation of natural knowledge. Indeed, attention to manipulation alone explains very little. One must also attend to the governing principles which lend order and meaning to those manipulations, which explain to what end those manipulations are directed and for what end they are performed. Put otherwise, efficient causes must be married to final causes if we are to shed adequate explanatory light on the rule-governed dynamics which we observe in the worlds of both nature and art. Coherent operational explanations presuppose that for the sake of which those operations occur, because, without this presupposition, those operations would not be intelligible as anything more than random activity.

7. Conclusion

Tongue planted firmly in cheek, Shapin begins his 1996 book, *The Scientific Revolution*, with the following declaration: 'There was no such thing as the Scientific Revolution, and this is a book about it.'[120] Indeed, although historians of science have, in recent years, become increasingly circumspect about the notion of a Scientific Revolution, few have been willing to abandon it entirely. In particular, the idea that early-modern science was inaugurated by a sudden, radical, and complete — in short, a *revolutionary* — epistemic break with the past has been largely abandoned by historians, with the historiographic emphasis shifting to more nuanced considerations of both the continuities and discontinuities which exist between early-modern science and its predecessors.

The Heideggerian account of the Scientific Revolution which I have begun to outline in this chapter shares this circumspective stance. Recall that my explication of Heidegger's account of modern science as *mathēsis* began with Heidegger's insistence that facts, measurement, and experiment, broadly construed, figure as continuous threads running from modern science all the way back through medieval to ancient science. Moreover, according to Heidegger, the Scientific Revolution was sparked not by the rejection of tradition, but instead by

---

120 Shapin (1996), *The Scientific Revolution*, p. 1.

the consolidation of an existential possibility which had, until that time, lain relatively dormant within that tradition. In this way, Heidegger's use of the term 'revolution' [*Umwälzung*] harkens back to an older meaning of that term: not a sudden and radical break with the past, but instead a transformative, as opposed to an atavistic, return to it.

The extent to which Heidegger viewed the Scientific Revolution as powerfully tied to the past is clearly evinced by his claim that the Aristotelian concept of final cause was not, after all, abandoned by early-modern scientific practitioners. In this, as we have seen, Heidegger argues for greater historical continuity than do both Dear and Shapin. Yet, this concept was retained only to then undergo transformation in a way which profoundly altered the extant practices of fact-stating, measuring, and experimenting. Heidegger identified this transformation with the mathematisation of scientific knowledge. At its root, this mathematisation had to do, not with numerical practice, but with the development of a uniform measure, a basic blueprint, circumscribing the thingness of scientific things. It was the mathematisation of final causes, rather than their rejection, which allowed physical apparatus to flood into the knowledge-making practices of early-modern natural philosophy. It was how these apparatuses were used, and the end to which they were put, which mark a decisive transformation in the meaning of those practices. Furthermore, it was the concentration of scientific experience under a uniform measure which allowed numerical practice to likewise flood into, and give new form to, the early-modern natural philosophical experience of nature. Hence, Heidegger's account of science as *mathēsis* may be read as charting a neutral course between two competing historiographic schools, each of which favours an explanation of the Scientific Revolution in terms of either the rise of experimental or mathematico-numerical practice. His account is meant to illuminate the common soil in which these two aspects of modern science are rooted, and from which they both sprang. It thus allows us an opportunity to reconnect recent historical studies of early-modern experimental philosophy with the theory-oriented studies which have typically been the brief of historians more concerned with early-modern mathematical practice. Chapter Six will give concrete, micro-historical attention to the ways in which the mathematicisation of final causes transformed the fortunes of early-modern experimental practice.

Chapter Six

Mathematics, Experiment, and the Ends of Scientific Practice

1. Introduction

In Chapter Five, we familiarised ourselves with Heidegger's concept of *mathēsis*. This concept lies at the heart of his attempt to understand the emergence of early-modern science in terms of mathematisation. In particular, *mathēsis* — as a kind of learning or studying wherein we learn what we, in a general and indeterminate way, already know — is meant to capture the mathematisation of the pre-modern Aristotelian notion of final cause. As we saw, Heidegger grounds his analysis in our experience of the thing as ready-to-hand within a work-world, that is, as something with which we are at work. According to Heidegger, such a thing has an end-directedness. When we work with (rather than against) it, we let it be what it is in its directedness towards some end; we let it, so far as is possible, fulfill that end.

I suggested that this end is the final cause of the thing, that for the sake of which we let the thing, in the course of working with it, be what it is. Heidegger argues that, in the case of early-modern scientific work, the end of the thing is a ground plan, a basic blueprint. As the final cause of the thing, this basic blueprint provides the measure against which our work with the thing makes sense. It guides our judgement

 http://dx.doi.org/10.11647/OBP.0129.06

about what is, and what is not, relevant as we work with the thing, and thus also our basic understanding of the thingness of the thing.

According to Heidegger, the measure given by the ground plan applies to the scientific thing *as such*. All potential scientific things will be circumscribed within the experiential realm laid out according to this measure. This circumscription forms the basis of mathematisation. The totality of end-directednesses within this realm — and thus also the ends of the corresponding scientific practices — becomes consolidated under one uniform measure — a single, final end. Hence, the scientific thing is what it is insofar as it conforms to this measure. This conformity is not, however, a strict determination. Recall that those things which the ground plan renders knowable are initially known in an only general and indeterminate way. Through *mathēsis*, this knowledge may then be developed into something more specific and determinate, and this determination may take a number of different shapes. All of these shapes will, however, still share the same general form. For example, individual plants and animals may refer to two different categories — the plant-like and the animal-like — but those two categories, in turn, refer to a single, more general category — the thing-like. Similarly, we may refer to things either in quantitative or operational terms, but we must first be able to refer to them as things. Hence, even though all scientific things conform to the same general measure, they are nevertheless amenable to specification and determination in a variety of different ways.

As a consequence, *mathēsis*, as the mathematical projection of the thingness of things, as a basic projective structure in the subjectivity of the subject, served to facilitate the influx of both numerical and instrumental practices into early-modern natural philosophy. By allowing for the consolidation of all potentially scientific things under one general measure, *mathēsis* expanded the range of application for concrete mathematical techniques. If a thing could be drawn within the scope of scientific experience, then one could also quantify it. *Mathēsis* likewise enabled and expanded the range of application for concrete instrumental techniques. Hence, scientific things came to be increasingly viewed as a legitimate subject matter for artful manipulation. Quantification and manipulation are thus two different possibilities for specifying the uniform thingness of the scientific thing. Thus, as was claimed at the close of Chapter Five, Heidegger's introduction of *mathēsis*, as a central and collective existential impulse behind the rise of early-modern science, charts a middle course between the historiographic Scylla and Charybdis of mathematics and experiment.

This is not so much a rejection as it is a resolution of the historical difference between what Thomas Kuhn identified as mathematical and experimental traditions in the physical sciences.[1] Each tradition represents a different way of specifying and determining the same general and indeterminate knowledge of things. Kuhn's traditions may thus be better described as sub-traditions within a broader tradition in which scientific thinking is guided by a uniform measure against which scientific things are experienced as intelligible. The shared root of these two traditions in mathematical projection furthermore helps to explain why these two traditions could eventually come together, especially, as Kuhn observes, in the case of nineteenth-century physics.[2] Like the hills on either side of a valley, a bridge can be built to join them. However, Heidegger's account reminds us that, despite the distance between them, each hill nevertheless forms one side of the same valley.

Peter Dear also challenges Kuhn's distinction between mathematical and experimental practices, arguing for an 'intimate relationship' between the two, manifested most spectacularly in the late-seventeenth-century mathematico-experimental work of Isaac Newton.[3] Yet, as we saw in Chapter Five, Dear locates the heart of experimental practice not, as Kuhn did, in the tradition represented by Robert Boyle's mid-seventeenth-century experimental philosophy and the Royal Society of London (of which Boyle and then Newton served as President), but instead in the physical apparatuses and artful manipulations of the Aristotelian tradition of mixed mathematics. Indeed, Dear argues that 'Boylean experimental philosophy was not the high road to modern experimentation; it was a detour.'[4] Since Kuhn's experimental tradition was precisely the Boylean one, it would seem that Dear has not really,

---

1   Thomas Kuhn (1977), 'Mathematical versus Experimental Traditions in the Development of Physical Science,' in *The Essential Tension: Selected Studies in Scientific Tradition and Change,* by Thomas Kuhn (Chicago: University of Chicago Press), pp. 31–65.

2   Kuhn (1977), 'Mathematical versus Experimental Traditions,' p. 63.

3   Peter Dear (1995), *Discipline & Experience: The Mathematical Way in the Scientific Revolution* (Chicago: University of Chicago Press), p. 246.

4   Dear (1995), *Discipline & Experience,* p. 2.

after all, discovered an intimate relationship between the two sides of Kuhn's distinction.

Steven Shapin, in contrast, preserves Kuhn's distinction, and emphasises the importance of Boyle's experimental philosophy as a forebear of the Newtonian programme.[5] In fact, Shapin even strengthens Kuhn's distinction by carefully outlining Boyle's resolve in insulating experimental philosophy from techniques of mathematical demonstration. According to Shapin, Newton would later both adopt Boylean experimental practice and marry it to those same mathematical techniques.[6] Yet, as we will see later in this chapter, although Boyle eschewed mathematical forms of persuasion, his experimental philosophy was not incompatible with *mathēsis*. Indeed, I will argue that Boylean experimental philosophy was, in the Heideggerian sense, strongly mathematical. Although Boyle rejected the concrete techniques of specification typical of mathematical practice, his understanding of the thingness of scientific things was nevertheless guided by a uniform measure. This uniform measure influenced, in turn, the way Boyle worked with, or manipulated, those things.

Shapin notes that Boyle's suspicion of mathematics included the rejection of an 'ontology' contending that 'physical qualities were uniform.'[7] Instead, Boyle allowed that 'substances like air and water varied in their physical properties from one locale to another and from one time to another.'[8] In other words, Boyle rejected an ontology which insisted that the properties of token instances of the same type always be specified in the same way. But Boyle's tolerance of context-dependency in the specification of physical properties is compatible with my claim that he was a mathematical philosopher in the Heideggerian sense.

---

5   Steven Shapin (1988), 'Robert Boyle and Mathematics: Reality, Representation, and Experimental Practice,' *Science in Context* 2(1), 23–58; Steven Shapin (1994), *A Social History of Truth: Civility and Science in Seventeenth-Century England* (Chicago: University of Chicago Press), pp. 310–54.

6   Steven Shapin (1996), *The Scientific Revolution* (Chicago: University of Chicago Press), p. 116. Note that, while Kuhn made the tentative psychologistic suggestion that the experimental/mathematical distinction may be 'rooted in the nature of the human mind,' Shapin's account is more sociological, attributing to Boyle the conviction that 'mathematical means of persuasion were embedded within an improper, even immoral, social order' (Kuhn (1977), 'Mathematical versus Experimental Traditions,' p. 64; Shapin (1988), 'Robert Boyle and Mathematics,' p. 33).

7   Shapin (1988), 'Robert Boyle and Mathematics,' p. 47.

8   Shapin (1988), 'Robert Boyle and Mathematics,' p. 48.

Boyle could have both understood the thingness of things according to a single uniform measure, and allowed that this thingness be specified and determined in different context-sensitive ways. What appears to have been important to him was the existence of a common ontological measure against which these localised differences could then be meaningfully judged.

In fact, Boyle's belief that the 'extension of experimental culture' could be achieved through the imposition of 'universal metrological standards' may be viewed as a concrete manifestation of his tacit, or unconscious, commitment to a uniform metaphysical standard according to which experimental experience, in general, was to be organised.[9] I use the term 'unconscious' in acknowledgement of Kuhn's observation that, because seventeenth-century experimental philosophers like Boyle typically decried metaphysics and celebrated experiment, the interaction which did occur between the two was 'usually unconscious.'[10] Consequently, as an explicit concept by which to make sense of Boyle's experimental practice, *mathēsis* figures into the present account as an analyst's category rather than an actor's category.

My method is thus not historicist in the currently prevailing sense, defined by Shapin as a 'practice devoted to interpreting historical action in historical actors' terms.'[11] Nor is it presentist, in the sense of using present-day terms to understand the past. *Mathēsis* is not an established present-day term. It is a term originating in ancient Greek discourse, as well as the etymological source of the present-day words 'mathematical' and 'mathematics.' Heidegger returns to the ancient term in order to re-introduce a meaning which no longer forms an explicit part of present-day usage. Indeed, this ancient meaning also appears not to have figured in Boyle's use of the term 'Mathematicks.' According to Shapin, especially in the context of the experimental philosophy, Boyle understood mathematics largely as a set of techniques for producing

---

9   Shapin (1988), 'Robert Boyle and Mathematics,' pp. 30, 29.
10 Kuhn (1977), 'Mathematical versus Experimental Traditions,' p. 44.
11 Shapin (1994), *A Social History of Truth*, p. xvi n. 1. Shapin offers no programmatic defence of historicism, adding: 'nor do I believe that historicism is without its problems and proper limitation' (Shapin (1994), *A Social History of Truth*, p. 328). See also Steven Shapin (1992), 'Discipline and Bounding: The History and Sociology of Science as Seen through the Externalism-Internalism Debate,' *History of Science* 30(4), 333–69 (pp. 353–59).

deductive certainty and numerical precision. By addressing the etymology of the word, Heidegger sought to recover a meaning which had become sedimented in subsequent culture, thus continuing to influence participants without figuring into their active vocabulary. Hence, *mathēsis* might be viewed as a *tacit* actor's category, with its influence on Boylean experimental culture being inferred through its explicit effects on the linguistic and non-linguistic practices of that culture.

In what follows, I will argue that Boyle was, in the Heideggerian sense of *mathēsis*, a mathematical philosopher. My focus will be on Boyle's dispute with the natural philosopher and mathematician Francis Line. A key point in this controversy was the legitimacy of Aristotelian final causes in experimental discourse. Boyle decried final causes as metaphysical, and condemned Line's use of them. He commented, 'I am not very forward to allow acting for ends to bodies inanimate, and consequently devoid of knowledge.'[12] This implies an understanding of final causes which links them to sentience and knowledge. Yet, as we saw in Chapter Five, neither Aristotle's nor Heidegger's concept of final cause necessarily ascribes sentience or knowledge to the thing whose movements it was meant to help explain. Hence, I will argue, the term 'final cause' included a sedimented meaning, one which Boyle neglected in his dispute with Line. Furthermore, the dynamics of Boyle's argument against Line reveal his own tacit, or unconscious, reliance on final causes in this sedimented and neglected sense.

This notion of final causes ties directly into *mathēsis*. Indeed, the final cause is the mathematical measure which allows us to make sense of the operations of both nature and art. Heidegger's account of seventeenth-century science especially focuses on the mathematisation of Aristotelian final causes, that is, their consolidation under a single uniform measure. This account gives particular attention to the roles of Galileo and Newton. I will first review Heidegger's brief discussion of these two figures, before teasing out the implications of his comments

---

12　Robert Boyle (1662 [1966]), *A Defence of the Doctrine Touching the Spring and Weight of the Air, Proposed by Mr. R. Boyle, in his New Physico-Mechanical Experiments; Against the Objections of Franciscus Linus. Wherewith the Objector's Funicular Hypothesis is also Examined*, in *The Works of Robert Boyle*, vol. 1, by Robert Boyle, ed. by Thomas Birch (Hildesheim: Georg Olms), pp. 118–85 (p. 143).

for our understanding of the early-modern experiment. Then I will move on to Boyle's dispute with Line, showing that the historical line Heidegger traces from Galileo to Newton runs directly through Boyle. In the penultimate section, I will bring Heidegger's concept of the mathematical into dialogue with David Bloor's concept of social imagery, and briefly explore some general implications of this combination for the historiographic method of the sociology of scientific knowledge (SSK).

Before we leap into the main current of the chapter, it bears emphasising that what follows is not an exercise in intellectual history. Recall, once more, that Heidegger defined *mathēsis* in terms of a twofold reciprocal relation between a mathematical projection of nature, on the one hand, and work experiences, on the other. Hence, in explicating and expanding on Heidegger's argument, I will give significant attention to the materiality of early-modern scientific and technological experience, in general, and Boyle's experimental manipulations of nature, in particular. My goal is to capture, if only roughly, the way in which material practice and metaphysical project came to mutually reinforce one another, each giving strength to the other, until a powerful new way of understanding and intervening in natural processes emerged, a novel intellectual and material culture, a new way of being in the world.

2. The Galilean First Thing and the Aims of Experiment

In her 2004 book, *The Body of the Artisan: Art and Experience in the Scientific Revolution*, Pamela Smith argues that 'the methods, goals, and *episteme* of art' are 'central to an understanding of the Scientific Revolution.'[13] In particular, she emphasises the crucial contribution of an 'artisanal epistemology' to the emergence of early-modern science. This epistemology was widely manifest in the artistic activity of the sixteenth and seventeenth centuries, and it provides evidence for a 'form of cognition' unique to the craft operations of the period.[14] In step

---

13   Pamela H. Smith (2004), *The Body of the Artisan: Art and Experience in the Scientific Revolution* (Chicago: University of Chicago Press), p. 19.

14   Smith (2004), *The Body of the Artisan*, p. 21.

with Dear's emphasis on the seventeenth-century preoccupation with operational explanations, discussed in Chapter Five, Smith argues that these artisans were more concerned with understanding the processes of nature than with creating representations of nature. To this end, they developed a deep and sophisticated expertise about the way matter behaves under a diverse range of conditions, with a particular interest in cases involving the generation and transformation of matter.[15] This expert knowledge was not abstract and theoretical, but embodied and practical: 'in the sixteenth and seventeenth centuries the pursuit of natural knowledge became *active* and began to involve the body; that is, one had to observe, record, and engage bodily with nature.'[16] Echoing the earlier 'craftsman thesis' of Edgar Zilsel, Smith argues that communication by these expert artisans with physicians, scholars, princes, and city governors created a social dynamic from which 'new attitudes toward nature and a new discourse about it emerged.'[17] Furthermore, like Zilsel, who claimed that these social developments provided only 'necessary conditions,' not sufficient conditions, for the rise of early-modern science, Smith also cautions that her account of artisanal practice offers an only 'partial answer' to the question of what brought about the Scientific Revolution.[18] Although she places more stress on the contribution of artisans, Smith nevertheless emphasises that this period was marked by a 'mutual influence' between artisans and humanists, between practitioners and theorists, an influence which was 'reciprocal and dialectical,' an influence which 'runs both ways.'[19] Her book, then, is not a reductionist account of science in terms of the skilled manipulation of matter by elite artisans, but instead a powerful reminder of the necessary and profound role played by those artisans in a larger social transformation. In so arguing, she reinforces a point against intellectualist accounts of the Scientific Revolution represented by, among others, Alexandre Koyré's reductionist emphasis on theoretical imagination as both necessary and sufficient for the emergence of

---

15 Smith (2004), *The Body of the Artisan*, pp. 7, 14.

16 Smith (2004), *The Body of the Artisan*, p. 18.

17 Edgar Zilsel (1942), 'The Sociological Roots of Science,' *American Journal of Sociology* 47(4), 544–62; Smith (2004), *The Body of the Artisan*, p. 151.

18 Zilsel (1942), 'The Sociological Roots of Science,' p. 547; Smith (2004), *The Body of the Artisan*, p. 19.

19 Smith (2004), *The Body of the Artisan*, p. 22.

early-modern science. Indeed, Koyré explicitly rejected Zilsel's claim for an artisanal influence on Galileo's work.[20] John Randall was likewise dismissive of Zilsel's craftsman thesis, which he viewed, incorrectly, as necessarily rooted in a Platonic metaphysics.[21]

It is important to recognise that Heidegger's account of the Scientific Revolution in terms of *mathēsis* predates the mid-twentieth-century swing towards theory-dominant explanations of early-modern science. As noted in Chapter Two, Heidegger was a reductionist champion of neither theory nor practice, but was instead concerned with gaining a better understanding of the complex relationship between the two. Recall his exclamation that 'it is by no means patent where the ontological boundary between "theoretical" and "non-theoretical" behaviour really runs!'[22] However, when it comes to historical analysis, Heidegger, like Smith, puts his emphasis on one side of this dynamic: in his case, giving more attention to intellectual rather than to pragmatic factors. Yet, just as Smith urges us, despite the one-sidedness of her account, to keep the reciprocal nature of the dynamic in mind, so too should we approach Heidegger's own historical comments in the same fashion. Heidegger gives scarcely any attention to the material practices of early-modern science, but his account does not preclude such attention. In fact, as we have already seen in the last chapter, he explicitly styles *mathēsis* as a reciprocal relation between ways of working with things, on the one hand, and the metaphysical projection of the thingness of things, on the other.

In this section, I will first gloss Heidegger's brief historical comments on the emergence of early-modern science, and then suggest how these comments may relate to the more practice-based accounts proffered by historians like Smith. This will further strengthen the point, made in Chapter Five, that an analysis of manipulation alone cannot properly explain the profound historical changes we are hoping to understand. In order to make sense of early-modern accounts of nature, we must

---

20 Alexandre Koyré (1943a), 'Galileo and Plato,' *Journal of the History of Ideas* 4(4), 400–28 (p. 401 n. 6).

21 John Herman Randall, Jr (1961), *The School of Padua and the Emergence of Modern Science* (Padova: Editrice Antenore), p. 130.

22 Martin Heidegger (1962a [1927]), *Being and Time*, trans. by John Macquarrie and Edward Robinson (Oxford: Blackwell), p. 409 [358]. (Following scholarly convention, page numbers in square brackets refer to the original 1927 German edition of *Being and Time*.)

give attention to both efficient and final causes, to both operations and ends. In order to understand what early-modern natural philosophers were up to, we should focus not just on the fact that they manipulated nature using experimental equipment, but also on the way they selectively ordered those manipulations with the intention of producing reliable natural knowledge. I have proposed that Heidegger's account of the Scientific Revolution be interpreted as a story of change in the prevailing conception of final causation in respect of natural processes. I consider this interpretation compatible with Zilsel's and Smith's respective macro- and micro-historical attention to the transformation in relations between humanists and artisans, practitioners and theorists, during the same period. This is a proposed interpretation, rather than a straightforward gloss, of Heidegger's account, because this is not how he explicitly described matters. I believe, however, that what follows is consistent with his overall approach in the 1920s and 1930s, though I will also note some consequential points of critical departure from a few of his more specific claims.

Recall from Chapter Five that, for the Aristotelian natural philosopher of the late Renaissance, the natural movement of a thing was understood as movement to proper place, or, to form. This natural movement was regulated by the thing's final cause. From this it follows that a fundamental change in natural philosophical conceptions of motion should also involve a fundamental change in corresponding conceptions of final cause. As the understanding of natural motion fundamentally changes, so too does the understanding of the rules or guidelines which give order and meaning to that motion. Although Heidegger did not have much to say about changes in early-modern conceptions of final causation, he does give brief but crucial attention to the corresponding changes in early-modern conceptions of natural motion. A review of Heidegger's observations on this point will help us to render more explicit the implied corresponding changes in prevailing conceptions of final causation during that same period.

Heidegger's discussion focuses on Newton's First Law of Motion, his principle of inertia, which reads: 'Every body [*corpus omne*] continues in its state of rest, or uniform motion in a straight line, unless it is compelled to change that state by a force impressed upon it.'[23] Heidegger points

---

23   Martin Heidegger (1967 [1962]), *What Is a Thing?*, trans. by William B. Barton, Jr., and Vera Deutsch (Chicago: Henry Regnery), p. 78.

out that the phrase *corpus omne*, or 'every body,' indicates Newton's rejection of the fundamental Aristotelian distinction between terrestrial and celestial bodies.[24] Indeed, under this law, the thingness of things is no longer subject to any kind of qualitative division: Newton's First Law treats all things as being basically the same, as being circumscribed by the same basic blueprint. Furthermore, according to Heidegger, Newton's law is a formal articulation of an understanding of thingness which had already been expressed in Galileo's earlier account of free fall. For Heidegger, Galileo's key innovation was to explain the difference in the time it takes for distinct bodies to fall to earth 'not from the different inner natures of the bodies or from their own corresponding relation to their particular place,' but from the external forces acting on them, for example, the resistance of the air.[25] With this, Galileo rejects the Aristotelian attempt to contrastively explain a difference in natural motion between two bodies by reference to a difference in their inner orientations towards distinct places within a qualitatively differentiated cosmos. This latter kind of explanation is no longer valid because all things are now conceived of as belonging to qualitatively identical places in the cosmos. Heidegger writes that with Galileo the cosmos is now understood to be 'the realm of the uniform space-time context of motion.'[26] In qualitative terms, the heterogeneity of places within the cosmos has now become a homogeneity of positions within a uniform spatial realm.

There is a clear affinity between Heidegger's account of Galilean motion and his earlier discussion of place in the context of the change-over from practical immersion in a work-world to a theoretical understanding of things as objects. In *Being and Time*, Heidegger writes that:

> In the 'physical' assertion 'The hammer is heavy' we *overlook* not only the tool-character of the entity we encounter, but also something that belongs to any ready-to-hand equipment: its place. Its place becomes a matter of indifference. This does not mean that what is present-at-hand loses its 'location' altogether. But its place becomes a spatio-temporal position, a 'world-point,' which is in no way distinguished from any other.[27]

---

24   Heidegger (1967), *What Is a Thing?*, p. 86.
25   Heidegger (1967), *What Is a Thing?*, p. 90.
26   Heidegger (1967), *What Is a Thing?*, p. 92.
27   Heidegger (1962a), *Being and Time*, p. 413 [361–62].

Here, Heidegger is describing the nascent experience according to which early-modern mathematical physicists began to make sense of nature. In his later discussion of Galileo's free-fall experiments, he emphasises the sharp contrast between this way of experiencing nature as intelligible and the way distinctive of orthodox Aristotelians of the same period.

> Both Galileo and his opponents saw the same 'fact.' But they interpreted the same fact differently and made the same happening visible to themselves in different ways. [...] Both thought something along with the same appearance but they thought something different, not only about the single case, but fundamentally, regarding the essence of a body and the nature of its motion.[28]

In contrast to his more orthodox contemporaries, Galileo's work experiences were shaped by a projection of nature in which the qualitative differences between things, so central to the prevailing Aristotelian image of the cosmos, had now become a matter of indifference. He presupposed uniformity vis-à-vis the thingness of things. At a basic phenomenological level, he experienced all things as the same. Methodologically, by drawing attention to the phenomenology of this basic Galilean experience, Heidegger seeks to uncover the ontological core of Galilean natural philosophy. Indeed, as Heidegger argued, '[o]nly as phenomenology, is ontology possible.'[29]

For Heidegger, an ontological commitment to the qualitative uniformity of things is clearly evident in Galileo's final book, the 1638 *Discourses and Mathematical Demonstrations Relating to Two New Sciences*, from which he paraphrases Galileo as stating '*Mobile mente concipio omni secluso impedimento*': 'I think in my mind of something moveable that is left entirely to itself.'[30] This generic moveable body will become the *corpus omne* of Newton's First Law. Free from external influence, it exists in an autonomous state of uniform and perpetual motion (or rest). The social field of intelligibility in which the existence of such a thing can make sense is what Heidegger dubbed the mathematical projection of nature. He now identifies this shared projection with the

---

28   Heidegger (1967), *What Is a Thing?*, p. 90.

29   Heidegger (1962a), *Being and Time*, p. 60 [35].

30   Heidegger (1967), *What Is a Thing?*, p. 91.

mente concipio of Galileo: 'There is a prior grasping together in this *mente concipere* of what should be uniformly determinative of each body as such, i.e., for being bodily. All bodies are alike. No motion is special. Every place is like every other, each moment like any other.'[31] This is, for Heidegger, the missing mathematical element which marks the emergence of early-modern science: '[t]he mathematical is the *"mente concipere"* of Galileo.'[32] Thus the key feature of early-modern *mathēsis*, what crucially distinguishes it from Renaissance *regressus*, discussed in Chapter Five, is the uniformity or univocality in its projection of the thingness of the things. As Heidegger argues, '[a]ll determinations of bodies have one basic blueprint (*Grundriss*).'[33] In other words, as we saw in Chapter Five, all bodies — and, more broadly, the thingness of all things — are experienced and understood according to one basic and uniform measure. Against this measure, 'the structure of every thing and its relation to every other thing is sketched in advance.'[34] Where the community of orthodox Aristotelian natural philosophers experienced a qualitatively differentiated, hierarchically ordered cosmos, Galilean philosophers now see a uniformly ordered world in which every place, and every thing, is qualitatively like every other.

On the basis of his conception of the mathematical projection in terms of a basic blueprint, Heidegger concludes that '[n]ow nature is no longer an inner capacity of a body, determining its form of motion and place.'[35] He seems here to have in mind a definition of nature in terms of final cause, that is, in terms of a principle regulating a thing's natural movement and place in the cosmos. If, as Heidegger argues, this final cause can no longer be viewed as something internal to the thing, then it must be viewed as external to that thing. However, this does not seem to be Heidegger's conclusion. Instead, he argues that '[n]ature is now the realm of the uniform space-time context.'[36] But this seems like a non sequitur. It is difficult to see how this space-time 'realm' is supposed to determine the form of motion and the place of a thing. It would seem, rather, to be the arena in which such a determination may occur. Hence,

---

31   Heidegger (1967), *What Is a Thing?*, p. 91.
32   Heidegger (1967), *What Is a Thing?*, p. 116.
33   Heidegger (1967), *What Is a Thing?*, p. 91.
34   Heidegger (1967), *What Is a Thing?*, p. 92.
35   Heidegger (1967), *What Is a Thing?*, p. 92.
36   Heidegger (1967), *What Is a Thing?*, p. 92.

a definition of nature in terms of this realm does not replace the inner capacity of a thing to move towards its proper place in the cosmos. Instead, one may say that the thing retains an inner capacity to follow specific rules of motion, rules which guide it to its natural place. But this place is not unique to the thing's kind, or category; it is rather a generic place, qualitatively identical to every other place, in the uniform space-time realm. It would thus be more consistent with Heidegger's overall account to describe the mathematical, not as replacing the inner capacities of things, but rather as rendering those capacities uniform in a way which mirrors the uniformity of the post-Aristotelian cosmos. In a uniform world, a differential, or pluralistic, account of final causes becomes unnecessary. Because this world admits of no differences, there is no need to reckon difference in the fundamental rules governing the natural movements of things. Indeed, now only one basic set of rules is needed in order to understand these movements at their most primitive level. In this way, the mathematical projection opens up an experiential space in which the concept of a universal physical law first becomes intelligible, and then realises its formal articulation in Newton's First Law: every body continues in its state of rest, or uniform motion in a straight line, unless it is compelled to change that state by a force impressed upon it. In the Newtonian universe, it is an implicit assumption that every physical thing possesses a capacity to consistently follow this basic and universal law. This law now serves as the final cause of a thing's most primitive form of motion: it is what regulates that motion; it is that for the sake of which the thing moves, elementally, through a uniform space-time realm.[37] In achieving pure conformity to this universal law — and so escaping the contingent effects of external, localised causal forces — the thing assumes its proper place, or state,

---

37   Trish Glazebrook calls this the early-modern 'homogenization' of natural place. She
      correctly observes that now '[n]o distinction is made between things on the basis
      of motion toward an end.' However, from this she then incorrectly concludes that
      'things are apprehended not in terms of [internal] essence or *telos* [i.e., end],' 'but on
      the basis of external force' (Trish Glazebrook (2000b), 'From φύσις to Nature, τέχνη
      to Technology: Heidegger on Aristotle, Galileo, and Newton,' *The Southern Journal of
      Philosophy* 38, 95–118 (pp. 109, 110; my brackets); see also, Trish Glazebrook (2001b),
      'The Role of the Beiträge in Heidegger's Critique of Science,' *Philosophy Today* 45(1):
      24–32 (p. 25)). But, if this were true, then the First Law would be unintelligible.
      Explanations in terms of external force entail a conception of internal essence, even
      when that essence is homogenous across entities.

in the post-Aristotelian cosmos, that is, a place where it is left entirely to itself, to be at rest, or to move uniformly in a straight line, without interference from without.

Yet, Heidegger argues that Newton's First Law 'speaks of a thing that does not exist. It demands a fundamental representation of things which contradicts the ordinary.'[38] The problem is that, although every thing may be capable of following this law, there is no basis in ordinary experience to accept this because we do not ordinarily encounter things as being unaffected by external forces. Put in Galileo's terms, we do not ordinarily experience a thing 'left entirely to itself.' What is more, Heidegger argues that '[t]here is also no experiment which could ever bring such a body to direct perception.'[39] Hence, the new mathematical projection of nature puts at its centre an elemental understanding of the thing which is closed off not just from ordinary experience, but from experimental experience as well. We can encounter it neither in ordinary life nor in the laboratory.

Heidegger argues that this mathematical understanding of the thing, which places it beyond the scope of experimental observation, also makes possible the experiment as 'a *necessary* and prime *component* of [scientific] knowledge.'[40] Because the mathematically projected thing is a necessary condition for the possibility of experimental knowledge, it cannot also be something discoverable by experimental means. Heidegger writes that '[i]t is precisely the projecting-open of nature in the *mathematical sense* that is the presupposition for the necessity and possibility of "experiment."'[41] The idea seems to be that the art of experiment, as a set of operations for investigating nature, is necessarily guided by a mathematical — more precisely, an axiomatic — understanding of things as primitively autonomous, as left purely to themselves in a qualitatively uniform world. One may picture the cosmic demiurge, forging the 'First Thing' into existence, and then sending it out, alone, into the empty expanse of space, where it sails unobstructed under the rule-governed constancy of its own internal inertia.

---

38 Heidegger (1967), *What Is a Thing?*, p. 89.
39 Heidegger (1967), *What Is a Thing?*, p. 89.
40 Martin Heidegger (1999 [1989]), *Contributions to Philosophy (From Enowning)*, trans. by Pravis Emad and Kenneth Maly (Bloomington: Indiana University Press), p. 113; my brackets.
41 Heidegger (1999 [1989]), *Contributions to Philosophy*, p. 113.

But the experimental practitioner will never discover this First Thing, because she works in a world populated by an untold number of things, each impinging on the autonomy of the others. What the axiomatic image of the First Thing gives the experimentalist is an understanding of things which permits her to meaningfully distinguish their fundamental thingness, their whatness or essence, from the effects wrought on them by the external contingencies of their local environment. Hence, a thing falls to the earth, not because it is composed essentially of earth, but because the earth exerts an external influence on it, accelerating it downward. The earth thus interferes with the autonomy of the thing, accelerating it out of its indigenous state of uniform rectilinear motion (or rest). In doing so, it allows the thing to become an object of experimental observation. More specifically, the experimentalist can now empirically investigate the acceleration of the thing as caused by the earth. Furthermore, because the thing's relationship to the earth is not unique, one can predict that the earth will affect other things in the same way. Hence, a heavier or lighter body should also accelerate towards the earth at the same rate: the effect should be universal, as determined by the mathematically projected basic blueprint of nature.

As Heidegger observes in respect of Galileo's free-fall experiment, this prediction was not confirmed, because two simultaneously released bodies, of differential weight, hit the earth's surface at different times. Yet Galileo was able to reason that this was not evidence of a difference in the respective relations of the two bodies to the earth, but rather of the differential effect of a further external influence, air resistance, on the bodies. Again, the axiomatic image of the First Thing allows this further effect to be explained away as irrelevant to the phenomenon under test. In a vacuum, or near vacuum, this effect should not be present, as was famously confirmed by David Scott when he let a hammer and a feather fall freely from an equal height during the 1971 Apollo 15 moonwalk. The point is that the thing, as mathematically projected, allows the experimentalist to identify certain causal influences as irrelevant to the effect under study, in the interests of isolating just those causes responsible for that effect. Because the project establishes the *a priori* uniformity of all bodies, including their uniform receptivity to fixed rules of motion, an observed difference in the behaviour of two bodies must indicate, not an essential difference in the bodies themselves, but rather a difference in the external causal nexuses which act on those bodies. To artificially achieve the same observed effect in the behaviour of the two bodies is to have successfully eliminated the difference in their respective causal nexuses — as Scott dramatically demonstrated on the moon's surface. In this way, the experimental practitioner isolates the cause of a particular behaviour, or effect, and then demonstrates the regularity of this causal relation.

This artificially facilitated isolation of causes, and the accompanying determination of the relevant causal relation, is a process of discrimination which presupposes a stable standard of judgement. According to Heidegger, this standard is provided by the mathematical projection of the thingness of things, which stipulates that no observable effect on the thing is essential to the nature of that thing, and so all such effects can, in principle, be eliminated without compromising the indigenous nature of the thing. The thingness of the thing is thus defined in terms of the purity of its conformity, in the absence of external influence, to what Newton formalised as the First Law of Motion. The experimentalist cannot empirically demonstrate the existence of this pure state, but she can employ it as a norm by which to justify the elimination of noise from an experimental system, that is, the externally generated effects interfering with the specific causal relations which the experiment is meant to isolate and test. In principle, all observable effects might be eliminated from an experimental system, thus leaving the thing entirely to itself, in a purified state of mathematically projected thingness. But this marks the limit of possibility for experimental practice, because it makes scientific knowledge of things impossible. In practice, the experimentalist is rather more concerned with isolating and stabilising specific causes, so as to reliably demonstrate, or explain, specific effects.

This artificially facilitated act is a demonstration of the kind which Paduan medical Aristotelians called *propter quid*, that is, a demonstration of *why* an effect is. Recall, from Chapter Five, Agostino Nifo's four stages of scientific method. First, one observes an effect. Second, one discovers the cause of that effect. Third, one gains precise knowledge of the cause through *negotiatio*, or definition. Fourth, on the basis of this precise knowledge, one now knows the effect *propter quid*, that is, in a well-reasoned, or well-grounded, manner. Nifo argued that this well-grounded knowledge of an effect carries the certainty of demonstrative knowledge, but only in an empirical, not a mathematical, sense. *Negotiatio* is not a purely rational, or intellectual, operation, but also includes a material element. The isolation and definition of a relevant cause requires the physical manipulation of the causal system under investigation. As argued in Chapter Five, these manipulations require a principle to guide them, to give them order and meaning, otherwise they would amount only to random behaviour. In the case of early-modern science, this principle was the missing mathematical element enabling the transition from medieval *regressus* to early-modern *mathēsis*. In Aristotelian terms, this element supplies the final cause guiding experimental manipulations, the organising principle, or source of intelligibility, for the new experimental philosophy. Crucially, this element includes an *a priori* conception of the thingness of the thing, a conception which is not discovered inductively through observation of the thing under investigation — as Jacopo Zabarella had argued in the case of *regressus* — but which is instead already known before the investigation has even begun. It serves as a basic axiom upon which that investigation becomes possible. According to Heidegger, this is a uniform conception of the thing as law-abiding and autonomous, a thing which, when left to itself, conforms absolutely to an inherent and universal principle of inertia, a principle which would become formally articulated in Newton's First Law. This uniform conception — the Galilean First Thing — provides a basic blueprint for the metaphysical projection of the thingness of things in general, a projection which Heidegger uses to explain the emergence of early-modern science.

More narrowly, this mathematical conception renders intelligible the technical stabilisation of the thing in an artificial set-up where most of the causal forces acting on it are systematically eliminated, or rendered irrelevant, in the interests of isolating specific, observable, and contingent causal effects, effects the artificially induced repetition of which evince a regularity in the thing's relation to its artificially controlled environment. In this way, the art of experiment may be viewed, not as a violation of natural processes, but instead as an attempt to complement or complete them. Viewed from within the mathematical project, the new experimental philosophy does not seek to impose a final cause on the thing from without, so as to override its natural tendencies, but instead to organise and employ artificial operations in accordance with an image of the thing as it is when left to its own indigenous, rule-governed nature.[42] The aims of experimental practice are thus twofold. First, and more familiarly, the experiment aims to disclose specific regularities in an artificially controlled experimental system, a system comprised of two or more interacting things: for example, the uniform acceleration of a body vis-à-vis the earth. Second, and more provocatively, the experiment employs artificial means to move, or manipulate, the thing in a way which brings it closer to being what it is, to achieving its proper place in the uniform space-time realm as an internally disciplined and wholly autonomous thing. It seeks, in other words, to treat things in accordance with the one basic blueprint of the mathematical projection. The second aim provides the phenomenological background against which the first aim can be intelligibly pursued. The regularity of a relation between two things presupposes the constancy and rule-conformity of each thing on its own, and this presupposition is provided by the mathematical projection of the Galilean First Thing. As a consequence of this projection, the practitioner understands the goal of her artificial manipulations to be in conformity with nature's own indigenous dispositions. Experimental art thus not only imitates nature, but also, to repeat a passage from Aristotle first cited in Chapter Five, 'art partly completes what nature cannot bring to a finish.'[43] As we will see in the next section, within this mathematically structured experience, experimental manipulations acquire an aura of transparency, they are experienced as granting neutral access to an independently existing and stable natural order. They are experienced, in other words, as disclosing objective matters of natural fact.

---

42   On this point, my interpretation differs from that of Glazebrook, who has Heidegger arguing that '[t]he experiment [...] is violent in that it sets beings up to behave in ways they would not when left to themselves' (Trish Glazebrook (2000a), *Heidegger's Philosophy of Science* (New York: Fordham University Press), p. 104; see also Trish Glazebrook (1998), 'Heidegger on the Experiment,' *Philosophy Today* 42(3), 250–61). She supposes that, for Heidegger, the final end towards which a thing is naturally directed is necessarily violated by the modern experiment. This may have been the view of orthodox Aristotelian natural philosophers, but it was not, on my reading, Heidegger's view.

43   Aristotle (1941c), *Physica*, trans. by R. P. Hardie and R. K. Gaye, in *The Basic Works of Aristotle*, ed. by Richard KcKeon (New York: Random House), pp. 213–394 (p. 250 [lines 199a14–17]).

3. Releasing Experimental Things

Historian of science Robert Kohler has argued that the scientific laboratory is governed by a 'logic of placelessness': 'Laboratory workers eliminate the elements of place from their experiments.'[44] This claim is stronger than Heidegger's claim that the mathematical projection of nature 'overlooks' place, rendering it a matter of 'indifference.'[45] Yet the difference between elimination and indifference has been lost on some critics of Heidegger's existential account of science. Karin Knorr-Cetina, for example, writes that Heidegger's account of science 'is founded on nothing more than the *decontextualization* from which it was derived.'[46] William Blattner likewise argues that, for Heidegger, natural science 'releases' things from situatedness, 'decontextualizes them.' Dimitri Ginev follows in step, presenting Heidegger as asserting the 'disappearance of any situatedness (or place) in what becomes mathematically projected.' However, although Heidegger does argue that the thing is 'released' in the mathematical projection, this means neither that it becomes decontextualised, nor that its place is eliminated. It means, instead, that the thing comes to be experienced differently. Its situatedness in an everyday environment becomes unimportant for the subject's understanding of what it is. Its context has not been eliminated, but instead replaced by the artificially constructed and controlled environment of the laboratory. Laboratory practitioners still experience the thing as situated, but its situation has changed dramatically. The thing is now encountered in a different context, an artificial context designed to move the thing closer to its natural place, as determined by the mathematical projection, the place to which it is now understood to properly belong.[47]

---

44 Robert E. Kohler (2002), *Landscapes and Labscapes: Exploring the Lab-Field Border in Biology* (Chicago: University of Chicago Press), pp. 9, 6.

45 Heidegger (1962a), *Being and Time*, p. 413 [361].

46 Karin D. Knorr-Cetina (1981), *The Manufacture of Knowledge: An Essay on the Constructivist and Contextual Nature of Science* (Oxford: Pergamon Press), p. 143; William Blattner (1995), 'Decontextualization, Standardization, and Deweyan Science,' *Man and World* 28, 321–39 (p. 321); Dimitri Ginev (2011), *The Tenets of Cognitive Existentialism* (Athens OH: Ohio University Press), p. 4.

47 The claim that Heidegger embraced a logic of placelessness in respect of scientific things is further challenged by his 1931 statement that an explanation of the whatness of things in terms of their 'going towards their place [...] until today is not in the least refuted, in fact not even grasped' (Martin Heidegger (1995b

The claim that the experiment helps move the thing to its proper place in a mathematically projected world challenges the more common view that the experiment serves to control the thing. On the present account, the experiment does not control the thing, but rather the causal nexus in which the thing is situated, and it does this in the interest of letting the thing be what it is. This is the meaning of Heidegger's statement that the thing is 'released' [*entschränkt*] by natural science from the confines of its everyday situation.⁴⁸ In German, the noun *Schrank* can mean 'bound,' in the mathematical sense of the upper or lower bounds of a finite number set. But it also carries the more general meaning of a limit, fence, or barrier. Hence, the phrase *etwas in Schranken halten* means to restrain something, to hold it back, to rein it in. When a thing becomes *entschränkt*, it is 'dis-bounded,' released from the bounds, bonds, or barriers which had restrained it — it is set free. Recall, from the last section, the mythical image of the cosmic demiurge, forging the First Thing in her workshop, sending it off to sail, alone and unhindered, across the empty expanse of space, ruled only by the constancy of its own internal inertia. Now imagine that thing suddenly plunging into the blooming and buzzing causal nexus of the world as we ordinarily experience it. It is from this latter situation that the experimentalist seeks to release the thing, stripping away the causal interference which prevents it from being what it is, and thus turning it back towards the purity of its own mythical beginnings.⁴⁹

[1985]). *Aristotle's* Metaphysics Θ 1–3: *On the Essence and Actuality of Force*, trans. Walter Brogan & Peter Warnek (Bloomington: Indiana University Press), p. 67). Indeed, Heidegger sought to understand the radical transformation of this kind of explanation in the early-modern period, rather than to affirm its rejection.

48 Heidegger (1962a), *Being and Time*, p. 413 [362].

49 Heidegger's account of the experiment appears to share some characteristics with Nancy Cartwright's more recent account. She writes that the experiment provides 'the circumstances where the feature under study operates, as Galileo taught, without hindrance or impediment, so that its nature is revealed in its behaviour' (Nancy Cartwright (1999), *The Dappled World: A Study of the Boundaries of Science* (Cambridge: Cambridge University Press), p. 84). Furthermore, she argues that we must already know, in some general sense, what this nature is in order to design an experiment which will successfully reveal it: '[w]ithout the concept of *natures*, or something very like it, we have no way of knowing what it is we are testing' (p. 90). For Cartwright, these are 'Aristotelian-style natures' (p. 81). Hence, she concludes: 'The empiricists of the scientific revolution wanted to oust Aristotle entirely from the new learning. I have argued that they did no such thing' (p. 103). A detailed comparison of the respective accounts of Heidegger and Cartwright

The question now is: how did late Renaissance practitioners come to understand the thing in this way? What caused this profound shift in the phenomenology of their scientific experience of things? The short answer is: the emergence of the mathematical as an increasingly powerful influence on scientific cognition. Yet Heidegger argues that 'this mathematical must [...] be grasped from causes that lie even deeper.'[50] With this, he points us towards the conditions of possibility for the early-modern mathematical experience of things.

Heidegger's own attempt to shed light on these conditions is not especially helpful, but he does hint at the direction in which a further explanation may be pursued. Sticking close to his own philosophical preferences, Heidegger looks for the 'deep' causes of the mathematical in the work of Descartes. In particular, he focuses on what was probably Descartes's first philosophical work, *Rules for the Direction of the Mind*, an unfinished manuscript which he may have written in 1628, and which was already mentioned in Chapter Four. In this work, declares Heidegger, 'the modern concept of science is coined.'[51] The *Rules* provide substantial evidence for the formal articulation of the mathematical, particularly in respect of its tendency to 'explicate itself as the standard of *all* thought and to establish the rules which thereby arise.'[52] Accordingly, as we saw in Chapter Four, Descartes develops a concept of thinking in terms of the *res cogitans*, the thinking thing, which he defined in the first-person singular. The thinking thing thus becomes an individual 'I'; thinking is always an 'I think.' As a consequence, through Descartes's efforts, the mathematical becomes grounded in the 'I,' which itself becomes the basis for all thought, all the rules which govern reason. Recall, furthermore, Descartes's claim that all knowledge statements, in addition to positing the 'I' principle as their necessary ground, also always posit an equally fundamental principle of non-contradiction. This second principle secures the purity of the 'I,' and so also the purity of reason. The individual 'I' is the condition of

---

might be worthwhile, but this is not the place for it. Suffice here only to note that, on Heidegger's account, Cartwright's natures will ultimately refer back to a possibility in the subject's shared historical existence, and will thus be amenable to sociological investigation.

50 Heidegger (1967), *What Is a Thing?*, p. 95.
51 Heidegger (1967), *What Is a Thing?*, p. 101.
52 Heidegger (1967), *What Is a Thing?*, p. 100.

possibility for thinking, as such, and the principle of non-contradiction is the condition of possibility for the purity of that thinking.

There is a striking similarity between this early-modern understanding of the thinking thing, on the one hand, and the early-modern understanding of the physical thing, on the other. The *res cogitans* shares an important homology with the *res extensa*, the extended, or corporeal, thing. Just like the Galilean First Thing, introduced above, the Cartesian 'I' is individual and autonomous, governed only by the constancy, the purity, of its own inner principle. This similarity should not come as a surprise. As we saw in Chapter Four, Heidegger insisted on the unity of things and thinking. Because the concept of each is dialectically bound to the other, the historical development of one will reflect that of the other. According to Heidegger, this dialectical relation was located by Aristotle in the 'is' of the proposition, but subsequently obscured when Descartes buried it in the 'I' of the subject. Kant then re-discovered it in the individual subject's feeling of 'respect' [*Achtung*] for the law, its phenomenological understanding of itself as a law-abiding being, as a being capable of being affected by the law. Heidegger then construes Kant's notion of respect as self-respect, and argues that, for Kant, the individual subject is bound by rules which it gives to itself. In opposition to this, Heidegger argues that rule-following is not individualistic, but presupposes one's participation in a tradition, an awareness of one's historical being-with-others. Recapping the argument of Chapter Four, the principle of unity which binds together things and thinking in the coherent production of scientific knowledge is, for Heidegger, rooted in tradition, a historical and social phenomenon, which has been subject to a string of philosophical articulations over the generations, tracing back through Descartes's 'I' to Aristotle's 'is,' and then to its beginnings in Plato's mythical image of the demiurge. Heidegger identifies this principle of unity with the for-the-sake-of-which, his version of the Aristotelian final cause. Hence, he outlines a philosophical history of the notion of final cause from Plato's demiurge, through Aristotle, Descartes and Kant, to his own concept of the subject, Dasein. For Heidegger, 'Dasein's very Being [i]s the sole authentic "for-the-sake-of-which."'[53] In other words, the source of the principles which

---

53   Heidegger (1962a), *Being and Time*, p. 116 [84].

give order and meaning to our experience of things is our own social and historical existence.

This is a radical reinterpretation, but not a complete rejection, of the ancient Greek craft analogy. The mythical world of the demiurge is now reformulated as the social world. The art which unifies things and thinking, which organises and thus renders intelligible our experience of things, the art which helps things by letting them become what they are, is a finite and social human art. More specifically, the rules which guide that art are finite, social rules. Hence, Heidegger argues that the intelligibility of Newton's First Law, as universally applicable to all corporeal things, is possible only on the basis of a particular historical projection, namely, a mathematical projection of the uniform thingness of things. This projection must, furthermore, be understood as a finite and historical feature of human social existence. Heidegger locates its distal cause in the ancient image of the cosmic demiurge, but his account of this mythical image in historical and social terms provides an opportunity to also locate its causes in more proximal socio-historical human actions, including the actions of the late Renaissance artisanal class.

The influence of this artisanal class on Galileo's natural philosophy was a key feature of Zilsel's craftsman thesis, which placed at its centre the macro-sociological transition from feudal to capitalist patterns of social organisation in that period. But more micro-sociological explanations are also possible. Smith, for example, points to recent research suggesting that late Renaissance craft methods were the result of a carefully thought-out technology, one which depended crucially on the maintenance of a strict consistency in the physical materials to which those methods were applied.[54] Such practical concerns with uniformity in craftwork would seem to manifest themselves also in the experimental practitioners' desire to maintain the stability and uniformity of experimental things in order to reliably reckon the regularities in their causal relations with one another. As argued above, this uniformity finds its ultimate expression in Galileo's principle of inertia, where a thing left to itself is just like every other thing, moving or resting with a constancy determined by a single, universal rule. The

---

54   Smith (2004), *The Body of the Artisan*, p. 7.

differences between things are thus to be judged by differences in the external causes which play on them in their bumpy passage through the heavy traffic of the cosmos. Likewise, the differences between artisanal products, when each is wrought from the same uniform material, are to be judged by differences in the operations to which they have been subjected in the intensive manufactory of the terrestrial workshop.

Based on the present interpretation of Heidegger, I suggest that the proximal sensibility of the artisanal workshop may have provided early-modern experimental practitioners with a powerful analogy by which to conceptualise the thing in terms of material uniformity and reliable responsiveness to standardised methods of manipulation. The point is that artisans provided natural philosophers with more than just an example of embodied knowledge-making, that is, knowledge won through the manipulation of natural materials; they also demonstrated an attitude towards embodied knowledge-making which prized the uniformity of both methods and materials. They put much effort into developing a sophisticated material diagnostics according to which the response of a particular material to standardised techniques of manipulation and manufacture could be reliably predicted. For example, Smith notes Paracelsus's sixteenth-century observation that 'carpenters of his day knew how to choose, cut, and prepare wood panels, so that even after centuries very little warping and twisting takes place.'[55] Likewise, the use by Renaissance sculptors of 'consistent alloys [...] indicates that they knew their materials and went to some trouble to procure what they needed.'[56] Late Renaissance artisans were thus compelled by principles of control and prediction in their work with materials. These materials were controlled for consistency, so that their behavioural responses to standardised forms of manipulation could be anticipated with confidence. Establishing the uniformity of material dispositions was thus a key value of artisanal practice.

Ursula Klein and Emma Spary argue that early-modern laboratories emerged as sites where this intense interest in the stabilised connection between manual techniques and material responsiveness flourished: '[l]aboratories [...] represented a strand of early-modern expertise developing around the manufacture of materials, as well as chemical

---

55   Smith (2004), *The Body of the Artisan*, p. 86.
56   Smith (2004), *The Body of the Artisan*, p. 86.

techniques such a distilling, smelting, or dissolving.'[57] Indeed, the metallurgical practice of smelting is especially suggestive in this regard. As Christoph Bartels has argued, the 'beginnings of both early-modern science and technology are closely linked to mining and metal production.'[58] The efficient manufacture of base metals was both a physically and an intellectually demanding undertaking, one which required 'hands-on knowledge and skill as well as text-based and mathematical knowledge.'[59] Base metals like copper, mercury, tin, and lead do not ordinarily occur in their elemental forms, but must be extracted from natural ores through a complex process known as smelting. This process frees the base metal from the ore, using intense heat in a controlled environment to induce a chemical transformation in which unwanted substances are driven off in the form of gases and slag. It bears emphasising that the smelting process does not simply melt the base metals out of the ores in which they are present. Instead, smelting is a transformative process in which base metals are 'released' from the chemical bonds which bind them together with sulfur, oxygen, or silicon in their ordinarily occurring ore state. According to Bartels, the depletion of metal-rich ores in Europe's leading mining districts in the second half of the sixteenth century, which increased the difficulty of metal production, led to a sudden rationalisation of smelting techniques, marked by innovations in 'precision measurement, data collection, the use of mathematics, attempts at standardization, the writing of technical instructions, [and] the writing of technical books to be published.'[60]

There is thus a striking operational homology between smelting, on the one hand, and experimental practice as described by Heidegger, on the other. Just as Heidegger argued that the experimentalist seeks to release the thing from the ordinary causal constraints which prevent

---

57 Ursula Klein and Emma C. Spary (2010), 'Introduction: Why Materials?,' in *Materials and Expertise in Early Modern Europe: Between Market and Laboratory*, ed. by Ursula Klein and Emma C. Spary (Chicago: University of Chicago Press), pp. 1–23 (p. 6).

58 Christoph Bartels (2010), 'The Production of Silver, Copper, and Lead in the Harz Mountains from Late Medieval Times to the Onset of Industrialization,' in *Materials and Expertise in Early Modern Europe: Between Market and Laboratory*, ed. by Ursula Klein and Emma C. Spary (Chicago: University of Chicago Press), pp. 71–100 (p. 100).

59 Bartels (2010), 'The Production of Silver, Copper, and Lead,' p. 97.

60 Bartels (2010), 'The Production of Silver, Copper, and Lead,' p. 97.

it from being what it is, so too does the metallurgist seek to release the metal from the ordinary causal constraints which prevent it from being what it is. Moreover, there was a rapid increase in the rationalisation and publication of smelting techniques during the period just prior to the emergence of the modern experiment in the early seventeenth century. In the case of both smelting and experiment, the object of pursuit is a product of art, and, in both cases, the artful production of this object seeks to return it to its pure, indigenous state. In neither case, however, can one speak of an opposition between art and nature. Instead, the operational dynamic is one in which art comes to the aid of nature by progressively stripping away the external causal constraints which prevent the thing from being what it is, thereby allowing it to more fully realise its proper place in the cosmic order. Because these artful manipulations are experienced, within the mathematical projection, as facilitating natural processes, they carry with them an aura of transparency; they are understood to provide practitioners with neutral access to an independently existing natural order, that is, to objective matters of fact.

Heidegger's account of the early-modern experiment thus lends conceptual support to recent historiographic attempts to explain the emergence of early-modern science, in part, by reference to the skilled artisanal manipulation of materials. In particular, Heidegger's account throws light on that for the sake of which these manipulations were often performed: a cultural image of the First Thing. These historical studies, in turn, lend support to Heidegger's philosophical account, most strikingly the studies of late Renaissance and early-modern metallurgy. If, as historians like Bartels, Klein, Smith, and Spary argue, the emergence of early-modern science was consequentially linked to mining and metal production, then it is possible that the mythical image of the First Thing found concrete, existential support in the artisanal experience of manufacturing pure metals. The pursuit of pure metals through increasingly sophisticated and rationalised metallurgical practices offers a powerful analogy for the natural philosophical pursuit of the Galilean First Thing through the practices of experiment. In this way, early-modern work experiences, on the one hand, and the mathematical projection of nature, on the other, may be viewed as reciprocally related, as serving to mutually reinforce one another.

Heidegger writes that '[t]he more exactly the ground plan of nature is projected, the more exact becomes the possibility of experiment.'[61] In other words, the more precise the projected ground plan becomes, the more legitimate and unequivocally compelling experiment becomes as a method by which to render nature intelligible. The mathematical projection is the condition of possibility for the experiment, and so the more secure it becomes, the more powerfully it comes to shape our experience of nature, the more apt we will be to regard the experiment as a preferred means of investigation. Yet, it is also the case that, the more effectively the experiment verifies a mathematical conception of nature, the more secure the mathematical projection becomes as the existential basis for that conception. An effective experiment not only verifies a theory, it also reinforces the authority of the projection in which both experiment and theory become possible as a means by which to make sense of nature. According to Heidegger, the modern physical experiment does this by producing an ever more exact account of the thing, through ever more precise acts of numerical measurement.[62] These rigorous acts of measurement serve to increase the precision of our understanding, and hence to reduce vagueness, or equivocation, in our descriptions of nature. In rigorously disciplining things by experimental means, we facilitate increased rigour in our thinking and theories about those things. Numerical measurement thus offers an effective and efficient means by which to underwrite an unambiguous, determinate, and logically consistent account of natural order. Numerical measurement furthermore becomes compelling as an investigative technique to the extent that the thingness of things is itself projected as inherently unambiguous, as pure and naturally amenable to precise determination.

This emphasis on exactitude allows for a further specification of the argument, made in the previous section, that the aims of experimental practice are twofold. First, the experiment discloses specific regularities in the relation of one thing to another. For Heidegger, this disclosure, as we can now see, is one which tends towards precision in its specification

---

61   Martin Heidegger (1977a [1952]), 'The Age of the World Picture,' in *The Question Concerning Technology*, by Martin Heidegger, trans. by William Lovitt (New York: Harper & Row), pp. 115–54 (p. 122).

62   Heidegger (1977a), 'The Age of the World Picture,' p. 122.

of the thing's behavioural regularities. Second, the experiment moves the thing closer to what it is, to being an internally disciplined and autonomous thing. This movement, as we can also now see, involves the material simplification and control of the thing's causal environment, thus disambiguating the thing's behavioural responses, allowing them to be unequivocally individualised, such that their regularities may then be exactly specified. As noted earlier, the second aim provides the phenomenological background against which the first aim may be intelligibly pursued. The material reduction of causal ambiguity in an experimental system allows the experimental practitioner to specify natural processes as precisely defined regularities, or even as exact laws. Heidegger argues that only through the specification of such regularities, 'only within the purview of rule and law[, do] facts become clear as the facts that they are.'[63] This is so because the physical matter to which a statement of natural fact refers has itself been isolated from the multifarious and competing causes of its ordinary environment, thereby allowing the regularities of its specific environmental relations to be clearly individuated and precisely measured. In the course of this process, the matter in question thus becomes a *matter of fact*, a discrete thing to which an unambiguous propositional *statement of fact* may then be applied. In the next section, we will apply Heidegger's account of the early-modern experiment in a concrete, micro-historical case study of one dramatic dispute between two seventeenth-century natural philosophers: Robert Boyle and Francis Line.

4. Boyle versus Line: A Study in Experimental Fact-Making

In their 1985 book, *Leviathan and the Air-Pump*, a classic of the SSK literature, Steven Shapin and Simon Schaffer argue that seventeenth-century experimental practice depended crucially on the emergence of a clearly defined community of experimental philosophers. Admission into this community required one to accept the view that artificially produced matters of fact could provide a legitimate basis for reliable

---

63   Heidegger (1977a), 'The Age of the World Picture,' p. 120.

knowledge of nature.[64] This social restriction helped to define the boundaries of the experimental community, determining those who were, and those who were not, to be counted among its members. According to Shapin and Schaffer, the establishment of the epistemic rule that reliable natural knowledge must rest on experimental matters of fact depended, in important part, on the prior definition of an experimental community through social rules of inclusion and exclusion. Hence, as they see it, the epistemic problem of how best to acquire knowledge of nature was contingent on the social problem of how best to define the community of natural philosophers: 'the effective solution to the problem of knowledge was predicated on a solution to the problem of social order.'[65]

This insight from SSK can help us to expand on Heidegger's claim, cited at the end of the last section, that scientific facts become clear as the facts they are only within the purview of rule and law, that is, only within a domain of practice in which causal regularities are disambiguated and thus more precisely specified. Shapin and Schaffer's argument that solutions to problems of knowledge are based on solutions to problems of social order suggests that effective epistemic techniques for the causal disambiguation of matters of fact are (necessarily but not sufficiently) contingent on the deployment of effective social techniques for the specification of membership in the relevant epistemic community. By demanding commitment to the epistemic primacy of experimentally produced matters of fact, early-modern experimental philosophers effectively eliminated, as irrelevant, explanations based on causes which could not be experimentally produced. On first blush, this may suggest that only efficient, or operational causes, were meant to play a role in the new experimentalists' explanations of natural phenomena. However, as we have already seen, explanations in terms of operational causes, if they are to explain causal *regularities*, must make at least tacit use of final causes. Hence, the social dynamics of exclusion which specified the early-modern experimental community could not have rejected final causes, as such, without threatening the intelligibility of the experimental enterprise. It should follow, then,

---

64 Steven Shapin and Simon Schaffer (1985), *Leviathan and the Air-Pump: Hobbes, Boyle, and the Experimental Life* (Princeton: Princeton University Press), p. 24.

65 Shapin and Schaffer (1985), *Leviathan and the Air-Pump*, p. 282.

that these dynamics functioned instead to disambiguate between the legitimate and illegitimate uses of final causes in explanations of experimentally produced phenomena. The burden of this section will be to demonstrate that the social logic of early-modern experimental subjectivity served to disambiguate final causes by relying, if only tacitly, on the mythical image of the First Thing. This image functioned as the uniform measure by which to differentiate between proper and improper causal explanations of experimental phenomena. This differentiation was, at base, a social one: the image of the First Thing also enabled a discrimination between authentic and inauthentic members of the experimental community. In other words, it also guided judgements about the proper social organisation of seventeenth-century experimental philosophy.

I wish to support these claims by examining a case of controversy in seventeenth-century experimental philosophy over the correct causal explanation of the 'Torricellian effect,' named for mathematician and physicist Evangelista Torricelli, who first produced it in 1644. Shapin and Schaffer observe that, in the decade following its production, this experimental effect 'was associated with [...] questions of immense cosmological importance.'[66] Given the diverse and dramatically contradictory causes which were proposed to explain the Torricellian effect, Shapin and Schaffer name it as 'a key example of scandal in natural philosophy.'[67]

To perform a reasonable approximation of the Torricellian experiment, do the following. Take a thin glass tube, about one metre long, and hold it vertically with the lower opening of the tube blocked with one of your fingers. Then have the tube filled right to the top with mercury (wear gloves!), being careful to eliminate any air bubbles. Seal the upper opening of the tube with a finger of your other hand, making sure that there is no air trapped between your finger and the mercury. Now immerse the lower end of the tube into a beaker half-filled with mercury, and remove your finger from this lower opening while keeping the upper opening of the tube firmly covered. What happens? One might expect the mercury to drop out of the tube, pulled down by the earth's gravity, since your finger is no longer holding it up by blocking

---

66   Shapin and Schaffer (1985), *Leviathan and the Air-Pump*, p. 41.

67   Shapin and Schaffer (1985), *Leviathan and the Air-Pump*, p. 41.

the bottom opening of the tube. In fact, the height of the mercury in the tube does drop as some of the mercury runs out of the tube and into the beaker. However, the mercury level in the tube does not drop all the way down to the level of mercury in the beaker. Indeed, most of the mercury remains in the tube, suspended at a height of about seventy-five centimetres above the level of mercury in the beaker (assuming the experiment is being performed near sea level). Even in the absence of your finger sealing the bottom of the tube, something still seems to be holding the mercury up, thus interfering with the effects of gravity. This suspension of the column of mercury in the tube, significantly above the mercury level in the beaker below, is the main experimental effect produced in the Torricellian experiment. In what follows, I will refer to this suspended column of mercury as the 'Torricellian effect.' Seventeenth-century natural philosophers disputed with one another over the cause of the Torricellian effect, that is, over the legitimate answer to the question of what holds the column of mercury suspended at a height of about seventy-five centimetres.

There are two secondary features of the Torricellian experiment also worth noting. First, because the mercury level in the metre-long tube has dropped, a space of about twenty-five centimetres has now opened up in the upper portion of the tube, between the surface of the suspended column of mercury and the finger sealing the upper end of the tube. This space is known as the 'Torricellian space.' Second, the appearance of this space in the tube is accompanied by a feeling, in the flesh of the finger sealing the upper end of the tube, that it is now being forced down into the tube. This experience is a secondary effect of the Torricellian experiment, and this effect is normally called 'suction.' As we will see, however, the term 'suction' is problematic, as it presupposes that the flesh of the finger is being *pulled* rather than *pushed* into the tube. In other words, it presupposes that the relevant physical cause of this effect is located *inside* the tube rather than *outside* of it. In early-modern debates over the meaning of the Torricellian experiment, much depended on the position one took in respect of this fine distinction.

I will focus on one specific seventeenth-century dispute over the natural philosophical meaning of the Torricellian experiment, one which occurred between Robert Boyle, on the one side, and Francis Line, who published under the Latin name Franciscus Linus, on the other. This dispute was occasioned by the publication in 1660 of Boyle's book *New Experiments Physico-Mechanical, Touching on the Spring of the Air, and its Effects; Made, for the Most Part, in an New Pneumatical Engine.* Two aspects of the title of Boyle's book deserve emphasis. First, the title indicates that Boyle has undertaken an investigation of the cause of certain effects, the cause in question here stipulated as the 'spring of the air.' Second, this investigation was conducted, for the most part, using a new experimental instrument which Boyle called a 'pneumatic engine,' but which is usually referred to by historians of science, including Shapin and Schaffer, as an 'air-pump.' Stated simply, the air-pump was a sealed glass vessel from which air was evacuated by means of an attached pump.

Boyle performed several experiments within this glass vessel, investigating the results produced on an observed phenomenon when air was removed, consequentially if not entirely, from the causal nexus within which that phenomenon was embedded. For example, in Experiment Seventeen of his book, Boyle lowered the Torricellian experiment through an opening in the top of the glass vessel of his air-pump, such that the open beaker of mercury was completely inside the vessel but the vertical Torricellian tube still protruded up through the vessel opening.[68] (The tube used in this case had a closed upper end, so it was unnecessary to seal it with a finger.) Boyle then sealed the vessel opening though which the tube protruded, so as to render the vessel as air-tight as possible. With everything in place, he then used the pump to evacuate as much air as possible from the glass vessel.

By removing air from the space surrounding the lower portion of the Torricellian apparatus, particularly from around the open beaker of mercury, Boyle introduced dramatic changes to the Torricellian effect. Specifically, in the absence, or near absence, of air in the glass vessel, the suspended column of mercury in the tube dropped to almost the same height as the mercury in the beaker (which itself rose slightly). In other words, the Torricellian effect had been almost entirely eliminated. On the basis of this result, Boyle argued that it must have been the

---

68   Robert Boyle (1660 [1966]), *New Experiments Physico-Mechanical, Touching on the Spring of the Air, and its Effects; Made, for the Most Part, in an New Pneumatical Engine,* in *The Works of Robert Boyle,* vol. 1, ed. by Thomas Birch (Hildesheim: Georg Olms), pp. 1–117 (pp. 33–39).

presence of ambient air, pushing down on the surface of the mercury in the beaker, which counteracted the effects of gravity, thus causing the mercury in the tube to be suspended at a height of about seventy-five centimetres. Boyle's experiment purportedly served to reduce ambiguity and increase rigour in the natural philosophical understanding of the Torricellian effect by more definitely specifying the cause of that effect, narrowing it down to the pressure of the ambient air on the surface of the mercury in the beaker.

However, Boyle was not content to simply argue *that* ambient air was the principal cause of the Torricellian effect. He also claimed to know *what* about the air produced this effect. He identified this 'what' as the 'spring' of the air. In Experiment One of *New Experiments*, Boyle wrote that he finds it 'not superfluous nor unreasonable' to 'insinuate' the air's spring as a 'likely' cause for the experimental effects produced in the glass vessel of his air-pump.[69] However, Boyle's initial diffidence regarding the epistemic status of the air's spring quickly disappeared. As Shapin and Schaffer observe, by the early 1660s, Boyle viewed the spring of the air as an established matter of fact rather than as a hypothesis. However, Boyle did not spell out the reasons for his growing confidence. As Shapin and Schaffer write, '[v]iewed naively, or as a stranger might view it, it is unclear why the spring of the air, as the professed cause of the observed results, should be treated as a matter of fact rather than as a speculative hypothesis.'[70] In response to this puzzle, they make the sociological argument that 'Boyle's criteria and rules for making his preferred distinctions between matters of fact and [hypothetical] causes have the status of *conventions*.'[71] These conventions, in turn, drew their legitimacy from the 'total pattern of activities' constitutive of the experimental culture which Boyle and his colleagues were struggling to establish.[72] Shapin and Schaffer support their sociological explanation for Boyle's discrimination between facts and speculative causes with a detailed examination of the way Boyle responded to his critics, including the Jesuit mathematician and natural philosopher Francis Line.

---

69   Boyle (1660), *New Experiments Physico-Mechanical*, p. 11.
70   Shapin and Schaffer (1985), *Leviathan and the Air-Pump*, p. 51.
71   Shapin and Schaffer (1985), *Leviathan and the Air-Pump*, p. 52.
72   Shapin and Schaffer (1985), *Leviathan and the Air-Pump*, p. 52.

Before turning to Boyle's dispute with Line, it will be useful to first briefly consider what Boyle meant when he attributed a spring to the air. At the most general level, Boyle had in mind a corpuscular conception of the air as being made up of individual microscopic particles. He thus understood the spring of the air as a disposition of individual air particles which was activated when those particles came into contact with one another, or with any other body. Hence, he writes that

> our air either consists of, or at least abounds with, parts of such a nature, that in case they be bent or compressed by the weight of the incumbent part of the atmosphere, or by any other body, they do endeavour, as much as in them lieth, to free themselves from that pressure, by bearing against the contiguous bodies that keep them bent; and, as soon as those bodies are removed, or reduced to give them way, by presently unbending and stretching out themselves.[73]

Put briefly, air particles endeavour to free themselves when placed under pressure by an adjacent particle. Each particle possesses a 'power or principle of self-dilation,' which becomes actualised as an 'endeavour outward' when the particle is under compression.[74] Boyle conceived of the earth's atmosphere as composed of innumerable particles heaped up on top of one another, such that air particles near the earth's surface where maximally compressed under the great weight of the atmosphere above them. In respect of the Torricellian experiment, Boyle argued that the pressure of the atmospheric air on the surface of the mercury in the beaker was the sole cause of the Torricellian effect. By placing the Torricellian apparatus in the glass vessel of his air-pump, and then exhausting most of the air from the vessel, Boyle removed the cause of this effect. As air particles were removed from the vessel, the remaining particles were able to progressively exercise their power of self-dilation, unbending themselves until they exerted scarcely any more pressure on the mercury in the beaker. In other words, Boyle used his air-pump to 'release' the air corpuscles remaining within its chamber from the external influence of neighbouring corpuscles. This, in turn, released the surface of mercury in the beaker from the weight of the atmosphere. As a consequence, the column of mercury in the tube descended under

73   Boyle (1660), *New Experiments Physico-Mechanical*, p. 11.
74   Boyle (1660), *New Experiments Physico-Mechanical*, p. 11.

the force of gravity, draining out into the beaker. A key consequence of Boyle's explanation is that it rendered the Torricellian space, between the suspended column of mercury and the top of the tube, causally inert. In Boyle's view, the Torricellian effect could be sufficiently explained by reference to the removal of air pressure from the surface of the mercury in the beaker.

Boyle was careful to distinguish this particular explanation of the air's spring from the general fact that the air has a spring. He emphasised this distinction by noting the contrasting explanation of René Descartes, who conceived of the air particles as lifted by heat and the 'restless agitation of that celestial matter' surrounding the earth to the point where they begin to whirl about, 'each corpuscle endeavour[ing] to beat off all the others from coming within the little sphere requisite to its motion about its own centre.'[75] Yet, despite the differences between these two explanations, they both assume that the air is composed of particles possessing an 'endeavour outward,' a disposition to push away neighbouring bodies in order to achieve for themselves a state free from external causal interference. Boyle's and Descartes's divergent explanations for the air's spring thus stand as different specifications, or articulations, of the same general idea. Put in Aristotelian terms, this disposition plays the role of efficient cause in the movements or operations of these particles. Moreover, as Boyle's own words suggest, the meaning of the particles' outward movement, that for the sake of which they so move, appears to be a state of freedom from external influence, in other words, a state of autonomy. This autonomy may be understood as the final cause of the particles' outward movement, as the principle according to which one is able to make sense of those minute mechanical operations, to experience them as intelligible.

Let us now turn to the explanation Francis Line gave for the Torricellian effect. Line presented his arguments in a 1661 book titled *Tractatus de Corporum Inseparabilitate*, or 'A Treatise on the Inseparable Nature of Bodies.'[76] Unlike Boyle, Line did not treat the Torricellian space

---

75   Boyle (1660), *New Experiments Physico-Mechanical*, p. 12.
76   Linus, Franciscus (1661), *Tractatus de Corporum Inseparabilitate in quo Experimenta de Vacuo, tam Torriculliana, quam Magdeburgica, et Boyliana examinantur, veraque eorum causa detecta, ostenditur, vacuum naturaliter dari non posse: unde et Aristotelica de Rarefactione sententia tam contra Assertores Vacuitatum, quam Corpusculorum demonstratur* (London).

as irrelevant for a causal explanation of the Torricellian effect. Indeed, he suggested that this space contains a fine thread of matter — like Boyle's springs, too fine to be seen by the eye — which connects the upper surface of the suspended column of mercury to the flesh of the finger sealing the top end of the tube. This thread, or *funiculus* in Latin, helped to prevent the mercury column from falling below seventy-five centimetres, and thus contributed causally to the Torricellian effect. Hence, unlike Boyle's corpuscular springs, which possess a power of self-dilation, Line's thread possesses a power of self-contraction. This was manifest as an endeavour inward, in contrast to the endeavour outward of Boyle's springs; where Boyle's springs push things away, Line's thread pulls them together. Line argued, in particular, that it pulls the mercury and the finger, at either end of the Torricellian space, towards one another. I will refer to this as Line's 'thread hypothesis.' Like Boyle's spring hypothesis, the thread hypothesis sought to explain the Torricellian effect in efficient, or operational, causal terms.[77]

Furthermore, Line's thread hypothesis also sought to explain a secondary feature of the Torricellian experiment, namely, the experience of the flesh of one's finger being forced down into the Torricellian space. He argued that the thread attached to the finger's flesh, by its power of self-contraction, was what pulled the flesh into the tube. This explanation was also meant to explain why one's finger was not only forced into the tube, but also why it stuck to the tube end. Indeed, much as effort is required to pull one's palm from the hose end of an operating vacuum cleaner, so too is effort required to pull one's finger from the top end of the Torricellian tube, that is, in Line's terminology, to break the thread attaching one's finger to the top of the mercury column. Line argued that Boyle's spring hypothesis failed to sufficiently explain this secondary effect of the Torricellian apparatus.

Line also thought that his thread hypothesis had another advantage over Boyle's spring hypothesis. As a Jesuit scholar, he was obligated to uphold the Scholastic doctrine that nature abhors a vacuum. Line thus also argued that the thread forms in the Torricellian space in order to

---

77   Conor Reilly notes that Line's thread hypothesis was not original: '[s]uch a concept had already been suggested under a number of different forms, by several scholars' (Conor Reilly (1969), *Francis Line S. J.: An Exiled English Scientist 1595–1675* (Rome: Institutum Historicum), p. 65).

fill the void left by the falling mercury, thereby preventing a vacuum. Thus, in addition to explaining the adhesion of one's finger to the tube, as well as contributing to an explanation of the Torricellian effect itself, Line argued that his thread hypothesis also confirmed a widely held belief about nature: 'And hence is confirmed that common axiom used in the schools for so many ages past, that nature doth abhor a vacuum.'[78] In Line's view, then, the prevention of a vacuum — that is, the absolute separation of bodies — was that for the sake of which the thread forms in the tube and pulls inward. The prevention of a vacuum, of the separation of the constituents of matter, was the final cause, or reason, explaining why the thread is disposed to form and contract itself. In contrast to Line, Boyle declined to give an explanation for why his corpuscular springs are disposed to dilate themselves. Although he sought to explain the Torricellian effect in terms of an efficient, or operational cause — the spring's endeavour outward — Boyle did not likewise offer an explanation in terms of a final cause, that is to say, he did not provide a reason for why the springs always move themselves outward as opposed to in some other direction.

In 1662, Boyle published a response to Line in which he used his earlier silence about final causation as a strength in promoting his spring hypothesis over Line's alternative proposal. Furthermore, he exploited Line's reference to final causation in order to attack the intelligibility of the latter's proposed efficient cause, the thread hypothesis. Boyle writes:

> It seems also very difficult to conceive, how this extenuated substance should acquire so strong a spring inward, as the examiner all along his books ascribes to it. Nor will it serve his turn to require of us in exchange an explication of the air's spring outward, since he acknowledges, as well as we, that it has such a spring.[79]

Here, Boyle argues that, because Line has accepted the legitimacy of Boyle's spring hypothesis, Boyle is under no obligation to provide a further explanation of spring in terms of final causation. This implies

---

78   Cited in Boyle (1660), *New Experiments Physico-Mechanical*, p. 135.
79   Robert Boyle (1662 [1966]), *A Defence of the Doctrine Touching the Spring and Weight of the Air, Proposed by Mr. R. Boyle, in his New Physico-Mechanical Experiments; Against the Objections of Franciscus Linus. Wherewith the Objector's Funicular Hypothesis is also Examined*, in *The Works of Robert Boyle*, vol. 1, ed. by Thomas Birch (Hildesheim: Georg Olms), pp. 118–85 (p. 143).

the view that an explanation in terms of operational causes can be accepted as legitimate independently of considerations of final cause. Yet, although Boyle allows himself the benefit of this supposition, he is not willing to extend this same benefit to Line. He uses Line's appeal to the *horror vacui* as a final cause to undermine the legitimacy of Line's appeal to a self-contracting thread as an efficient cause. Yet, as his own work makes clear, there is no necessary connection between a particular efficient cause and a specific final cause. Hence, Boyle might have instead recognised the possible legitimacy of Line's thread hypothesis in isolation from the *horror vacui*, and he might even have reinterpreted that hypothesis in a way more in step with his own principles.

In fact, this appears to have been Line's strategy towards Boyle. As Boyle notes, Line claimed that '[i]t cannot be conceived [*concipi non posse*]' on a corpuscular model how Boyle's spring could take up more space when it dilates itself.[80] Line seems to have had in mind the Aristotelian idea that 'things can increase in size not only by the entrance of something but also by qualitative change; e.g. if water were to be transformed into air.'[81] Hence, a sponge will increase in size by absorbing water, but a volume of water will increase in size by qualitatively changing from a liquid to a gas. In the second case, nothing has been added to the substance; instead the substance itself undergoes a change. On the corpuscular model, in contrast, there is no qualitative change in the basic material of water when it changes from a liquid into a gas. The corpuscles of water remain the same, but more space is opened between them. Likewise, there is no qualitative change to a corpuscular spring when it self-dilates. The more successfully it can beat back its neighbours, the more space it will take up, even as its physical size remains unchanged. It thus seems clear that Line and Boyle drew on radically different conceptions of matter in their respective interpretations of the phenomenon of spring. And both claimed to find the other's conception unintelligible. Yet Line was willing to accept the air's spring as a matter of fact so long as it was explained in accordance with his own non-corpuscular principles. In contrast, Boyle was reluctant to do the same for Line's thread hypothesis. Indeed, he sought to reject that hypothesis by arguing, in part, that the principles Line

---

80   Boyle (1662), *A Defence of the Doctrine Touching the Spring and Weight of the Air*, p. 126.
81   Aristotle (1941c), *Physica*, p. 282 (lines 214a36–214b2).

used to explain it — above all, the *horror vacui* — were unintelligible. Given that Boyle was happy, in other contexts, to distinguish between operational hypotheses and their explanation, his attempt here to erase that distinction seems less than fully compelling.

In addition to unintelligibility, Boyle offered three other reasons to reject Line's thread hypothesis: 'this hypothesis of our author's [...] seems to be partly precarious, partly unintelligible, partly insufficient, and besides needless.'[82] He immediately notes, however, that 'it will not be so convenient' to prove each of these claims in isolation from the others.[83] Hence, for example, Boyle combines his argument from precariousness with charges of unintelligibility when he argues that Line's thread hypothesis relies on a belief in the *horror vacui*, which has not been 'cogently prove[d].'[84] This lack of proof, concludes Boyle, 'may help to shew his doctrine to be precarious.'[85] But, if the merits of Line's thread hypothesis can be judged independently of his belief in the *horror vacui*, then this argument from precariousness lacks adequate force.

Perhaps the most promising reason Boyle gave to reject Line's thread hypothesis was that it is needless, or unnecessary. One of Line's strongest empirical arguments for his thread hypothesis was that it explains both why one feels one's finger being forced into the Torricellian tube, as well as why the end of the tube sticks to one's finger. In addressing the former case, Boyle writes that

> the finger that stops the tube being exposed on the upper parts and the sides to the external air, has the whole weight and pressure of the atmosphere upon it. [...] [C]onsequently that part of the finger that is within the tube will have much less pressure against it from the dilated air without. By which means the pulp of the finger will be thrust in (which our author is pleased to call sucked in).[86]

In addressing the latter case, Boyle argues that the 'pressure of the ambient air' will 'thrust in the pulp of the finger at the upper orifice of the tube, and make it stick closely enough to the lip of it.'[87] Boyle immediately

---

82   Boyle (1662), *A Defence of the Doctrine Touching the Spring and Weight of the Air*, p. 134.
83   Boyle (1662), *A Defence of the Doctrine Touching the Spring and Weight of the Air*, p. 134.
84   Boyle (1662), *A Defence of the Doctrine Touching the Spring and Weight of the Air*, p. 135.
85   Boyle (1662), *A Defence of the Doctrine Touching the Spring and Weight of the Air*, p. 137.
86   Boyle (1662), *A Defence of the Doctrine Touching the Spring and Weight of the Air*, p. 126.
87   Boyle (1662), *A Defence of the Doctrine Touching the Spring and Weight of the Air*, p. 129.

recognised that Line would not be convinced by this latter explanation, writing that 'I know the examiner affirms, that no thrusting or pressure from without can ever effect such an adhesion of the finger to the tube. But this should be as well proved as said.'[88] Surely, however, the onus lay with Boyle to prove that his hypothesis sufficiently explains this adhesion, rather than with his critic to prove that it does not. In any case, Line was not alone in finding Boyle's explanation unconvincing. The powerful Dutch natural philosopher, Christiaan Huygens, commented at the time that he could 'not see that either Linus' hypothesis or Mr. Boyle's is satisfactory, that is, why the siphon sticks to the finger, so that one must use some little force to draw it off.'[89] It seems clear that, at the time of the dispute, Boyle's claim to have given a sufficient explanation of 'suction' was not uncontroversial. Hence, it could not have provided adequate grounds for arguing that Line's thread hypothesis was not needed for an explanation of this phenomenon.

It should be emphasised that Line, in order to provide space for his own explanation of the Torricellian effect, did not need to likewise argue that Boyle's explanation was unnecessary, but only that it was insufficient. This would have been enough for him to justify the introduction of his thread hypothesis as necessary, but not sufficient, for a comprehensive explanation of the effect. Line's position was consistent with the view that both his thread hypothesis and Boyle's spring hypothesis were necessary for an explanation of the Torricellian effect, but that neither was, on its own, sufficient for such an explanation. On this basis, Line could have agreed with Boyle's claim that his thread hypothesis was 'partly insufficient' without revising his original argument.[90]

It seems that Boyle did not fully understand Line's position with respect to this point. Although he recognised in several passages that Line was criticising his spring hypothesis for being insufficient, in other passages he reacted as if Line were rejecting it as wholly unnecessary. In the first instance, Boyle's observation that Line 'denies it not, that the air has some weight and spring, but affirms, that it is very insufficient to perform such great matters' was one which he

---

88   Boyle (1662), *A Defence of the Doctrine Touching the Spring and Weight of the Air*, p. 129.

89   Cited in Reilly (1969), *Francis Line S. J.*, p. 85.

90   Elizabeth Potter also makes this observation (Elizabeth Potter (2001), *Gender and Boyle's Law of Gases* (Bloomington: Indiana University Press), pp. 36, 133).

repeated several times throughout his response to Line.[91] Yet he also described Line as arguing that 'the things we ascribe to the weight and spring of the air are really performed by neither,' and also of trying to 'invalidate the hypothesis of weight and spring of the air.'[92] In other words, despite explicitly recognising evidence to the contrary, Boyle ultimately treated Line as employing a zero-sum strategy, whereby only one of their respective hypotheses could be explanatorily relevant. However, it was, in fact, Boyle who employed this strategy, not Line. Boyle argued that his spring hypothesis provided not just a necessary, but also a sufficient, explanation for the Torricellian effect. Hence, he viewed Line's proposed explanation as not only insufficient, but also unnecessary, or 'needless.' But this appears to misrepresent the dispute. Whereas Boyle treated the choice between the spring and thread hypotheses in exclusive either/or terms, Line seems to have taken a more inclusive approach in which both hypotheses could contribute to a comprehensive operational explanation of the Torricellian effect.

In sum, the four reasons Boyle gave for his rejection of Line's thread hypothesis — that it is 'partly precarious, partly unintelligible, partly insufficient, and besides needless' — do not seem to adequately justify that rejection. Two further reasons may help us to more fully understand Boyle's response to Line. The first reason has to do with Boyle's desire to ·neutralise the explanatory significance of the Torricellian space. This space had been a flashpoint for metaphysical disputes over the existence of a vacuum. The Aristotelian doctrine of the *horror vacui* played a central role in this dispute. As Shapin and Schaffer show, Boyle was not trying to resolve this dispute, but instead to sideline it: '[w]hat he was endeavouring to create was a natural philosophical discourse in which such questions [about the existence of a "metaphysical" vacuum] were inadmissible.'[93] By neutralising these disputes, Boyle and his colleagues hoped to provide a solution to the problem of social order made pressing by the intense civil conflict of their period. Their solution

---

91 Boyle (1662), *A Defence of the Doctrine Touching the Spring and Weight of the Air*, p. 156; see also pp. 124, 162, 178.

92 Boyle (1662), *A Defence of the Doctrine Touching the Spring and Weight of the Air*, pp. 134, 178.

93 Shapin and Schaffer (1985), *Leviathan and the Air-Pump*, p. 45.

involved creating a new mode of discourse in which metaphysics was supposed to have no legitimate place. This discourse embraced the language of 'experimental "physiology"' in opposition to the language of 'metaphysics.' This distinction set up a powerful boundary within discourse about the natural world, and '[d]issension involving violations of this boundary [...] was deemed fatal.'[94] Boyle rejected Line's thread hypothesis because it violated this boundary by invoking the metaphysical doctrine of the *horror vacui*, and thereby threatened the social order which Boyle and his colleagues were struggling to establish. According to Shapin and Schaffer, this boundary was a social convention rooted in collective efforts to reestablish social and political order. Hence, Boyle's epistemological reasons for rejecting Line's thread hypothesis cannot be neatly separated from considerations of the social context in which the debate took place. By invoking the *horror vacui* — the inseparability of matter — Line not only threated Boyle's professed empiricism, but also the conception of social order he shared with other members of the emerging community of experimentalists.

Yet, as I argued above, once Line's thread hypothesis is separated from his commitment to the doctrine of the *horror vacui*, this doctrine can no longer provide a compelling reason for rejecting the hypothesis. In this light, Line's thread hypothesis seems no more metaphysical than Boyle's spring hypothesis. Indeed, Boyle even interprets it as an inverse of his own hypothesis: whereas he posits a spring outward, Line posits a spring inward. It is the directionality of the two hypotheses which most clearly and consequentially distinguishes them. This brings us to the second of the two further reasons for Boyle's rejection of Line's thread hypothesis. As an efficient, or moving, cause, Line's thread moves itself in the wrong direction. It acts to move things together, rather than to push them apart. I have argued, in general, that the directedness of an efficient cause, as that which regulates or guides a thing's operations and thus gives them a clear and stable meaning, is what Aristotle had meant by the notion of final cause. In this sense, any interpretation of a thing which ascribes to it a regular pattern of behaviour will rely on this notion. Without it, the regularity of the thing's operations would appear to be a coincidence, a matter of chance, rather than the result of a

---

94   Shapin and Schaffer (1985), *Leviathan and the Air-Pump*, p. 80.

power or principle belonging to that thing. Hence, when Boyle ascribed a power of self-dilation, an endeavour outward, to his corpuscular springs, he was implicitly recognising that a final cause regulates the operations of those corpuscles.[95]

Boyle's tacit ascription of a final cause to the operations of his corpuscular springs was buried beneath his declaration that the spring of the air was a matter of fact. As observed above, Boyle insisted on a strict distinction between matters of fact, on the one hand, and speculative hypotheses, on the other. Indeed, first recognising the Torricellian effect to be a matter of fact, Boyle then speculated that its cause was the outward spring of the air. However, Boyle quickly dropped his speculative tone and declared this outward spring to also be a well-established matter of fact rather than a hypothesis. As Shapin and Schaffer note, Boyle did not explicate a reason for this change in his attitude, and they suggest that a sociological explanation can be given, one which makes sense of Boyle's apparently arbitrary distinctions as having been guided by social conventions. On this basis, we can treat Boyle's discrimination between metaphysical and empirical language as a social discrimination motivated by a specific conception of social order. This was a discrimination in which a power of outward spring could be accepted as an empirical matter of fact, while a power of inward spring was rejected as an unintelligible metaphysical hypothesis.

The credibility of Boyle's claim to have eschewed metaphysics was contingent on his success in rendering invisible the interpretative space

---

95   The outward endeavour of Boyle's microscopically insensible corpuscles was metaphysical. However, he also studied sensible phenomena manifesting a similar power of motion. In a 1685 treatise, Boyle remarks on thick pieces of glass which, when cooled after removal from a furnace, burst apart, their pieces 'fly[ing] to a not inconsiderable distance from one another' (Robert Boyle (1685 [1966]), *An Essay of the Great Effects of Even Languid and Unheeded Motion*, in *The Works of Robert Boyle*, vol. 5, ed. by Thomas Birch (Hildesheim: Georg Olms), pp. 1–37 (p. 24)). Likewise, when a bow string is cut, 'the bow will fly suddenly outwards, and the parts of the string will swiftly and violently shrink from one another' (p. 26). And gemstones, when extracted from the hard cement of an aggregate rock, 'quickly expanded themselves, as if it were by an internal and violently compressed spring, and would presently burst asunder' (p. 27). From this, Boyle concludes that 'bodies, which as to sense, are in a natural state of rest, may be in a violent one, as of tension, and may have [...] a strong endeavour to fly off or recede from one another' (p. 26). Boyle may have viewed these phenomena as empirical vindication for his metaphysical claim that corpuscular springs manifest an internal power of outward endeavour when released from compression.

between the Torricellian effect and his causal explanation of that effect in terms of outward spring. Hence, the most pressing threat posed to Boyle by Line's thread hypothesis was that it posited a power of inward spring which plausibly explained many of the same aspects of the Torricellian experiment. In this way, Line's intervention threatened to re-expose the contingency of the interpretative space existing between the Torricellian effect and Boyle's causal spring hypothesis, a space which Boyle had worked hard to render invisible. Building on Shapin and Schaffer's analysis, one may conclude that Line's thread hypothesis threatened the conception of social order which tacitly grounded Boyle's explanation of the Torricellian effect. However, contrary to what Shapin and Schaffer argue, this threat cannot be explained in terms of Line's appeal to metaphysics, in the form of final causes, because Boyle also shared this commitment to final causation. It would seem, instead, that Boyle viewed the threat more as directed at his particular, tacit conception of the final cause of an individual corpuscle's movement: a state of autonomy, of freedom from the things which surround it. Yet, as we have seen, Line did not challenge Boyle's conception of outward spring, as such. Line accepted the necessity of Boyle's proposed cause, but challenged its sufficiency as an explanation of the Torricellian experiment. Hence, what was ultimately at stake for Boyle was not the validity of his spring hypothesis, but its exclusivity. Insofar as his hypothesis provided a solution to the problem of social order, Boyle meant for it to serve as an exclusive solution, one that was both necessary and sufficient.

These two features of Boyle's explanation of the Torricellian effect — outward spring and exclusivity — can be fit together with two complementary features of Heidegger's explanation of the emergence of early-modern science in terms of the mathematical. These latter two features are what I have called the First Thing, and what Heidegger called the basic blueprint. The mythical image of the First Thing helps to further explicate Boyle's model of a spring-like corpuscle which endeavours to beat away all neighbouring bodies, thereby opening up around itself a free space in which it can realise its natural, autonomous state. This autonomous state is the final cause regulating the behaviour of Boyle's corpuscular springs. It is that place towards which things, by exercising their power of outward spring, endeavour to move themselves. Boyle's air-pump supported this endeavour, helping the corpuscles to realise their natural state. This end state or place is represented by the First Thing. Moreover, this end state is the only one towards which things may be legitimately directed. In other words, Boyle's corpuscular springs can be ultimately regulated by only one final cause. Hence, the image of the First Thing also functions as the one basic blueprint governing the indigenous operations of all air corpuscles. Just as Boyle's inward spring was meant to play an exclusive role in his explanation of the Torricellian experiment, so too was the image of the First Thing meant to serve as the one basic blueprint governing explanations of material phenomena in the new Galilean universe. The Galilean First Thing was the basic blueprint guiding Boyle's explanation of the spring of the air. It was the tacitly held mathematical image underpinning the social logic of inclusion and exclusion which enabled Boyle and his colleagues to discriminate between legitimate and illegitimate accounts of experimental phenomena, and hence also between authentic and inauthentic members of the emerging seventeenth-century community of experimental philosophers. The success of Boyle's refutation of Line's thread hypothesis was consequentially determined by this tacit image. In the next section, I will argue that this tacitly held, mathematical image of the First Thing was, at root, a *social* image.

5. Social Imagery and Early-Modern Science

The idea that knowledge is governed by an image — a *social* image — has had an important, if somewhat underdeveloped, place in SSK. Indeed, David Bloor considered it so central that he gave his influential 1976 introduction to SSK the title *Knowledge and Social Imagery*. In this section, I will discuss the relationship between Heidegger's mythical image of the Galilean First Thing and Bloor's notion of social imagery. This discussion will allow us to further explore, on the one hand, the sociological aspects of Heidegger's account of the emergence of early-modern science, and, on the other hand, the theoretical insights which SSK may draw from that account.

Bloor argues that 'we think about knowledge by manipulating images of society.'[96] A clear distinction must be drawn here between an image

---

96 David Bloor (1991 [1976]), *Knowledge and Social Imagery*, 2nd edn (Chicago: University of Chicago Press), p. 52.

of society and society itself. Bloor argues that society is too complex and overwhelming for us to understand it directly. As a consequence, we must rely on simplified social models: 'Immersed as we are in society we cannot grasp it as a whole in our reflective consciousness except by using a simplified picture, an image, or what may be called an "ideology."'[97] This image subsequently guides the way we come to think about knowledge in general. Crucially, our manipulation of social images need not be a conscious act. In other words, the connection between a theory of knowledge and a social image will often be a tacit one. This means that social imagery will often serve as an analyst's category without also being an actor's category. According to Bloor, an analyst should not limit her explanations to the consciously deployed categories and concepts of those actors whose behaviour she is seeking to explain. She should also seek to uncover the social categories and concepts which structure an individual actor's thoughts and actions at a tacit, or unconscious, level. One of the key targets of this method of excavation is the social imagery which may shape and guide actors' thoughts and actions without their being aware of it.

The claim that social imagery guides actors' practices suggests that it plays a role similar to that of Aristotelian final causes. In other words, Bloor's concept of social imagery appears to include a teleological element. This introduces a potential tension into the comparison between Bloor's social imagery and the Heideggerian notion of the Galilean First Thing. Bloor strongly repudiates teleological forms of explanation as being inimical to SSK: 'There is no doubt that if the teleological model is true then the strong programme is false.'[98] Yet, this apparent tension may be resolved once we recognise that the offending aspects of what Bloor calls the 'teleological model' are absent from the notion of the First Thing. These aspects are, first, that the teleological model entails the existence of 'purposive' behaviour, and, second, that this behaviour is 'naturally oriented towards truth.'[99] The worry, in the first case, is that science is being treated as a process of 'mind or "consciousness,"' rather than as a configuration of socially sustained

---

97   Bloor (1991), *Knowledge and Social Imagery*, p. 53.

98   Bloor (1991), *Knowledge and Social Imagery*, p. 12.

99   David Bloor (1973), 'Wittgenstein and Mannheim on the Sociology of Mathematics,' Studies in History and Philosophy of Science 4(2), 173–91 (pp. 178, 185).

material practices.[100] In the second case, the worry is that this allegedly mental process is directed towards a non-contingent (i.e., asocial and ahistorical), or absolute, truth. As we have seen, SSK entails neither of these claims. As we have also seen, the image of the First Thing likewise entails neither of these claims. As the basic blueprint of early-modern scientific practice, the First Thing can guide behaviour at a wholly tacit and non-deliberative level. Furthermore, the grounds for the First Thing are not absolute or necessary, but ultimately bottom out in the subject's shared, historically contingent tradition. Hence, Bloor's reservations with respect to teleology do not apply in the case of the Galilean First Thing.[101]

Having resolved this initial worry about a potential conflict between Bloor's social imagery and the Heideggerian image of the First Thing, let us now turn to the example Bloor offers to illustrate his explanatory

---

100 Bloor (1973), 'Wittgenstein and Mannheim,' p. 174 n. 4; citing Popper.

101 In respect of the second point, note that Bloor has more recently reiterated his identification of 'telos' with a non-contingent 'inner necessity' guiding scientific practice (David Bloor (2011), *The Enigma of the Aerofoil: Rival Theories in Aerodynamics, 1909–1930* (Chicago: Chicago University Press), p. 436). As for the first point, the issue is similar to the one addressed in Chapter Four, where I argued that Bloor's identification of intentionality with mental content does not touch a Heideggerian account of intentionality, which construes this in non-mental terms. That Heidegger did not espouse a teleological model, in Bloor's sense, is further attested in Michael Friedman's observation that '[Ernst] Cassirer's philosophy of symbolic forms [...] diverges from Heidegger's existential analytic of Dasein precisely in emphasizing a teleological development toward the genuinely "objective" and "universally valid" realm of scientific truth' (Michael Friedman (2000), *A Parting of the Ways: Carnap, Cassirer, and Heidegger* (Chicago: Open Court), p. 138). Friedman also notes the influence of Cassirer's anti-naturalistic teleological historiography on such historians of science as Edwin Arthur Burtt, Eduard Jan Dijksterhuis, and Alexandre Koyré (Friedman (2000), *A Parting of the Ways*, p. 88; see Edwin Arthur Burtt (1925), *The Metaphysical Foundations of Modern Physical Science: A Historical and Critical Essay* (London: Kegan Paul, Trench, Trubner & Co.); Eduard Jan Dijksterhuis (1961 [1950]), *The Mechanization of the World Picture*, trans. by C. Dikshoorn (London: Oxford University Press); Koyré (1943a), 'Galileo and Plato'; and Alexandre Koyré (1943b), 'Galileo and the Scientific Revolution of the Seventeenth Century,' *Philosophical Review* 52(4), 333–48). This historiographic tradition appears to be a central object of Bloor's criticism. Barry Barnes criticises a similar teleological method in György Lukács's *History and Class Consciousness* (Barry Barnes (1977), *Interests and the Growth of Knowledge* (London: Routledge & Kegan Paul), p. 48; see György Lukács (1971 [1923]), *History and Class Consciousness: Studies in Marxist Dialectics*, trans. by Rodney Livingstone (London: Merlin Press)). For a more recent comparison of Heidegger and Cassirer, see Peter E. Gordon (2010), *Continental Divide: Heidegger, Cassirer, Davos* (Cambridge, MA: Harvard University Press).

method. Drawing from Karl Mannheim's 1925 essay, 'Conservative Thought,' Bloor attempts to uncover the social imagery influencing the works of philosopher of science Karl Popper and historian and philosopher of science Thomas Kuhn.[102] In a nutshell, Bloor argues that Popper's account of scientific knowledge reflects an Enlightenment ideology, while Kuhn's account of scientific knowledge reflects a Romantic ideology. These two ideologies represent two different images according to which Popper and Kuhn organised, and thus made sense of, the empirical data regarding modern scientific knowledge production. As Bloor puts it, in the case of each thinker, 'the same facts are fitted together to form a different picture.'[103] In thinking about scientific knowledge, in developing an epistemology of scientific practice, Popper and Kuhn each manipulated a distinct image of society, and that social image powerfully influenced the different ways in which they each came to perceive the structure and dynamics of scientific knowledge.

Bloor gives considerable attention to the differences between Popper's and Kuhn's epistemologies, mapping them respectively onto the more general cultural categories of Enlightenment and Romanticism. Here, however, I want to focus on one strong similarity between these two otherwise divergent accounts of scientific knowledge, as well as a difference which then appears within this common perspective. According to Bloor, Popper and Kuhn shared a perception of science as the work of 'collectivities' or 'social wholes.'[104] In other words, both thinkers tended to view scientific practice in terms of bounded and coherent spheres of social activity. However, each then goes on to analyse social wholes in dramatically different ways. As Bloor writes, Popper treated social wholes as 'unproblematically equivalent to sets of individual units.'[105] Kuhn, in contrast, rejected such individualism. In his work, '[s]ocial wholes are not treated as mere collections of individuals but are seen as having properties of a special kind, e.g. certain spirits, traditions, styles and national characteristics.'[106] Hence,

---

102 Karl Mannheim (1953 [1925]), 'Conservative Thought,' in *Essays on Sociology and Social Psychology*, by Karl Mannheim, ed. by Paul Kecskemeti (London: Routledge), pp. 74–164.

103 Bloor (1991), *Knowledge and Social Imagery*, p. 60.

104 Bloor (1991), *Knowledge and Social Imagery*, pp. 62, 63.

105 Bloor (1991), *Knowledge and Social Imagery*, p. 62.

106 Bloor (1991), *Knowledge and Social Imagery*, p. 63.

while for Popper 'individuality is not bound up in society,' for Kuhn '[i]ndividuals can only be understood in context.'[107] At root, this is a disagreement over how to specify the subjectivity of the subject, over what it is to be a human being.

Popper and Kuhn each approached scientific knowledge in terms of social wholes, but Popper treated those wholes as reducible to their individual components, while Kuhn treated them as irreducible to those individual components. At stake in this difference is the location of the principle or principles according to which a social whole is organised, according to which it is viewed as a structured unit rather than an unruly shambles. For Popper, organising principles are uniformly located in the individual parts making up the whole.[108] For Kuhn, they cannot be located in the individual parts but must instead be a property of the whole itself. According to Bloor, these divergent ontologies guide the respective epistemologies of their authors, with Popper favouring a view of scientific knowledge as the property of individuals and Kuhn a view of scientific knowledge as the property of groups. For present purposes, however, what is more important than this purported disagreement is the shared belief of these two thinkers in the existence of identifiable and coherent social wholes. In this sense, when Popper and Kuhn thought about scientific knowledge, their thoughts were being guided by the same kind of social image, an image of the social group as an organised whole. As a consequence, they both also tended to view scientific knowledge in terms of organised wholes, whether of concepts, practices, or a combination of both.

The implication of this is that the Enlightenment and Romantic ideologies which Bloor used to develop his comparative study are, at least in one important sense, not so different from one another. Indeed, they appear to share a common root in an image of the social group as an organised and bounded whole. This conclusion is supported by Mannheim's argument in the essay mentioned above, the essay on which Bloor modelled much of his discussion.

---

107 Bloor (1991), *Knowledge and Social Imagery*, pp. 62, 63.

108 I have elsewhere questioned Bloor's reading of Popper, arguing that Popper is much closer to being a communitarian epistemologist than is allowed for by Bloor (Jeff Kochan (2009b), 'Popper's Communitarianism,' in *Rethinking Popper*, ed. by Zuzana Parusniková and Robert S. Cohen (*Boston Studies in the Philosophy of Science* 272) (Berlin: Springer), pp. 287–303).

Mannheim addresses the same macro-sociological phenomenon as did Zilsel — the historical transition from feudalism to capitalism — but he picks it up at a later stage. Whereas Zilsel focussed on social changes during the fifteenth and sixteenth centuries, which he argues led to the Scientific Revolution in the early seventeenth century, Mannheim focusses instead on the emergence of conservative thought in the nineteenth century, presenting it as a response to the growing cultural hegemony of Enlightenment rationalism from the late eighteenth century onward. Like Zilsel, Mannheim analyses the usurpation of feudalism by capitalism as a process of rationalisation.[109] The process was 'revolutionary and radical just because it want[ed] to rationalize the whole social order right from the beginning in a systematic manner.'[110] Nineteenth-century conservative thought grew out of a widespread and deliberate resistance to this process. In the face of rationalisation, conservative thinkers made a deliberate effort to maintain and develop the remnants of a pre-capitalist experience which had once been taken for granted but was now threatened with extinction.[111] This effort led to the rise of the counter-revolutionary movement of Romanticism, which 'seized on the submerged ways of life and thought, snatched them from oblivion, consciously worked them out and developed them further, and finally set them up against the rationalist way of thought.'[112] Thus, what Mannheim dubs 'romantic conservatism' included two steps. First, it sought to rescue and revitalise the remnants of a disappearing pre-capitalist way of life. Second, it sought to actively articulate those remnants and develop them into an explicit and comprehensive project of intellectual resistance to Enlightenment rationalism. This second step was above all motivated by a rejection of the perceived rigidity and linearity of Enlightenment thinking, which Romantic thinkers endeavoured to replace with a more dynamic and dialectical form of thought.[113] Crucially, Mannheim argues that this second move reiterated, rather than abandoned, the faith of Enlightenment rationalism in the power of reason to understand the world: 'romantic thought

---

109 Mannheim (1953), 'Conservative Thought,' p. 85.
110 Mannheim (1953), 'Conservative Thought,' p. 149.
111 Mannheim (1953), 'Conservative Thought,' p. 115.
112 Mannheim (1953), 'Conservative Thought,' p. 89.
113 Mannheim (1953), 'Conservative Thought,' pp. 150–51.

(unintentionally perhaps) merely continues, though more radically, and with new methods, the same process which the Enlightenment had already hoped to complete — the thorough rationalization of the world.'[114] Mannheim uses the following analogy to illustrate the fundamental similarity, as well as the subsequent differences, between Enlightenment and Romantic thought:

> The conservative picture of things as a whole is like the inclusive sort of picture of a house which one might get by looking at it from all possible sides, a concrete picture of the house in all its detail from every angle. But the progressive is not interested in all this detail; he makes straight for the ground plan [*Grundriss*] of the house and his picture is suitable for rational analysis rather than for intuitive representation.[115]

The fundamental similarity between the two positions is their shared view that the world in its entirety can be captured in a 'picture of things as a whole,' a comprehensive image of everything that is. On the basis of this shared view, each position then employs a different method to rigorously work out the content of this world picture. The Enlightenment rationalist uses abstraction and systematisation to uncover the fixed 'ground plan' of the world. The Romantic thinker emphasises concrete description and the dynamic balance of competing forces in order to disclose the organic unity of the world. By each pursuing a distinct ideal of intellectual rigour, an ideal determined by standards internal to their own methods and experience, both rationalistic progressives and romantic conservatives aim to identify and define a comprehensive picture of the world construed as a whole.

Yet not every kind of conservativism depends on a conception of the world as a whole. In addition to Romantic conservativism, Mannheim identifies another form of nineteenth-century conservativism, which he dubs 'feudalistic conservatism.' This form followed only the first of the two steps characterising Romantic conservativism: that is, it consciously sought only to preserve and explicate elements of a pre-capitalist

---

114 Mannheim (1953), 'Conservative Thought,' p. 153.
115 Mannheim (1953), 'Conservative Thought,' p. 111. The bracketed text has been inserted from the German original (Karl Mannheim (1984 [1925]), *Konservatismus. Ein Beitrag zur Soziologie des Wissens*, ed. by David Kettler, Volker Meja and Nico Stehr (Frankfurt: Suhrkamp), p. 121). Note that *Grundriss* is the original term for both Mannheim's concept of 'ground plan' and Heidegger's concept of 'one basic blueprint,' under which, I have argued, the First Thing may be placed.

tradition which, by that time, existed only on the periphery of society.[116]
Feudalistic conservativism focusses its attention on the specific details
of concrete practice, making little effort to integrate those details into an
overall, coherent picture of the world. It resists the 'subsumption of the
individual and particular under one general principle' and so 'does not
really trouble itself with the *structure* of the world in which it lives.'[117]
Unlike Romanticism, feudal conservativism did not identify itself as a
counter-revolutionary response to Enlightenment rationalism. Hence,
feudalistic conservatives felt little compulsion to adopt, so as to then
renegotiate, the Enlightenment conception of the world as an object
which could be accessed, as a whole, through the power of reason. In
addition, their suspicion of theoretical abstraction, and their attention to
concrete practice, were 'exceedingly sober.'[118] They displayed no traces
of the romantic tendency to conceptualise the world as a dynamic unity
of competing, often non-rationalisable, life forces.

One upshot of Mannheim's argument is that, in addition to the
Enlightenment and Romantic imagery Bloor discusses in his comparison
of Popper and Kuhn, one can also add feudalistic conservativism. Unlike
the former two positions, feudalistic conservativism was not motivated
by a need to identify and elaborate a picture of the world as a whole.
In this, it proves closer to a pre-Enlightenment intellectual tradition,
one which concerned itself with the description and explanation of the
concrete details of experience without attempting to integrate those
details into one comprehensive and consistent whole. Hence, although
Mannheim was mainly concerned with the nineteenth century, when
Enlightenment rationalism had become well-developed and culturally
dominant, his comments on feudalistic conservativism point to an
earlier period, in the seventeenth century, when proto-Enlightenment
rationalism posed a still relatively weak threat to the dominant
feudalistic tradition.

These insights may now be applied to the dispute between Boyle
and Line in the early 1660s. I want to suggest that this dispute can be
understood as a conflict between an ascendant rationalism and a still-
dominant feudalistic traditionalism. More to the point, I am suggesting

---

116 Mannheim (1953), 'Conservative Thought,' p. 88.
117 Mannheim (1953), 'Conservative Thought,' pp. 155, 103.
118 Mannheim (1953), 'Conservative Thought,' p. 159.

that Boyle represents the early impulses of a rationalism which pursued a systematic model of the world as a whole, while Line represents the late tendencies of a tradition which was intensely engaged with concrete detail but had little interest in systematic analyses of a universalising scope. I want to make two arguments: first, that Bloor's concept of social imagery may be applied, in an illuminating but imperfect way, to this episode in the history of science; and second, that Heidegger's account of modern science explains why the application of Bloor's concept must be imperfect.

The application of the concept of social imagery to the dispute between Boyle and Line is imperfect because it is not symmetrical. If, as Bloor argues, a social image functions as a simplified model of the world construed as whole, and if Line was a traditionalist who had no use for such a model, then the concept of a social image will not be useful in examining his case. In contrast, if Boyle represents, at least in this instance, an emerging culture of rationalism, then the concept should apply in his case. Indeed, this asymmetry picks out one of the key issues at stake in the dispute. This was the issue of what role a simplified picture of things as a whole should play in the production of reliable experimental knowledge. Since, as Bloor argues, social imagery often guides an actor's thinking at a tacit level, the question of whether the world can be systematically treated as whole need not have been something Boyle and Line deliberately addressed in their dispute. The point is rather that we, as analysts instead of actors, may usefully illuminate the dispute in terms of this question.

In fact, the relevant analysis was already done in the last section. What I wish to pick out here with Bloor's concept of social imagery was there picked out with Heidegger's concept of the one basic blueprint. In short, Heidegger's basic blueprint is an instance of Bloor's social imagery. Both concepts are meant to highlight the exclusivity of a style of thinking, the fact that the image or blueprint serves as a comprehensive picture of things as a whole. I argued in the last section that Boyle's thinking was guided by such an image, an *a priori* image in which the thingness of the thing was specified beforehand according to the Galilean First Thing. If this image was furthermore a social image, in Bloor's sense, then it should follow that Boyle, when thinking about knowledge, was manipulating images of society. Indeed, this was one of the main points of Shapin and Schaffer's study. They argued that Boyle's epistemological concerns were, at root, concerns about social order, especially, the problem of disorder caused by political controversy. To this, I added the suggestion that the image of the First Thing served Boyle as a model by which to construct a clearly defined community of experimental philosophers — a social whole — in which the dangers of controversy could be effectively managed.

As we have seen, Boyle, in his dispute with Line, defended the sufficiency of his own spring hypothesis against the challenge posed by Line's thread hypothesis. This defence appears to have been motivated by Boyle's assumption that the phenomena of the Torricellian experiment could be subsumed under one general explanatory hypothesis. This epistemological assumption reflected Boyle's commitment to a simplified model of society in which members took for granted a consistent base of clearly defined principles and matters of fact. Line, in contrast, appears to have taken a more eclectic view, allowing instead that both his and Boyle's hypotheses were necessary for an explanation of the Torricellian phenomena. Underpinning Line's assumption was, it appears, not a commitment to a competing model of the social whole, but rather a complacent disregard for the necessity of any such model. It seems plausible that Line's epistemology, such as it was, reflected the prevailing feudal tradition — with its indifference towards simplified models of society — which was only just then beginning to face the challenge of an ascendant, rationalistic conception of the world. Hence, the dispute was not a conflict between two competing images of the social whole, but between Boyle's perceived need for such an image, on the one hand, and Line's apparent indifference to that need, on the other.

However, it was not the case that every follower of the feudal tradition was indifferent to the threat posed by the new rationalism. In fact, at stake in this period were conceptions not just of society and nature, but also of religion. This should not be surprising, as it is a commonplace among contemporary historians of science that conceptions of society, nature, and religion were, in Boyle's period, not easily separated. As the historian Quentin Skinner has observed, Boyle insisted that his experimental activity supported his religious faith because it provided evidence for the 'design' of the world. By demonstrating that nature as a whole was created on the basis of a rational plan, Boyle hoped to provide empirical grounds for the existence of a divine creator. '*L'horloge, donc l'horloger,*' notes Skinner, and then immediately adds: 'During the seventeenth century, however, this familiar trope of the Enlightenment was still widely believed to carry alarming religious consequences.'[119]

One such expression of alarm came from the Anglican cleric and humanist Méric Casaubon. In a letter 'Concerning Natural experimental Philosophie, and some books lately about it,' published in 1669, Casaubon criticised members of the Royal Society for claiming that religious controversies could be settled on the basis of 'plain reason.'[120] The nub of this worry was Casaubon's resistance to the idea that God's creation could be understood in accordance with a basic blueprint, accessible to anyone possessing natural reason. As Michael Spiller remarks, Casaubon viewed this ambition as deluded and dangerous, because 'reality is too vastly complex, and human beings too corrupt, irremediably frustrated by recurring and permanent vices and frailties.'[121] Any attempt to know the world as a whole was bound to fail, as nature outstrips finite human understanding, making controversy inevitable.

In this dispute over the nature of knowledge, the conflict was not about what picture of the world to adopt, as it would be in the later feud between Enlightenment rationalists and Romantics, but instead about whether or not a world picture should be adopted at all as an instrument in the acquisition of knowledge. As Spiller notes, Boyle was, at best, an uneasy accomplice in the development of early-modern rationalism.[122] Yet, in his dispute with Line, Boyle appears to have been strongly motivated by a simple and consistent conception of nature structured in terms of the Galilean First Thing. Line, for his part, appears to have been

119 Quentin Skinner (2002), 'Hobbes and the Politics of the Early Royal Society,' in *Visions of Politics*, vol. 3: *Hobbes and Civil Science*, by Quentin Skinner (Cambridge: Cambridge University Press), pp. 324–45 (p. 336).

120 Reprinted in Michael R. G. Spiller (1980), *'Concerning Natural Experimental Philosophie': Meric Casaubon and the Royal Society* (The Hague: Martinus Nijhoff), pp. 149–189. According to Richard Serjeantson, Casaubon 'had a strong fascination with the natural world: as a young man he attempted some of the experiments in Francis Bacon's *Sylva sylvarum*' (Richard W. Serjeantson (2004), 'Casaubon, Méric,' in *Oxford Dictionary of National Biography*, vol. 10, ed. by H. C. G. Matthews and Brian Harrison (Oxford: Oxford University Press), pp. 464–66 (p. 466)).

121 Spiller (1980), 'Concerning Natural Experimental Philosophie,' p. 130.

122 Spiller (1980), 'Concerning Natural Experimental Philosophie,' p. 136.

guided by no one picture at all, but instead by an eclectic willingness to blend together a variety of doctrines. Put in Heidegger's terms, only Boyle comported himself in a mathematical manner. In his dispute with Line, Boyle's thinking and conduct exemplified the emerging mathematical projection which marked the historical and existential transition from a medieval to an early-modern understanding of nature.

Heidegger argues that '[t]he world picture does not change from an earlier medieval one into a modern one, but rather the fact that the world becomes picture at all is what distinguishes the essence of the modern age.'[123] For Heidegger, the modern age, the age of modern science, is, at base, the age of the world picture. He thus presents a theory of scientific knowledge which applies only to the modern age. According to this theory, modern science is possible only on the basis of a projected picture of the world as a whole. This projection is the *a priori* condition of possibility for such knowledge. The subsequent pursuit of modern scientific knowledge is conceived as the filling out, the rigorous articulation and refinement, of a picture of the world which has already been projected, albeit only generally and confusedly, in advance.[124]

Heidegger's argument that thinking in terms of a world picture is a specifically modern development provides an opportunity to refine Bloor's sociological claim that we think about knowledge by manipulating images of society. Bloor writes that this is 'a theory about how people think. The hypotheses are not alleged to be necessary truths. [...] Furthermore the range of application of the picture here presented has yet to be determined.'[125] On the basis of Heidegger's discussion of the world picture, as well as the above analysis of the dispute between Boyle and Line, we may now tentatively limit the

---

123 Heidegger (1977a), 'The Age of the World Picture,' p. 130.

124 Hans-Jörg Rheinberger suggests that Heidegger's term *Weltbild* ('world picture') 'might be more appropriately translated as "[...] Planetary Configuration"' (Hans-Jörg Rheinberger (1997), *Towards a History of Epistemic Things: Synthesizing Proteins in the Test Tube* (Stanford: University of Stanford Press), p. 25 n. 6). This has the benefit of emphasising the material aspect of the phenomenon, but it also seems to deflect attention from the projective role played by the subject in creating that phenomenon, not to mention the intersubjective (social) labour required to materially extend it to, and sustain it on, a planetary scale. See also Hans-Jörg Rheinberger (2010a), *An Epistemology of the Concrete: Twentieth-Century Histories of Life* (Durham: Duke University Press), p. 234.

125 Bloor (1991), *Knowledge and Social Imagery*, p. 154.

range of application of Bloor's theory to the modern era. SSK's method of analysing scientific controversies in terms of conflicting, often tacitly held, pictures of society seems most applicable to modern controversies. These conflicting pictures are images of society construed as a whole, images which, in turn, motivate conceptions of knowledge construed as a whole. As human relations come, in the modern era, to be viewed in terms of social wholes, human knowing likewise comes to be viewed in terms of epistemic wholes. Hence, the modern problem of organising knowledge into a structured whole can be understood as the modern problem of organising society into a structured whole. Whether actors view knowledge as a system of concepts — organised around basic principles of logic and definition — or as a system of practices — organised around basic principles of prediction and control, or credibility and power — in both cases, the SSK practitioner will seek to uncover the underlying social image held by the actors, along with the basic principles around which that image is organised. However, in cases where groups of actors genuinely treat knowledge in a non-systematic and eclectic way, the implication of Heidegger's argument is that there will be no underlying image of a social whole which the SSK practitioner may uncover and then use to explain the actors' epistemic activities. The sociologist may, in these latter cases, discover shared social interests, and use these to explain behaviour, but, if Heidegger's argument is correct, these interests will not be the components of a single, integrated social image, a shared ideology or blueprint determined by a single coherent set of basic principles.

6. Conclusion

Some readers may feel that there is an elephant in the room which I have ignored throughout this chapter. This alleged elephant is the mechanical philosophy of the early-modern period. However, while far from being irrelevant, the mechanical philosophy is, in fact, largely peripheral to the main concerns of this chapter. There is, then, no elephant in the room. So why not talk about it?

Mechanism is usually contrasted with organicism, with the attendant account of the Scientific Revolution then emphasising a transition in the prevailing image of the world from an organic to a mechanical one.

What should be immediately clear, however, is that this difference in philosophical orientation — organistic versus mechanistic — is underpinned by a shared impulse to grasp the world as a single, structured whole. It is this shared impulse, rather than the difference between these two philosophies, which has been a central concern of this chapter. On this basis, we may reasonably doubt, for example, Carolyn Merchant's claim that '[t]he rejection and removal of organic and animistic features and the substitution of mechanically describable components would become the most significant and far-reaching effect of the Scientific Revolution.'[126] On the contrary, according to the argument presented here, the most far-reaching and significant effect of the Scientific Revolution was the institutionalisation of a concept of the world as a thing which is ordered — *organised* — according to a uniform measure, a basic blueprint. Both models of the world as a machine and as an organism reflect this institutionalisation, because both machines and organisms are, so we tend to think, discrete organised units, structured individuals or wholes.[127]

Merchant describes the Scientific Revolution as 'the transition from the organism to the machine as the dominant metaphor binding together the cosmos, society, and the self into a single cultural reality — a world view.'[128] She furthermore argues that the ancient notion of '[t]he female earth was central to the organic cosmology that was undermined by the Scientific Revolution.'[129] This may give the impression that the organic cosmology was itself an ancient institution, supplanted by a mechanistic cosmology in the early-modern period. However, Merchant also argues that the doctrine of the world's 'organic unity' emerged in the late Renaissance, citing as evidence works published only in the second half of the sixteenth century, or later.[130] Hence, it would appear that an

---

126 Carolyn Merchant (1980), *The Death of Nature: Women, Ecology and the Scientific Revolution* (San Francisco: Harper & Row), p. 125.

127 In 1940, Heidegger claimed that 'the idea of "organism" and of the "organic" is a purely modern, mechanistic-technological concept' (Martin Heidegger (1976 [1967]), 'On the Being and Conception of φύσις in Aristotle's Physics B, 1,' *Man and World* 9(3), 219–70 (p. 234)). This claim is stronger, and more contentious, than is required for the present argument.

128 Merchant (1980), *The Death of Nature*, p. xxii.

129 Merchant (1980), *The Death of Nature*, p. xx.

130 Merchant (1980), *The Death of Nature*, pp. 103–126. Works cited include: Giovanni Battista della Porta's *Magiae naturalis* (published 1558); Bernardino Telesio's *De*

organic world picture may not have been a pre-revolutionary institution after all, but rather a revolutionary challenger subsequently replaced by its equally revolutionary rival, the mechanical world picture. The ancient notion of a female earth may well have been replaced along with its cosmological counterpart, but it does not follow from this that this ancient notion represents a cosmology, an image of the world conceived as a whole.

On these admittedly fragile grounds, I tentatively suggest that the conflict between organic and mechanical philosophies was not a cause, but rather a consequence, of the rise of early-modern science. This was a struggle between two emergent conceptions of the world as a whole. It thus presupposed that the world can be so conceived, that is, as a whole. The institutionalisation of this presupposition was, I suggest, one of the most significant and far-reaching effects of early-modern scientific culture. It entailed not just that things be conceived in terms of a uniform measure, but that the world itself also be so conceived, as a thing disciplined by a single measure, a uniform image, a basic blueprint.

We may thus question Elizabeth Potter's attempt to assimilate the dispute between Boyle and Line to the broader ideological dispute between mechanical and organic cosmologies. Potter attempts to slot Line's thread hypothesis into an organic philosophy, calling it a 'non-mechanistic assumption' which presupposes the idea of '[a] World Spirit or Nature having consciousness and feelings.'[131] The implication is that Line conceived of the world as possessing consciousness and feeling, that is, as being a living, sentient organism.

Potter's argument follows Boyle's own by forcing a necessary connection between Line's thread hypothesis and his affirmation of the metaphysical doctrine of *horror vacui*. Yet, as I have argued, this connection may also be viewed as contingent. Indeed, Boyle insisted on the contingency of the relation between his own spring hypothesis and any further metaphysical doctrine. So why deny the same treatment to

---

rerum natura iuxta propia principia (published 1565); Giordano Bruno's *Spaccio de la bestia trionfante* (published 1584); Tommaso Campanella's *De sensu rerum et magia* (published 1620). Note that all of these works appeared after Nicolaus Copernicus's *De revolutionibus orbium coelestium* (published 1543), with which the beginning of the Scientific Revolution is often credited.

131　Potter (2001), *Gender and Boyle's Law of Gases*, p. 160.

Line? Furthermore, it is worth repeating that Line was obliged, as a Jesuit, to affirm the *horror vacui*. What is more, one may doubt whether this doctrine, when affirmed by a Jesuit, was meant to slot into an organic cosmology which construed the world as a living, conscious being. This construal is much stronger than the more modest claim that nature acts to prevent a vacuum. In order to attribute this act to nature, one need not ascribe to it consciousness and feeling. Like human activity, the activity of nature may be non-deliberative — as when an acorn grows into an oak tree. When Boyle attributed an outward endeavour to corpuscles of air, he presumably had this non-deliberative notion of activity in mind. Line's thread hypothesis need not be treated any differently.

It is worth noting that Line was a skilled mechanician, with 'a reputation for ingenuity in the construction of timepieces,' especially sundials.[132] Indeed, in his *Brief Lives*, compiled in the late seventeenth century, John Aubrey described Line as having constructed 'the finest Dialls in the World,' and also as possessing 'great skill in the Optiques.'[133] On the basis of this reputation, Line was invited by Charles II to design and build a sophisticated sundial, ultimately comprised of 250 individual units, which was installed at Whitehall in 1669.[134] Hence, although Line was evidently no mechanical philosopher, he was certainly not unfamiliar with mechanical ways of thinking and working. Moreover, Line's thread hypothesis might even be viewed in mechanical terms, since strings, lines, and threads seem, on a basic level, no less mechanical than springs. For example, Walter Charleton, elected to the Royal Society in 1663, gave substantial attention to threads in his mechanical studies of attraction and adhesion. With attraction, he asked: 'Why therefore should we not conceive, that in every Curious and Insensible Attraction of one bodie to another, Nature makes use of

---

132 Peter Davidson (2009), 'Francis Line S. J., *Explicatio horologii* (1673),' in *Jesuit Books in the Low Countries (1540–1773): A Selection from the Maurits Sabbe Library*, ed. by Paul Begheyn, SJ., Bernard Deprez, Rob Faesen, SJ, and Leo Kenis (Leuven: Uitgeverij Peeters), pp. 187–90 (p. 187).

133 John Aubrey (2014), *Brief Lives, with an Apparatus for the Lives of Our English Mathematical Writers*, vol. 1, ed. by Kate Bennett (Oxford: Oxford University Press), p. 151.

134 Davidson (2009), 'Francis Line S. J., *Explicatio horologii* (1673),' p. 189. Line published a detailed description, with illustrations, of the Whitehall sundial in 1673, in both Latin and English. Davidson's entry includes reproductions of the Latin title page and one of Line's illustrations.

certain slender Hooks, Lines, Chains, or the like intercedent Instruments, continued from the Attrahent to the Attracted [...]?' With adhesion, he observed that when amber 'is no sooner Warmed by rubbing, but there rise out of it certain small *Lines* or *Threads*, which adhere to a mans finger that toucheth it, and such as may, by gentle abduction of the finger, be prolonged to considerable distance.'[135] It is not surprising, then, that Boyle sometimes treated Line's thread hypothesis as being inadequately mechanical rather than as not being mechanical at all. For example, he argued that Line had not adequately explained how his thread attaches itself to the surfaces inside the Torricellian space: 'For I farther demand, how the Funiculus [i.e. thread] comes by such hooks or grapple-irons, or parts of the like shape, to take fast hold of all contiguous bodies, and even the smoothest, such as glass, and the calm surface of quicksilver.'[136] Hence, Boyle does not appear to have judged Line's thread hypothesis as necessarily non-mechanical. In fact, there is compelling reason to think that Line was quite comfortable with mechanical ways of thinking, and very little reason to think that he was committed to an organic cosmology.

More generally, I have argued that the central dispute between Boyle and Line cannot be reduced to a contest between distinct world pictures or ideologies. Instead, this dispute focussed on two different ways of conceptualising the directedness of natural processes, particularly in the context of the Torricellian experiment: either inward or outward. However, even this difference was not at the core of the dispute. At its centre, the dispute had instead to do with exclusivity versus inclusivity vis-à-vis the question of directedness. Boyle favoured an exclusive explanatory role for his spring hypothesis. Line, in contrast, sought only to win a legitimate place for his hypothesis alongside Boyle's own. He appears to have followed a piecemeal and pluralistic approach, incompatible with exclusivity, whereas Boyle was guided by an exclusive conception, a basic blueprint and uniform measure, of the thingness of things — what I have called the Galilean First Thing. As a

---

135 Walter Charleton (1654 [1966]), *Physiologia Epicuro-Gassendo-Charltoniana: Or a Fabrick of Science Natural, Upon the Hypothesis of Atoms* (New York: Johnson Reproductions), pp. 344, 345.

136 Boyle (1662), *A Defence of the Doctrine Touching the Spring and Weight of the Air*, p. 142.

consequence, Boyle was, on my argument, a mathematical philosopher, while Line (a professor of mathematics in Liège) was not.

This argument entails a significant modification, but not a rejection, of Kuhn's distinction between early-modern mathematical and experimental traditions. While still describing distinct arenas of epistemic practice, these traditions should be viewed as the sub-traditions of an emerging, broader early-modern tradition wherein things came increasingly to be experienced according to a single, uniform measure. This homogenisation of thing-experience was marked by a consolidation of the final causes of things under this one basic measure. As a consequence, the proper place of the scientific thing — the end towards which it naturally moves in becoming what it is — was rendered qualitatively uniform. Early-modern experimental philosophers did not, in practice, reject final causes in their attempts to understand the things of nature. Instead, what made their manner of working new was its dependency on a basic blueprint, a uniform projection of the thingness of things. According to Heidegger, the existential growth of this dependency was a defining feature of the mathematisation of early-modern science. This was a process whereby experimental apparatus could now be reliably used to release the scientific thing from the external constraints which prevented it from realising its own indigenous end. In other words, the experiment was designed to let the thing be what it is. Experimental art supported and supplemented nature by helping the thing approach its proper place within a mathematically projected cosmos.

Chapter Seven

Conclusion: Subjects, Systems, and Other Unfinished Business

Heidegger delivered his lectures on Galileo, Newton, and *mathēsis* during the winter semester of 1935–1936. In the autumn of 1935, just before, or just as, this lecture course was getting started, Heidegger also hosted, over the course of several days, a discussion by two distinguished German scientists: the physiologist and physician Viktor von Weizsäcker, and the physicist Werner Heisenberg. The discussion took place in Heidegger's small cabin, perched on the slope of a meadow above the Black Forest village of Todtnauberg. Also present, among others, was Carl Friedrich von Weizsäcker, a nephew of the older von Weizsäcker and a student of both Heisenberg and Niels Bohr. The younger von Weizsäcker would later become a distinguished nuclear physicist in his own right. The meeting in Todtnauberg marked the beginning of the younger von Weizsäcker's long friendship with Heidegger, one which lasted until Heidegger's death in 1976.

According to von Weizsäcker's later recollection, the participants of the meeting crowded around the narrow table in Heidegger's tiny living room, with Heidegger seated at the table head, Viktor von Weizsäcker and Heisenberg on either side of him, and Carl von Weizsäcker at

    http://dx.doi.org/10.11647/OBP.0129.07

the foot, facing Heidegger.[1] The two scientists spoke at length with one another, 'each interested in the other, but, in fact, separated by a chasm.' Heisenberg was deeply preoccupied with the 'abyss opened by the unsought after discovery of the inseparability of subject and object in quantum theory,' but he felt one could not live in this abyss. He insisted on 'a precise explanation of the role of the subject as observer, as experimenter.' Viktor von Weizsäcker, in contrast, wanted just as badly to see the subject as a 'living object': 'He did not view the objective observer of quantum theory as a subject at all.' When the two scientists had anchored themselves deep in mutual misunderstanding, Heidegger intervened.

> 'You, Mr. von Weizsäcker, seem to mean the following.' Three crystal clear sentences. 'Yes, that's exactly what I wanted to say.' 'And you, Mr. Heisenberg, probably mean this.' Three sentences of the same kind, and the answer: 'Yes, that's how I picture it.' 'Then the relation could be the following': Four or five short sentences. Both agreed: 'It could be like that.' The dialogue continued until Heidegger had to help them out of the next impasse.

A year earlier, in 1934, when the young von Weizsäcker was working with Bohr in Copenhagen, he bought a copy of *Being and Time* and read it alongside Kant's work. Although he had trouble understanding the writings of both, he felt Heidegger came closer than any other philosopher to the 'unsolved problems' he faced in his own work.[2] Later, in a 1949 essay on the 'Relations of Theoretical Physics to Heidegger's Thinking,' von Weizsäcker observed that 'the substantive separation of the *res cogitans* and *res extensa* is the methodological presupposition of the classical natural sciences in their entirety.' The methodological purity of these classical sciences forbids the mixing of the two. Furthermore, the ideal of classical physics had been to speak only of one of the two Cartesian substances: 'It should speak of the object as if there were no subject whose object it is or can be.' With the new physics, in contrast, a physicist who asks what the atom *is* 'has every reason not to repeat Descartes's error' by determining this 'is' in terms of the properties of

---

1   Carl Friedrich von Weizsäcker (1977a), 'Begegnungen in vier Jahrzehnten,' in *Erinnerung an Martin Heidegger*, ed. by Günther Neske (Pfullingen: Verlag Neske), pp. 239–47 (pp. 239–40); my translations.

2   von Weizsäcker (1977a), 'Begegnungen in vier Jahrzehnten,' pp. 240–41.

a discrete substance, but instead 'to follow Heidegger in questioning what "being" itself means.'[3]

Despite Heidegger's impartial adjudication of the debate between the older von Weizsäcker and Heisenberg, it is clear that the younger von Weizsäcker viewed Heidegger's philosophy as directly (though not entirely) supportive of the new physics, and as directly critical of the classical understanding exemplified by his uncle Viktor. Indeed, as we saw in Chapter One, Heidegger traced the original impulse of modern treatments of the subject as a 'living object' not just back to Descartes, but still further back to Aristotle's definition of the human being as *zōon logon echon*, later interpreted to mean 'animal rationale,' a living thing capable of reason. In Chapter Four, we then saw how, according to Heidegger, this ancient view was transformed (among other things, Christianised) within the framework of Descartes's substance ontology. The living thing becomes a discrete substance, an object with the properties of life and reason. By insisting that the subject be understood reductively as a living object, Viktor von Weizsäcker appears to have been reproducing the modern entanglement of Cartesian ontology and the methods of classical science. The Cartesian subject continues to circulate beneath the elder von Weizsäcker's commitment to an object ontology, asserting its presence through those deliberate methodological acts which, paradoxically, seek to erase it from the realm of legitimate scientific discourse. At least, this is how I imagine it would have looked from the perspectives of Heidegger, Heisenberg, and Carl von Weizsäcker. On the other hand, Heisenberg's insistence on the need for a precise explanation of the scientific subject would have presumably looked to the older von Weizsäcker like a regressive move threatening the basic methodological presuppositions of established scientific method, and hence the prevailing norms of objective knowledge production.[4]

---

3   Carl Friedrich von Weizsäcker (1990 [1949]), 'Beziehungen der theoretischen Physik zum Denken Heideggers,' in *Zum Weltbild der Physik*, 13th edn, by Carl Friedrich von Weizsäcker (Stuttgart: S. Hirzel Verlag), pp. 243–45 (pp. 244–45); my translations.

4   On the relationship between Heidegger and Heisenberg, see: Cathryn Carson (2010), 'Science as Instrumental Reason: Heidegger, Habermas, Heisenberg,' *Continental Philosophy Review* 42(4), 483–509; Bernard Freydberg (2002), 'What Becomes of Science in "The Future of Phenomenology"?,' *Research in Phenomenology* 32(1), 219–29; Trish Glazebrook (2000a), *Heidegger's Philosophy of Science* (New York: Fordham University Press), pp. 247–51; Werner Heisenberg (1959), 'Grundlegende

One central goal of this book has been to provide a more precise account of the subjectivity of the scientific subject — in a nutshell, to elaborate an account of science as social existence. This has involved the combination of SSK with Heidegger's existential conception of science. This combination is based, among other things, on the phenomenological deconstruction of the Cartesian subject-object distinction in its entirety, rather than on the reductive collapse of the subject-side of the distinction into its object-side.

Yet, from a certain perspective, this method may appear lamentably misguided, as it insists on a necessary role for the subject. The first three chapters of this book considered a number of criticisms which appear to have been motivated by this perspective. Looking back, I would like to suggest (not without mischief) that these criticisms may be labelled as either conservative or liberal. Chapter One briefly addressed the conservative criticism, and Chapters Two and Three addressed variations of the liberal criticism. On my reading, the conservative feels threatened by the thematisation of the subject because she believes that the prevailing methods of science prohibit a place for subjects. Instead, she emphasises the strictly object-based nature of knowledge: the subject must be suppressed in order to conserve the propriety of established scientific norms. The liberal, in turn, feels threatened by attention to the subject because she views this as a sly attempt to reassert the authority of the subject-object distinction, and thus to re-impose an unwelcome and atavistic constraint on the recently won autonomy of the post-Cartesian scientific imagination. Despite their differences, both criticisms mistakenly believe that a methodological orientation to the subject must be underpinned by an essentialist commitment to Cartesian substance ontology.

The conservative critics mentioned at the beginning of Chapter One accuse SSK practitioners of handling their alleged Cartesian

Voraussetzungen in der Physik der Elementarteilchen,' in *Martin Heidegger zum Siebzigsten Geburtstag. Festschrift*, ed. by Günther Neske (Pfullingen: Verlag Günther Neske), pp. 291–97; Werner Heisenberg (1977), 'Brief an Martin Heidegger zum 80. Geburtstag,' in *Dem Andenken Martin Heideggers. Zum 26. Mai 1976* (Frankfurt: Vittorio Klostermann), pp. 44–45; Hans-Peter Hempel (1990), *Natur und Geschichte. Der Jahrhundertdialog zwischen Heidegger und Heisenberg* (Frankfurt: Hain); and Hans Seigfried (1990), 'Autonomy and Quantum Physics: Nietzsche, Heidegger, and Heisenberg,' *Philosophy of Science* 57(4), 619–30.

commitment with incompetence. For these critics, the object-based method is meant to capture the independently existing object without interference from the subject. SSK practitioners argue that this is not possible, because knowledge of the object necessarily implicates the knowing subject. The critics interpret this as scepticism about the existence of an independent object of knowledge, or, put otherwise, scepticism about the fundamental separability of object from subject. This is more generally framed as scepticism about the existence of the external world. This scepticism, so the critics say, results from SSK's incompetent failure to protect the object-side from corruption by the subject-side of the Cartesian ontological schema. In other words, it results from an incompetent failure to follow the established methods of classical science.

I have argued that this criticism misjudges SSK in both motive and method. SSK practitioners deliberately seek to follow the methodological conventions of science by affirming the existence of an external world. At the same time, however, they wish to radically reform the theory of knowledge arising from the Cartesian schema. This attempted reform, I have suggested, has not been radical enough, and this has led to significant tensions in SSK's attitude towards external-world scepticism, tensions which have been effectively exploited by critics, conservative and liberal alike. I argued that SSK can resolve these tensions by adopting Heidegger's phenomenological deconstruction of the Cartesian schema, a move which explains the subject-object distinction in terms of a more basic subject-world relation. Now the subject is no longer seen as a social substance gaining access to an external world, but as an entity whose basic modes of existence include being-in-the-world and being-with-others. With this move, an affirmation of the existence of an external world becomes meaningless, because the subject is not something to which a world can be external. This move may be viewed as a preliminary answer to Heisenberg's call for a more precise explanation of the scientific subject, the experimenter, following the collapse of the orthodox subject-object distinction in the early twentieth century.

The liberal critic of Chapter Two was Joseph Rouse, who accuses both Heidegger and SSK practitioners of espousing a theory-dominant account of science. I challenged this criticism in both cases, while acknowledging that, in the case of SSK, Rouse makes some good points. Indeed, he exploits the tensions just mentioned in order to charge SSK with a 'vestigial commitment to epistemology.' By this he means a commitment to the Cartesian schema. According to Rouse, SSK practitioners seek to separate social and natural phenomena into autonomous domains, and they must then face the sceptical problem of how the constructive social domain achieves access to the natural domain in the production of knowledge. Rouse reads this as an atavistic recapitulation of the subject-object distinction, where the subject is a relatively rigid social substance and nature the discrete object of its epistemic attentions. Yet, without claiming that SSK practitioners have slipped entirely free from this schema, I argued that Rouse dramatically suppresses the not inconsiderable ways in which they have recognised the instability and dynamism of the social realm.

Rouse's argument that Heidegger, too, espouses a theory-dominant account of science is based on his claim that Heidegger's mathematical projection of nature is a theoretical phenomenon. In particular, Heidegger's emphasis on this allegedly theoretical projection purportedly leads him to dismiss experimental practice as merely incidental to scientific activity. In response, I have argued, over several chapters, that Heidegger's mathematical projection is the existential condition which makes modern scientific activity — both theoretical and experimental — possible. In his alternative proposal, Rouse turns this explanatory model on its head, arguing instead that meaning-constitutive scientific practices are, in fact, the condition of possibility for existence. The existence of subjects and objects is to be explained by practices, and not the other way around. Hence, the explanatory role of Heidegger's mathematical projection — as a historical mode of scientific subjectivity — is neutralised. In his proposed alternative to SSK, Rouse likewise seeks to dissolve the distinction between social subject and natural object in an ever-circulating field of heterogeneous scientific practices. Here, too, the explanatory role of the social subject has been neutralised.

In my view, such attempts to erase — rather than to deconstruct and transform — the traditional distinctions of subject and object, society and nature, mind and body, and theory and practice serve a particular conception of freedom, one which regards such distinctions as placing an unnecessary and unwelcome constraint on the free play of intellectual analysis. Underpinning this conception is a particular idea of the human being. This is an idea which credits the human being with an intrinsic ability to throw off the cognitive constraints of its social and historical circumstances, so as to achieve autonomy. It is, in other words, an idea which directly challenges the finitude of human reason and imagination.

This brings us to the liberal critic of Chapter Three, Bruno Latour. My analysis focussed narrowly on Latour's debate, in the 1990s, with David Bloor. In inviting his readers to take 'one more turn after the social turn,' Latour argued that the symmetry principle, a central tenet of SSK, uncritically reproduces the subject-object schema as formulated by Immanuel Kant. According to Latour, SSK is grounded in a distinction between a 'subject pole' and an 'object pole,' with all of SSK's explanatory resources being gathered around the subject pole, and none at all being drawn to the object pole. He declares this asymmetrical, and proposes that all explanatory resources should be removed from the subject pole as well, leaving both the subject and the object equally useless in explanations of scientific knowledge production. Under this new, and allegedly more thoroughgoing, symmetry, the subject and object, society and nature, themselves become topics requiring explanation. As with Rouse, they are to be explained in terms of fields of circulating scientific practice. Thus the explanatory significance of the social subject has once again been neutralised.

I argued that Latour, like Rouse, exploits genuine tensions in the SSK literature, but likewise exaggerates their significance, consequently suppressing the extent to which SSK practitioners have sought to radically reformulate the orthodox subject-object distinction. In my view, to acknowledge such efforts would have compromised Latour's main goal: to promote an unrestrained constructivity, stripped of its social and historical constraints, as the fundamental explanatory resource for studies of science.

I suggested that we can better understand what is at stake in the debate between Bloor and Latour by considering Heidegger's own analysis of the Kantian version of the subject-object distinction. Kant construes the object as a thing-in-itself. On Heidegger's reading, the thing-in-itself marks the extreme limit of human understanding. It is that aspect of nature which continually outstrips our correspondingly finite ability to make sense of it, to draw it into the snare of our constructive power. The thing-in-itself thus represents independently existing nature, and so figures as a basic presupposition in the position I have called 'minimal realism.' Accordingly, I argued that Latour's attack on the thing-in-itself amounts to a rejection of both minimal realism and the finitude of human understanding. As a consequence, Latour may be viewed as embracing a concept of unrestrained constructivity, as well as a position which I have dubbed 'pragmatic idealism.'

My main goal in these concluding reflections is not to relaunch my earlier critique of the positive doctrines of these liberal critics, but instead to reemphasise the inadequacy of their alleged refutations of SSK and Heidegger's existential conception of science. They have not succeeded in refuting these positions because they have not properly understood them. I will, nevertheless, offer the opinion that these critics have also failed in their positive attempt to radically liberate themselves from the subject-object distinction. Rather than superseding the informal, existential root of this distinction (that is, our feeling of directedness towards independently existing things), they have instead succeeded only in suppressing it as a topic of analysis, and hence as a potential basis for understanding science. As a result, these critics seem to grant free licence to the subject's activities while simultaneously withholding the conceptual tools by which to properly analyse and explain those activities. This unfortunate circumstance may be viewed as a form of — perhaps unwitting — intellectual dissimulation: the concealment of a concept which, in fact, plays a central role in the position being promoted. This concealed concept is, of course, the concept of the subject.

In contrast to this, Chapters Four, Five, and Six investigated the scientific subject as one side of a phenomenologically grounded distinction between thinking and things. One central consequence of this investigation has been to improve our understanding of the concept of scientific practice. The liberal dissimulation sinks the subject beneath the surface of a smoothly functioning system of practices, thereby masking the subject's crucial role in the constitution of those practices. The reintroduction of the subject allows us to topicalise practices in a way discouraged by those who, paradoxically, promote practice as one of their primary explanatory resources. By topicalising practices, we can develop a better understanding of what practices are, and this, in turn, will enable us to better control and apply the concept of practice as a central explanatory resource.

These considerations now allow us to segue back to Carl von Weizsäcker's reflections on Heidegger and post-classical physics. Von Weizsäcker felt that Heidegger's ruminations on the question of being could guide him in his attempt to overcome the classical Cartesian image of the atom as a property-bearing substance. Heisenberg, in kind, was searching for a new post-Cartesian understanding of the experimenter-subject's role vis-à-vis the atom and other experimentally detectable objects. As we have seen, Heidegger's question of being may be treated as a question about the meaning of the word 'being.' He argues that 'being' is a polysemous word, with connotations of both existence and essence, that-being and what-being. Hence, to ask about the being of the atom is to ask both whether it is and what it is.

I have argued that to assert the independent existence of a thing to which an experimental term refers is to be a minimal realist about that thing; it is to assert *that* the thing is, that it exists. This assertion formally articulates, in propositional terms, our experience of that thing's independent existence, an experience grounded in our receptivity towards, our ability to be affected by, that thing. On the basis of this experience, we know *that* the thing is, but not *what* it is; we can state whether it is, but not what it is. In contrast, being able to say what the thing is depends on our being able to experience it in its whatness as well as its thatness. This experience is grounded in our projective understanding of the thing, our ability to make sense of it, to construct it as meaningful. This experience, too, may be formally articulated in propositional statements about what the thing is, about its essence. Such acts of articulation may, however, also be non-linguistic, as, for example, in the case of a laboratory set-up, constructed to let the thing be what it is for the disciplined experience of an attentive observer. That the experimenter's experience is disciplined points to the sociological — in addition to the psychological, biological, and technological — conditions which give shape to that experience. For this reason, topicalising the scientific subject invites the application of psychological, biological, technological, and sociological explanatory resources.

We can now see how von Weizsäcker's question about the being of the atom and Heisenberg's question about the role of the experimenting subject come together. The that-being and what-being of the atom correlate, respectively, to the receptivity and constructivity of the scientific subject. Hence, in order to understand the being of the atom, we must also investigate the structure of scientific experience. This is why Heidegger argued that the question of being points necessarily towards the existential structure of the subject (Dasein), towards a phenomenological analysis of the subjectivity of the subject. As we have seen, Heidegger's analysis picks out at least four basic, interrelated characteristics of that subjectivity: (1) being-in-the-world; (2) being-with-others; (3) understanding; and (4) affectivity.

In Chapter One, the first characteristic was used to help SSK practitioners in their realist response to the external-world sceptic. According to Heidegger, being-in-the-world is a basic state of subjectivity when experienced in more immediate non-Cartesian and non-Kantian terms. But it is not the only basic state. The second characteristic — being-with-others — is an equally basic aspect of subjectivity. Indeed, Heidegger writes that '[t]he world of Dasein is a *with-world* [*Mitwelt*]. Being-in is *Being-with* Others.'[5] However, Heidegger did not develop this second characteristic of subjectivity as thoroughly as he did the first characteristic. I have used the insights and empirical studies of SSK to help mitigate this deficiency in Heidegger's work, first with respect to the social foundations of logic, in Chapter Four, and then with respect to the emergence of early-modern experimental practice, in Chapter Six. Certainly a good deal more work could still be done in this area.

The third basic characteristic of subjectivity — understanding — has received ample attention throughout the preceding chapters. This characteristic picks out the fundamentally cognitive aspect of subjectivity, an aspect underpinning and enabling all knowledge, whether it be tacit or explicit, non-propositional or propositional, practical or theoretical. Much attention has also been given to the changes-over from tacit to

---

5   Martin Heidegger (1962a [1927]), *Being and Time*, trans. by John Macquarrie and
     Edward Robinson (Oxford: Blackwell), p. 155 [118]. (Following scholarly convention,
     page numbers in square brackets refer to the original 1927 German edition of *Being
     and Time*.) Cf. Heidegger (1962a), *Being and Time*, p. 163 [125]: 'So far as Dasein *is*, it
     has Being-with-one-another as its kind of Being.'

explicit, non-propositional to propositional, and practical to theoretical forms of cognition. For Heidegger, all of these changes-over occur against the continuous backdrop of the subject's elemental state of understandingly being in the world with others. As we have seen, a key structure of this understanding is projection. Accordingly, Heidegger viewed the emergence of early-modern scientific practice and theory against the existential backdrop of a mathematical projection of the thingness of things. Understanding may thus be viewed in terms of the projective constructivity of the subject.

The fourth basic characteristic of the subjectivity of the subject — affectivity — has received comparatively little attention in this book. Indeed, I directly addressed it only in Chapter Three, and again, more briefly, in Chapter Four, as part of a discussion of the receptivity of the subject. The argument in Chapter Three was that Heidegger's construal of receptivity as affectivity grounds his reinterpretation (not rejection) of the Kantian thing-in-itself. Affectivity is thus a core concept of minimal realism. Things affect us, therefore they exist. I drew from clinical studies of anxiety to empirically support this claim.

In Chapter Four, we saw that Kant, in Heidegger's reading, recognised the affective nature of receptivity in his phenomenology of respect for rules. The compulsive character of this respect provides an existential condition of possibility for thinking and doing, for understanding, as such. Our affectivity vis-à-vis rules for understanding helps to explain their normative power as guides for our thinking and doing. However, as we also saw, in Heidegger's account it is not towards the rules themselves that our respect is directed, but towards the persons, the community, generating and sustaining those rules. On this basis, I developed the implications of Heidegger's phenomenology of rule-following by joining it with the more empirical work of SSK, summed up in Bloor's observation that '[w]e are compelled by rules in so far as we, collectively, compel one another.'[6]

However, unlike in the case of being-with-others, SSK can offer few resources for developing Heidegger's meagre account of the affective dimension of scientific subjectivity. Indeed, against the backdrop of Heidegger's claim that 'Dasein-with is already essentially manifest in a

---

6   David Bloor (1997b), *Wittgenstein, Rules and Institutions* (London: Routledge), p. 22.

co-affectivity and a co-understanding,' we may judge SSK to have been far more concerned with co-understanding than with co-affectivity.[7] This is a serious deficiency, one also reflected in the pages of this book. Despite the substantial attention which has already been given to changes-over in understanding, these changes-over will never be properly understood until comparable attention has been also given to the corresponding changes-over in affectivity. Although Heidegger argued that '[u]nderstanding always has its mood,' and although he recognised '[t]he fact that moods can [...] change over,' he gave scarcely any attention to the affective structure of cognitive changes-over in understanding.[8] For their part, SSK practitioners have likewise failed to produce studies systematically focussed on the affective aspect of scientific knowledge production.

I view this deficiency as 'unfinished business,' and thus as an invitation for future research. Accordingly, the remainder of this chapter will point forward, towards areas where valuable scholarly work might still be done. In moving forward, then, let us begin by first looking back, one last time, to Carl von Weizsäcker's reflections on Heidegger and twentieth-century science. Despite his high regard for Heidegger, von Weizsäcker also points to deficiencies in Heidegger's knowledge of the natural sciences. Von Weizsäcker's criticism hinges on Heidegger's later reflections, from the 1940s onward, on the relationship between modern physics and modern technology, and so carries us beyond the scope of Heidegger's earlier work, which has been the main focus of this book. Let us review some of Heidegger's later reflections before considering von Weizsäcker's criticism.

In 1953, Heidegger presented a lecture, published in 1954 as 'The Question Concerning Technology,' in which he argued that '[t]he modern physical theory of nature prepares the way [...] for the essence of technology,' and that it does so because it 'pursues and entraps [*stellt*] nature as a calculable coherence of forces.'[9] The implication is

---

7   Heidegger (1962a), *Being and Time*, p. 205 [162]; translation modified.
8   Heidegger (1962a), *Being and Time*, pp. 182 [143], 173 [134].
9   Martin Heidegger (1977b [1954]), 'The Question Concerning Technology,' in *The Question Concerning Technology and Other Essays*, by Martin Heidegger, trans. by William Lovitt (New York: Harper & Row), pp. 3–35 (pp. 22, 21); my brackets. Cf. Martin Heidegger (1954a), 'Die Frage nach der Technik,' in *Vorträge und Aufsätze*, by Martin Heidegger (Pfullingen: Verlag Günther Neske), pp. 9–40 (p. 25).

that modern physics paves the way for modern technology, because it 'entraps' nature. However, 'entraps' is a misleading translation of *stellen*, which typically means 'to place.' Indeed, just a few lines earlier, Heidegger had made clear that *stellen*, in his specialised usage, reflects two different meanings: first, an 'ordering which challenges forth' (*das herausfordernde Bestellen*); second, a 'producing which brings forth' (*das hervorbringende Her-stellen*).[10] He states that these two meanings are 'fundamentally different, and yet they remain related in their essence.'[11] The word 'entraps' emphasises the first meaning of *stellen*, and suppresses the second. To more succinctly express this first meaning, Heidegger proposes the word *Ge-stell*, typically translated, for this purpose, as 'enframing.'[12] As the essence of modern technology, enframing places nature in a frame, just as 'entrap' means to place something in a trap. To more succinctly express the second, contrasting meaning of *stellen*, Heidegger uses the ancient Greek word *poiēsis*, which is meant to signify the kind of non-modern technical production typical of art, poetry, and handicraft.[13] Heidegger thus denotes the double meaning he ascribes to *stellen* by the concepts of enframing and *poiēsis*.

As just mentioned, Heidegger states that these two meanings are fundamentally different, but he also says that they are related 'in their essence.' They are essentially related because '[b]oth are ways of revealing.'[14] For Heidegger, revealing means 'com[ing] into unconcealment.'[15] This may be glossed as a thing's being moved, metaphorically, from hiddenness or concealment, and placed in the light. Revealing thus means 'illuminating,' rendering a thing intelligible or understandable as what it is. According to Heidegger, enframing and *poiēsis* are two fundamentally different ways in which this is done, two fundamentally different ways in which a thing is 'placed' in the light of projective understanding.

---

10   Heidegger (1954a), 'Die Frage nach der Technik,' p. 24. Cf. Heidegger (1977b), 'The Question Concerning Technology,' p. 21.
11   Heidegger (1977b), 'The Question Concerning Technology,' p. 21.
12   Heidegger (1977b), 'The Question Concerning Technology,' p. 19. In ordinary usage, *Gestell* usually means 'frame,' 'rack,' or 'shelf.'
13   Heidegger (1977b), 'The Question Concerning Technology,' p. 10.
14   Heidegger (1977b), 'The Question Concerning Technology,' p. 21.
15   Heidegger (1977b), 'The Question Concerning Technology,' p. 11.

If placing a thing in the light of understanding means putting it in its proper place, the place where it may be encountered as what it is, then enframing and *poiēsis* each suggest fundamentally different conceptions of a thing's proper place, and hence also of its whatness. While this is not how Heidegger explicitly described the difference, the grounds for such an interpretation have already been established in Chapters Five and Six, with the discussions of *mathēsis* and final causation. In Chapter Five, I argued that we need the concept of a final cause in order to make sense of the artful manipulation of things. The final cause explains the order and meaning of such manipulations, grounds their intelligibility, thereby distinguishing them from random, unstructured behaviour. It thus plays a necessary role in any comprehensive account of practice. It explains how a practice realises its completion in a finished thing, how the practice lets the thing become what it is. In Aristotelian terms, to let a thing become what it is means to guide it to its proper place in the cosmos. A practice may thus be viewed as an act of placing, regulated by a final cause. There will be as many final causes as there are proper places in the cosmos.

In Chapter Six, I laid out Heidegger's account of the mathematisation of early-modern science in terms of *mathēsis*, a process whereby final causes become consolidated under a single, uniform measure. This uniform measure is the one final cause, the basic blueprint, regulating the early-modern experimental manipulation of nature. Under the guidance of this basic blueprint, the prevailing conception of the thingness, or whatness, of things is rendered uniform. As a consequence, the scientific thing now has only one proper place in the cosmos, and there is thus only one general way of properly 'placing' it in that cosmos, of letting it be what it is. Under these circumstances, the scientific thing is illuminated or revealed — experienced as intelligible — in the light of a single, homogenous understanding.

I want to now suggest that, as two different ways of placing things in the light of understanding, enframing and *poiēsis* are guided, respectively, by homogenous and heterogeneous pictures of place. Enframing is a homogenous revealing; *poiēsis* is a heterogeneous revealing. With enframing, then, the world is experienced in light of a single, uniform picture. With *poiēsis*, by contrast, the world is experienced in light of multiple diverse pictures.

Heidegger calls enframing a 'challenging revealing,' as well as an 'ordering revealing.'[16] His idea seems to have been that enframing challenges or orders things — forcibly places, or frames, them — into a single, homogenous picture, in light of which we may then experience them as what they are. In this way, '[e]nframing [...] blocks *poiēsis*,' prevents us from encountering things in the heterogeneous light of multiply different pictures.[17]

On the face of it, this seems like not such a bad thing when it comes to scientific practice. For example, standardised terminologies and methods allow for effective cooperation between differently situated laboratories. In this sense, homogenous revealing may be viewed as a strength of modern science. Natural scientists are able to coordinate with one another, and to build on one another's results, because they all inhabit, so to speak, the same general picture. Homogenous revealing allows for a productive co-understanding of nature among members of the scientific community. However, Heidegger also observes that this strength carries with it a danger, because the modern physical theory of nature 'prepares the way' for enframing: 'Modern physics is the herald of Enframing.'[18] His idea seems to have been that modern science proffers a standardised form of understanding which, when transported out of the realm of scientific practice, and into the broader realm of human society, poses a threat to the diverse range of understandings which are presumably integral to the well-being of that broader society. What stands as a strength in science proves to be a weakness in society in general. Modern machine technology provides the physical medium by which the homogenous revealing of modern science becomes suffused into society as a whole. However, as I suggested at the end of Chapter Six, the fact that machine technology should have provided this medium seems a matter of social and historical contingency. Whether one understands society in terms of a machine or in terms of an organism, one is, in both cases, conceptualising it as a whole, in light of a single, uniform picture. And it is this homogenisation of understanding which sits at the core of Heidegger's worries about scientific and technological modernity.

---

16   Heidegger (1977b), 'The Question Concerning Technology,' pp. 16, 19.
17   Heidegger (1977b), 'The Question Concerning Technology,' p. 30.
18   Heidegger (1977b), 'The Question Concerning Technology,' p. 22.

A significant question, which Heidegger leaves unanswered, is why this homogenisation should at all take place. Assuming Heidegger's account is correct, then how should society in general have become, or now be under threat of becoming, wholly enframed by the homogenous revealing peculiar to the comparatively small community of modern scientists? This question is perhaps best left to historians and sociologists of science and technology. From the perspective presented here, however, their answers should give consequential attention to the role of subjectivity in the process. In this regard, Heidegger offers a phenomenological clue when he refers to the 'irresistibility of ordering.'[19] Because the co-understanding giving shape to modern science is irresistible, its application beyond the scientific realm will also be experienced, increasingly, as irresistible or compelling. Hence, alternative experiences of nature will be blocked, among them, and most importantly, the experience enabled by the heterogeneous revealing of *poiēsis*. In some highly speculative comments, Heidegger suggests that the irresistibility of enframing may somehow lead to its own dissolution, by bringing to light an 'as yet unexperienced [...] saving power.'[20] This saving power will allow for a return to the heterogeneous revealing of *poiēsis*. Heidegger then wonders: 'Could it be that the fine arts [...] may expressly foster the growth of the saving power [...]?' This, then, is his highly tentative answer to the perceived threat of enframing. In concluding his speculations, Heidegger does not argue that, but instead wonders whether, the fine arts may furnish the necessary 'restraint' by which to counter the irresistibility of enframing. He thus leaves the question concerning technological enframing unresolved.[21]

---

19 Heidegger (1977b), 'The Question Concerning Technology,' p. 33.
20 Heidegger (1977b), 'The Question Concerning Technology,' p. 33.
21 Heidegger (1977b), 'The Question Concerning Technology,' pp. 35, 33. Andrew Feenberg has developed Heidegger's suggestion that the fine arts may facilitate resistance to enframing (Andrew Feenberg (2005), *Heidegger and Marcuse: The Catastrophe and Redemption of History* (London: Routledge). I have critiqued Feenberg's proposal from an SSK perspective (Jeff Kochan (2006), 'Feenberg and STS: Counter-Reflections on Bridging the Gap,' *Studies in History and Philosophy of Science* 37(4): 702–20). Feenberg has responded to this criticism (Andrew Feenberg (2006), 'Symmetry, Asymmetry, and the Real Possibility of Radical Change: A Reply to Kochan,' *Studies in the History and Philosophy of Science* 37(4), 721–27). Harry Collins and Trevor Pinch have thrown their hats into the debate as well (Harry M. Collins and Trevor Pinch (2007), 'Who Is to Blame for the Challenger Explosion?,' *Studies in History and Philosophy of Science* 38(1), 254–55). Wiebe Bijker

There is much to contend with here. For present purposes, however, let us limit ourselves to von Wiezsäcker's response to Heidegger.[22] Von Weizsäcker argues that Heidegger was right to identify enframing as 'the signature of our time.' Nevertheless, Heidegger's own reflections on enframing were handicapped, he writes, because 'Heidegger was unable to think the natural sciences through to their base.' In von Weizsäcker's view, the reason that an alternative to enframing has not yet presented itself 'lies in the fact that the path of science has not yet reached its end.'[23] As a physicist, von Weizsäcker felt himself obliged to follow scientific understanding to its conclusion, and he expressed mild optimism that an alternative would be thus disclosed, while he remained sensitive to the risk and uncertainty of the venture: 'The saving power is already intangibly present here in the middle of the world of tangibility. The prospecting and trespass of paths in the danger is accessible through planning and is therefore an obligation.'[24] The 'planning' to which von Weizsäcker refers seems to deliberately depend on empirical-causal analysis guided by the homogenising revealing of enframing. He describes enframing as 'the deconstruction of reality in conceptual acts of imagination, and the attempt to reconstruct the whole as the sum of interacting parts.'[25] For him, the

---

and Trevor Pinch cite this exchange as exemplifying debates over whether SSK (and its offshoot, the social construction of technological systems, or SCOTS) provide grounds for a political critique of science and technology (Wiebe E. Bijker and Trevor Pinch (2012), 'Preface to the Anniversary Edition,' in *The Social Construction of Technological Systems: New Directions in Sociology and History of Technology (25th Anniversary Edition)*, ed. by Wiebe E. Bijker, Thomas P. Hughes and Trevor Pinch (Cambridge, MA: MIT Press), pp. xi–xxxiv (p. xxvii n. 3)).

22 For my part, I have responded to Latour's Heideggeresque claim that modern science 'transform[s] society into a vast laboratory' by arguing that the field sciences offer an alternative model, one contrary to the epistemic imperialism implicit in Latour's laboratory model (Bruno Latour (1983), 'Give Me a Laboratory and I Will Raise the World,' in *Science Observed: Perspectives on the Social Study of Science*, ed. by Karin D. Knorr-Cetina and Michael Mulkay (London: SAGE), pp. 141–70 (p. 166)). I gloss the subjectivity of this alternative in terms of 'epistemic neighbourliness' (Jeff Kochan (2015d), 'Objective Styles in Northern Field Science,' *Studies in History and Philosophy of Science* 52, 1–12).

23 Carl Friedrich von Weizsäcker (1977b), 'Heidegger und die Naturwissenschaft,' in *Der Garten des Menschlichen: Beiträge zur geschichtlichen Anthropologie*, by Carl Friedrich von Weizsäcker (Munich: Carl Hanser Verlag), pp. 413–31 (pp. 431, 413–14, 427); my translations.

24 von Weizsäcker (1977b), 'Heidegger und die Naturwissenschaft,' p. 431.

25 von Weizsäcker (1977b), 'Heidegger und die Naturwissenschaft,' p. 431.

'whole' to be deconstructed and reconstructed includes the subject. He writes that '[t]he reflexive or phenomenological self-awareness of the subject can be supported or corrected by the causal insight of the natural sciences.'[26] The idea seems to be that subjectivity should be drawn within the frame of a world picture, where it will then be analytically deconstructed and reconstructed in a process which will, so von Weizsäcker hopes, ultimately disclose a mode of understanding which enables 'restraint' in the face of the 'irresistibility' of enframing. Where Heidegger saw the path from enframing back to *poiēsis* as travelling through the fine arts, von Weizsäcker saw it as travelling more deeply into the natural sciences, or, more specifically, more deeply into the dynamics of enframing itself. Indeed, he even named the scientific field most likely to yield this result: '[t]his conceptual reconstruction of a whole is today called systems theory.'[27] For him, an answer to the danger posed by modern technological thinking lies in the direction of systems theory, the scientific discipline of cybernetics.

Von Weizsäcker argues that in order to understand the conditions of possibility for modern science and technology, that is, in order to understand the origins of enframing, we must investigate 'the basic condition of possibility for conceptual thinking as such.'[28] In other words, the focus of his proposed system-theoretic investigation of enframing is squarely centred on an understanding of the subjectivity of the subject. The subject is to be viewed in cybernetic terms, as an entity existing within a self-regulating system. And this system, in turn, is to be viewed in ecological terms, as the physical surroundings in which conceptual thinking, as such, emerges, stabilises, and evolves. Von Weizsäcker's hope was that cybernetics will discover the 'fundamental laws formulat[ing] the way and manner by which a conceptual-empirical thinking of what is can be the case.' More specifically, these laws will describe the logic of objective experience, that is, the rules governing 'the correctness of behaviour [a]s *adaequatio ad rem*, adaptation to the conditions of the ecological niche.'[29] In a nutshell, then, von Weizsäcker proposed a cybernetic theory of normativity.

---

26   von Weizsäcker (1977b), 'Heidegger und die Naturwissenschaft,' p. 429.
27   von Weizsäcker (1977b), 'Heidegger und die Naturwissenschaft,' p. 431.
28   von Weizsäcker (1977b), 'Heidegger und die Naturwissenschaft,' p. 429.
29   von Weizsäcker (1977b), 'Heidegger und die Naturwissenschaft,' pp. 428, 429.

Although von Weizsäcker comments that Heidegger had never tried to dissuade him from his interest in cybernetics, Heidegger did, in fact, otherwise express clear scepticism about cybernetics in his later work, especially in respect of what he understood to be that theory's account of subjectivity. In an essay from the late 1960s, Heidegger writes that 'the new fundamental science that is called cybernetics [...] corresponds to the determination of man as an acting social animal [*handelnd-gesellschaftliches Wesens*].'[30] The translation of this passage could be misleading. The word 'acting' possesses a more neutral connotation than does *handelnd*. It simply means to take action, without specifying the nature of that action. Hence, it is closer to the German word *tätig*. The word *handelnd*, in contrast, also carries connotations of 'dealing' and 'trading,' and thus may carry a distinctly economic hue. The word 'social,' in turn, can mean 'gregarious' or 'sociable,' much like the German word *gesellig*. The word *gesellschaftlich*, on the other hand, may also be translated as 'societal' or 'corporate.' It thus suggests an organised body of individuals, rather than an individual who is simply disposed to engage with others. In Heidegger's view, then, cybernetics begins with an inappropriately narrow conception of the human being, one seemingly influenced by economic and corporate models of rationality and exchange. He was especially concerned about the influence of cybernetics on prevailing conceptions of language: 'Cybernetics transforms language into an exchange of news.'[31] In 1975, the year before his death, Heidegger wrote that '[t]he ascending dominion of linguistics and of the information sciences threatens to drive the efforts of thinking and poetizing and their great traditions out of human eyesight.'[32] Hence, for Heidegger, cybernetics offered little hope for salvation in the face of

---

30 Martin Heidegger (1993c [1969]), 'The End of Philosophy and the Task of Thinking,' trans. by Joan Stambaugh, in *Basic Writings*, revised and expanded edn, by Martin Heidegger, ed. by David Farrell Krell (New York: HarperCollins), pp. 431–49 (p. 434); my brackets. Cf. Martin Heidegger (1969), 'Das Ende der Philosophie und die Aufgabe des Denkens,' in *Zur Sache des Denkens*, by Martin Heidegger (Tübingen: Max Niemeyer Verlag), pp. 61–80 (p. 64).

31 Heidegger (1993c [1969]), 'The End of Philosophy and the Task of Thinking,' p. 64.

32 Heidegger, Martin (1987/88), 'Brief an Jean Beaufret / Letter to Jean Beaufret,' trans. by Steven Davis, *Heidegger Studies* 3/4, 3–6 (p. 5). Similar statements about cybernetics may also be found in Martin Heidegger (1998), 'Traditional Language and Technological Language,' trans. by Wanda Torres Gregory, *Journal of Philosophical Research* 23, 129–45.

the perceived threat of enframing. Indeed, by apparently eclipsing the poetical and contemplative aspects of human experience — by blocking *poiēsis* — cybernetics seemed, for Heidegger, to powerfully exemplify the very phenomenon of homogenous revealing characteristic of enframing.[33]

Whereas von Weizsäcker's theory of normativity puts the system at its centre, Heidegger's theory puts the subject at its centre. Nevertheless, the root difference between their respective theories lies in their contrasting interpretations of the subject. In my view, von Weizsäcker implicitly advances a theory of the subject which makes necessary a systems-theoretic account of normativity. Heidegger, in contrast, does not. In order to more clearly understand this difference, it will be useful to revisit two concepts introduced in Chapter Six, those of the Galilean First Thing and of the world picture.

---

33 Andrew Pickering consistently obscures Heidegger's identification of enframing with homogenous revealing, first by claiming that Heidegger actually draws a sharp distinction between the two, and then by identifying enframing with 'asymmetric domination' and revealing with 'performative and open-ended dances of agency' (Andrew Pickering (2009), 'The Politics of Theory: Producing Another World, With Some Thoughts on Latour,' *Journal of Cultural Economy* 2(1/2), 197–212 (pp. 205, 204)). On this infelicitous base, Pickering introduces labels for two distinct kinds of cybernetics: 'authoritarian enframing,' on the one hand, and 'liberal democratic revealing,' on the other (Andrew Pickering (2010), *The Cybernetic Brain: Sketches of Another Future* (Chicago: University of Chicago Press), passim). He then promotes the second as a politically attractive alternative to the first. This uneven conceptual terrain makes it difficult to properly address, in a short space, the relation of Pickering's account to the present discussion. Suffice to observe that, whatever the attractions of Pickering's alternative, his description of liberal-democratic cybernetics as the 'levelling of power relations' and the placement of all subjects 'metaphysically [...] on a single and level playing field' seems to suggest the homogenous revealing of enframing (Pickering (2010), *The Cybernetic Brain*, pp. 384, 393). On this view, swapping one uniform political blueprint for another, whatever the benefit, reshuffles the deck of enframing rather than rejects it. Note that the distinction between authoritarian and democratic technology has a long history in technology studies (Andrew Feenberg (1992), 'Subversive Rationalization: Technology, Power, and Democracy,' *Inquiry* 35(3/4), 301–22; Otto Mayr (1986), *Authority, Liberty & Automatic Machinery in Early Modern Europe* (Baltimore: John Hopkins University Press); Lewis Mumford (1964), 'Authoritarian and Democratic Technics,' *Technology and Culture* 5(1), 1–8; and Langdon Winner (1980), 'Do Artifacts Have Politics?,' *Daedalus* 109(1), 121–36). Carol Steiner has staged an imaginary dialogue between Pickering and Heidegger (Carol J. Steiner (2008), 'Ontological Dance: A Dialogue between Heidegger and Pickering,' in *The Mangle in Practice: Science, Society, and Becoming*, ed. by Andrew Pickering and Keith Guzik (Durham: Duke University Press), pp. 243–65).

Recall that the Galilean First Thing is a thing left entirely to itself, to be at rest, or to move uniformly in a straight line, without interference from without. Newton formalised this concept in the First Law, which interprets the thing, fundamentally, as a law-abiding thing. Note that this image, of a thing left entirely to itself, negates the necessity of a system to which that thing belongs as a constituent part. This situation changes, however, when seventeenth-century experimental philosophers begin to consider the thing as existing in a world also inhabited by other things. Robert Boyle and his colleagues, as we saw, viewed the experimental thing as endeavouring to beat back its neighbours, so as to open up around itself an autonomous space, a space in which to achieve its native independence. One puzzle which seems to arise from such a conception of things is that of the order among things: if things are fundamentally disposed to seek their independence from one another, then why are we not now faced with a world of disjointed chaos? I want to suggest that the concept of system emerged as an answer to this question. The order we observe is not to be explained in terms of individuals, but in terms of an organised whole in relation to which those individuals stand as constitutive parts. Order is a property of the whole, irreducible to the part. This ordered whole is a system. Insofar as this system possesses the principle of its own organisation, it is a self-organising system. While the individual strives to move itself towards autonomy, the system strives to move itself towards order, equilibrium. Autonomy and equilibrium thus figure as the respective final causes of the individual thing and the system. The latter explains the presence of order left unaccounted for by the former. The world pictured as a self-regulating system thus offsets the threat of world disorder implied in the thing conceived as an autonomous individual.[34]

If, however, one conceives of the thing in a different manner, one suggesting relative order among things rather than potential chaos,

---

34   Mayr notes that the idea that the 'discord of individuals [...] contribute[s] to a concord of the whole' was already much discussed in the seventeenth century, but only acquired an adequate explanation a century later with the concept of the self-regulating system (Mayr (1986), *Authority, Liberty & Automatic Machinery*, p. 185). According to Mayr, '[w]hat won the concept such popularity was its promise of linking the values of equilibrium and liberty' (p. 188). The concept would reach full development only in the twentieth century, with the arrival of cybernetics theory (p. 187).

one will then require no additional concept of system to explain order. Order among things will instead be the result of things regulating themselves, rather than of things being the constituents of a discrete, self-regulating system. Indeed, in 1950, three years before presenting 'The Question Concerning Technology,' Heidegger presented to the same audience a lecture entitled 'The Thing,' which was published together with the later lecture in 1954. He was evidently involved in a further reconceptualisation of the thing during the same period in which he was developing his critique of enframing.[35]

Heidegger's alternative conception of the thing is a strange one. He grounds it in an old Germanic meaning for 'thing' as 'gathering.'[36] Furthermore, when a thing 'does its thing,' that is, when it subsists as a gathering, we may say — employing an obsolete English usage, meaning 'to reconcile' — that the thing *things*, that it brings reconciliation in the form of a gathering. What does the thing, by thinging, gather together? According to Heidegger, it gathers together earth, sky, gods, and mortals in a unitary 'fourfold.' This gathered fourfold, in turn, makes a 'world' possible. Heidegger writes: 'The thing stays — gathers and unites — the fourfold. The thing things world.'[37]

In developing this self-consciously poetical account of the thing, the later Heidegger uses the example of a wine jug — a vessel or 'stowage' (*Gefäß*) in which to 'stow' (*fassen*) wine.[38] To pour wine for someone is to 'bestow' (*schenken*) it on them, and the result of this bestowing is a 'bestowal' (*Geschenk*, or 'gift'). This bestowal lets the jug be what it is *qua* jug. The bestowing of wine gathers together, in the same moment — in

---

35   See Martin Heidegger (1971b [1954]), 'The Thing,' in *Poetry, Language, Thought*, by Martin Heidegger, trans. by Albert Hofstadter (New York: Harper & Row), pp. 165–86. The audience for the two lectures was the Bavarian Academy of Fine Arts. In fact, earlier versions of both lectures were presented together, on the same day, in 1948, to the Club at Bremen. See the introductory note in Heidegger (1977b), 'The Thing,' pp. ix–x.

36   Heidegger (1971b), 'The Thing,' p. 177.

37   Heidegger (1971b), 'The Thing,' p. 181. Cf. the original German: '*Das Ding verweilt das Geviert. Das Ding dingt Welt*' (Martin Heidegger (1954b), '*Das Ding*,' in *Vorträge und Aufsätze*, by Martin Heidegger (Pfullingen: Verlag Günther Neske), pp. 157–79 (p. 173)). The normally intransitive verb *verweilen* is put to unconventional transitive use in this passage, thus exacerbating attempts at translation. In this context, the word also suggests 'temporalises' and, translating more freely, 'harbours.'

38   Heidegger (1954b), '*Das Ding*,' pp. 158–66; my translations. Cf. Heidegger (1971b), 'The Thing,' pp. 166–74.

one bestowal — the earth and sky, gods and mortals. In this moment, with the bestowed wine, one experiences the earth's nutrients and the sky's sun and rain. One also experiences the mortals whose thirst the wine slakes, and, when it is offered as a libation, the gods whom it honours. Crucially, Heidegger argues that the bestowing of wine as a libation is the 'true' (*eigentliche*) bestowal.[39] Through this act the jug is let be what it essentially is. The gods are apparently the fundamental for-the-sake-of-which of the wine jug in the context of use, the principal reason for the jug's being what it is.

This claim may be more fully unpacked by returning to 'The Question Concerning Technology,' where Heidegger discusses a ceremonial bowl. Drawing on Aristotle's doctrine of the four causes in order to explain the whatness of the bowl, Heidegger writes the following.

> But a third cause remains above all responsible for it. It is that which limits the bowl beforehand within a realm of dedication and offering. Through this, the bowl is defined [*umgrenzt*] as a sacrificial implement. That which defines brings the thing to an end. With this end the thing does not finish, but rather, from it, the thing begins to be what, following its manufacture, it will be. Bringing-to-an-end, consummation, in this sense, is what the Greeks called *telos*, which one all too often translates, and thus misconstrues, as 'goal' and 'purpose.'[40]

Just as the jug is truly let be what it is in the act of pouring a libation for the gods, so too the bowl is, above all, let be what it is in an act of dedication and offering, presumably also directed towards the gods. That for the sake of which a thing is enrolled in a practice is the final cause, the ultimate meaning and end, of that thing. As I argued in Chapter Five, this end is the norm governing the thing's proper usage. A thing is thus defined by its directedness towards that end.

Yet, as I furthermore argued in Chapter Six, the experimental thing is also defined by its directedness towards an end, namely, the state of pure autonomy represented by the Galilean First Thing. In my view, the difference between the two conceptions lies in the fact that this autonomy, as an end, stands as the only end, the singular basic blueprint, for the thingness of things in general. In contrast, on

---

39   Heidegger (1954b), 'Das Ding,' p. 165. Cf. Heidegger (1971b), 'The Thing,' p. 173.
40   Heidegger (1954a), 'Die Frage nach der Technik,' p. 13; my translation. Cf. Heidegger (1977b), 'The Question Concerning Technology,' p. 8.

Heidegger's alternative conception the final cause according to which the thingness of things is determined is presented as plural — 'the gods' — rather than as singular. It is, in other words, a heterogeneous rather than a homogenous conception. Furthermore, Heidegger's alternative is couched in terms which seem to resist formalisation, in much the same way as the term 'heap,' which we considered at the beginning of Chapter Five.[41] Although a plurality of gods may be thematised as a mytho-poetical pantheon, it will nevertheless resist formalisation as a system. Heidegger presumably felt that a Newtonian rationalisation of the fourfold is not possible, because rationalisation demands homogenisation (in this case, monotheism) in order to avoid contradiction. Viewed through the lens of Heidegger's late reflections, one might see the Galilean First Thing as a god which has successfully beaten back the other gods, thereby establishing itself as an autonomous and uniform measure for the mathematicisation of nature.

But let us bring Heidegger's mytho-poetical phenomenology back down to a more concrete level. I suggest that it may still be possible to proceed scientifically on the basis of Heidegger's comments. My proposal involves reading Heidegger's late reflections in light of his earlier views on final causation and the subjectivity of the subject, as discussed in Chapter Five. Recall Heidegger's claim that the final causes guiding and regulating practice ultimately bottom out in the existential possibilities of the subject, possibilities which comprise its historical tradition. I propose to view Heidegger's concept of the gods as a mytho-poetical thematisation of these basic existential possibilities, and I wish to argue that these possibilities — this tradition — may instead be thematised in terminology more suited to historical and sociological analysis.[42] In

---

41   Indeed, as Miles Burnyeat has shown, the paradox of the heap has been also used to undermine the attempted rationalisation of conceptions of the gods (Miles F. Burnyeat (1982), 'Gods and Heaps,' in *Language and Logos*, ed. by Malcolm Schofeld and Martha Craven Nussbaum (Cambridge: Cambridge University Press), pp. 315–38). Note, also, Heidegger's 1962 comment that 'a poem does not [...] let itself be programmed' (Heidegger (1998), 'Traditional Language and Technological Language,' p. 141).

42   For arguments that Heidegger's concept of the gods can be reinterpreted in sociological terms, see: Jeff Kochan (2010b), 'Latour's Heidegger,' *Social Studies of Science* 40(4), 579–98 (p. 592); Charles Spinosa (2000), 'Heidegger on Living Gods,' in *Heidegger, Coping, and Cognitive Science: Essays in Honor of Hubert L. Dreyfus*, ed. by Mark A. Wrathall and Jeff Malpas (Cambridge: Cambridge University Press), pp. 209–28 (p. 216); and Julian Young (2006), 'The Fourfold,' in *The Cambridge*

this regard, it is important to also recall the normative role played by final causes: such causes provide the measure by which practitioners distinguish proper from improper action, thereby giving order and direction to their activities. Hence, to respect the rules governing one's actions means to respect the existential possibilities comprising one's shared historical tradition. By respecting these possibilities, as manifest both in one's own actions as well as in the actions of others, one sustains and gives shape to the tradition. In this way, the subject *is* its possibilities. Moreover, as being-in-the-world, as one who historically constructs and experiences meaning in the world, the subject also always experiences and pursues its possibilities in conjunction with others, as being-with-others. Meaning-constitutive action is thus, necessarily, both social and historical action, because the source and cause of that action — the subject — is social and historical.

Heidegger's gods may thus be viewed in terms of the historical possibilities for meaning-constitutive action available to persons who share and sustain a tradition. Furthermore, these possibilities are experienced as more or less compelling. We feel compelled to follow certain possibilities for action, rather than others, because they have normative force for us, because we view them as good rather than bad, proper rather than improper. As I argued above, this feeling of compulsion is the existential condition of possibility for understanding as such. We understand because we can be affected, and thus also guided, by the rules which structure that understanding. As I also argued above, as well as earlier in Chapter Four, according to both Heidegger and SSK, our receptivity to rules, in particular, and a tradition, more broadly, is, at base, a receptivity to each other. As Bloor says, '[w]e are compelled by rules in so far as we, collectively, compel one another.'[43] Within this phenomenological constellation, which determines the subjectivity of the subject, being-in-the-world, being-with-others, and understanding combine with affectivity in order to produce an experience of intelligibility, or cognitive order. Heidegger's gods are thus a mytho-poetical marker for the affective component of human subjectivity. Indeed, Heidegger describes the gods as providing

---

Companion to Heidegger, 2nd edn, ed. by Charles B. Guignon (Cambridge: Cambridge University Press), pp. 373–92 (p. 375).

43   Bloor (1997b), *Wittgenstein, Rules and Institutions*, p. 22.

the basis for 'every affective disposition [*wesentliche Gestimmtheit*] from respect and joy to mourning and terror.'[44] The affectivity and the sociality of the subject combine in a mutual receptivity which enables it to projectively understand its environment as an ordered world.

For both Heidegger and SSK, cognitive order is grounded in social order, and social order is, in turn, produced in micro-social interactions between mutually receptive subjects who both compel and defer to one another in (and often by manipulating bits of) a common material environment. We might call this an 'interactionist theory of normativity,' in contrast to von Weizsäcker's cybernetic theory of normativity. The norms governing behaviour are not, finally, to be located in a system, of which each person is a constitutive part, but in the often mundane interactions between persons who, to a greater or lesser extent, share a history and culture. Whether or not these interactions achieve a level of consistency and stability allowing for their reification as a 'system' will be a matter for empirical investigation. But, if such were the case, then it would be important to remember that this system is, in fact, a simplification, a reification, a matter of conceptual art and convenience, rather than a fundamental statement about the nature of normativity, much less the ontology of groups.

Interactionism is SSK's social theory of choice, and has been discussed at length by Barry Barnes.[45] An interactionist social theory puts the individual subject at the methodological centre of explanations of social, and thus also of cognitive, order. According to this theory, the subject does not eschew interaction, beating away its neighbours in pursuit of its own autonomy, but instead seeks to interact with them as a matter of natural necessity. This is the sociological equivalent of Heidegger's claim that being-with-others is a fundamental characteristic of the subjectivity of the subject. A natural disposition to interact with others is, however, not yet an explanation of social order. Barnes argues that

---

44   Martin Heidegger (1992 [1982]), *Parmenides,* trans. by André Schuwer and Richard Rojcewicz (Bloomington: Indiana University Press), p. 106, my brackets. Cf. Martin Heidegger (1982b), *Parmenides* (*Gesamtausgabe,* vol. 54) (Frankfurt: Vittorio Klostermann), p. 157.

45   Barry Barnes (1995), *The Elements of Social Theory* (London: UCL Press), chpt. 3; Barry Barnes (2001), 'Practice as Collective Action,' in *The Practice Turn in Contemporary Theory,* ed. by Theodore R. Schatzki, Karin Knorr Cetina and Eike von Savigny (London: Routledge).

this disposition is accompanied by 'the readiness of the individual to align her cognition with that of others and co-ordinate her activities with theirs.'[46] This is an innate orientation towards cooperation with others, but not towards conformity to rules. According to Barnes, a person's natural gregariousness does not constrain her to a fixed social order, but rather facilitates her participation in a 'form of life.'[47] This is the sociological near equivalent of Heidegger's claim that being-with-others is a necessary (but not a sufficient) condition of the subject's being able to make sense of things, to project a space of intelligibility — a world — in which things may be encountered as what they are. Put another way, gregariousness plays a necessary role in revealing, in opening up a world of understanding.

From this it follows that enframing — homogenous revealing — may be experienced as the opening up of a world within which action becomes possible, rather than as an external constraint on action. However, so Heidegger worries, the existential basis for this freedom to act, its condition of possibility, is ontologically limited. From an interactionist perspective, in such a world, the subject's natural gregariousness is increasingly guided into an intersubjective alignment of cognition and co-ordinated action which may allow for considerable freedom, but only within a limited cognitive and material domain, a domain in which alternative existential possibilities have been blocked or suppressed. Heidegger's later work thematises, or topicalises, the phenomenology of this situation. SSK provides the resources by which to explain it.

On an interactionist account, the homogenising impulse of enframing is to be explained by reference to the micro-social interactions through which enframing is produced and sustained. Enframing does not have

---

46   Barnes (1995), *The Elements of Social Theory*, p. 102.
47   Barnes (1995), *The Elements of Social Theory*, p. 103. In a welcome step beyond his earlier work, discussed in Chapter Two, Joseph Rouse now defends an account of normative practice which appears, in its essentials, very similar to Barnes's older and sociologically more detailed account: 'a practice is [...] held together by the interactions among its constitutive performances' (Joseph Rouse (2015), *Articulating the World: Conceptual Understanding and the Scientific Image* (Chicago: University of Chicago Press), p. 163). Note, however, Rouse's reference to interacting performances, not performers, a difference in focus reflecting his continuing attempt to neutralise the subject as an explanatory resource. It is a small intellectual tragedy that Rouse's earlier misreading of SSK should now blind him to the benefits of positively enrolling SSK in the elaboration of his own interactionist theory of normativity.

a life of its own, existing independently of people, and exercising an external power over them, but must instead be explained reductively in terms of the interactions of those people. As Barnes notes, this methodological point may be difficult to square with immediate personal experience.

> For a single involved individual, norms may still be experienced as having a fixed meaning beyond her own discretion and as pressing on her judgement like an external force. This is just one of many instances of the fact that what is created and controlled collectively may be experienced as given and coercing individually, and what is actually the power of many other people may be experienced individually as a power external to people altogether.[48]

The basic idea is that enframing 'does not carry us along but rather [...] we carry it along.'[49] The same insight may also be applied to systems: the system does not carry us along, we carry it along. We are compelled by the system only insofar as we, collectively, compel one another.[50]

On this basis, an interactionist theory of normativity turns a systems theory of normativity on its head. Von Weizsäcker proposed that a systems-theoretic deconstruction and reconstruction of 'reality' would yield the 'fundamental laws' which govern the ecological conditions of possibility for conceptual thinking. He thus characterised the subject as being carried along in a system whose law-like regularities enable conceptual understanding. The interactionist account reverses this, arguing that the aligned cognition and co-ordinated action — that is, the co-understanding — of subjects may enable a shared experience of the world as a single, integrated system, a coherently organised reality. If a person experiences this world picture as compelling, it is not because the picture itself is compelling, but because the person is aligned with others who deem it a good thing to feel compelled by that picture, and so actively encourage the person to feel so compelled. The community

---

48   Barnes (1995), *The Elements of Social Theory*, p. 75.

49   Barnes (1995), *The Elements of Social Theory*, p. 112.

50   In a discussion of Heidegger and science studies, Steiner reads the later Heidegger as claiming that 'when we understand Being as the *nil source* of our knowledge, [...] then what we know is no longer under our control. [...] All that remains to move on to the path to primordial Being is to *embrace the mystery*' (Carol J. Steiner (1999), 'Constructive Science and Technology Studies: On the Path to Being?,' *Social Studies of Science* 29(4), 583–616 (p. 589)). I will reserve judgement on whether this fairly represents Heidegger's later views. My task here is to use SSK to demystify explanations of this kind, a task I view as compatible with Heidegger's earlier work.

compels the person to feel compelled by the picture. In so doing, the community carries the picture along.

This attention to feelings of compulsion evinces the deep relationship between co-understanding and co-affectivity. Von Weizsäcker viewed the systems-theoretical reassembly of subjectivity as an 'obligation.' For him, the path of science *must* be pursued to its end. Such statements tell us something about the particular mode of co-affectivity characteristic of von Weizsäcker's own scientific community. In these final paragraphs, I want to suggest that much exciting work could still be done investigating such modes of scientific co-affectivity.

Here, too, SSK can offer guidance, with Barnes's interactionist elaboration of Max Weber's concept of 'status groups,' an elaboration which puts at its centre the emotional susceptibility to one another of group members.[51] A status group arises when a number of people come together on the basis of a characteristic, or 'good,' which they share in common, and which they simultaneously use to exclude from membership those who fail to uphold that good. This dynamic of inclusion and exclusion defines a field of legitimate competition: group members may compete with one another for status within the group, while those excluded are disqualified from such competition. The criterion which determines inclusion in and exclusion from the group will be a matter of historical contingency. However, the dynamic holding group members together, in accordance with the relevant, contingent criterion, is, Barnes suggests, a matter of necessity.[52] Indeed, this fundamental dynamic forms the basis of human sociality; it is a necessary characteristic of the subjectivity of the subject, comparable to Heidegger's being-with-others.

Moreover, also in step with Heidegger, Barnes argues that the sociality of the subject necessarily combines with affectivity: 'we convey to each other signals of praise and blame, approval and disapproval, recognition or rejection, honour or contempt, socially organised to sustain the collective good, and our susceptibility to these signals is what

---

51   Barry Barnes (1992b), 'Status Groups and Collective Action,' *Sociology* 26(2), 259–70; see also Barnes (1995), *The Elements of Social Theory*, chpt. 5. Cf. Max Weber (1968), *Economy and Society*, trans. by Günther Roth and Claus Wittich (Berkeley: University of California Press). Note that there is an established and growing body of literature addressing the topics of scientific emotion, in particular, and epistemic emotion, more generally. See the Appendix at the end of this chapter for an incomplete overview (p. 381).

52   Barnes (1992b), 'Status Groups and Collective Action,' pp. 260, 266, 263.

encourages collective action.' Our conveyance of, and susceptibility to, these signals — our co-affectivity — 'is the basis of our sociability, and may be taken as given in accounting for any particular manifestation or consequence of that sociability.'[53] This is a description of social order in terms of the emotionally infused interaction of individuals who are biologically so predisposed. Barnes credits Thomas Scheff with having 'systematized' this description under the rubric of a 'deference-emotion system.'[54] This is an interactive system in which 'the individual constantly monitors how he or she stands in the eyes of others, and experiences feelings of pride or shame accordingly.' However, it is difficult to thematise and articulate these interactions: 'cultural taboos often prohibit explicit recognition of the system in everyday discourse, because the system is everywhere operative and hence not totally under the conscious, calculative control of members themselves.'[55] Once we accept the existence of this emotion-based dynamic — including its cognitive importance — then a largely uncharted research area opens itself up for exploration by science studies scholars. This is an opportunity not only to discover new facts about the way science is and has been done, but also to develop new sociological and historiographic methods by which to discover those facts.

Status groups are formed and sustained by members' elicitation of feelings of pride and shame in one another. One's behaviour will attract deference, on the one hand, or disregard, even disdain, on the other, depending on whether or not it promotes the collective good. This collective good will vary between groups on the basis of their contingent social and historical circumstances. In Chapter Four, we saw that, historically, the idea of the good has played an essential role in organising understanding, enabling intelligibility, by allowing for normative distinctions to be made between epistemically good and bad phenomena. We saw, furthermore, that Heidegger reinterprets Plato's idea of the good in terms of the for-the-sake-of-which, that is, the

---

53   Barnes (1992b), 'Status Groups and Collective Action,' p. 263.
54   Barnes (1995), *The Elements of Social Theory*, p. 72. Cf. Thomas J. Scheff (1988), 'Shame and Conformity: The Deference-Emotion System,' *American Sociological Review* 53(3), 395–406. Both Barnes and Scheff acknowledge their debt in this context to Goffman's work on 'interaction rituals' (Erving Goffman (1967), *Interaction Ritual* (New York: Anchor)).
55   Barnes (1992b), 'Status Groups and Collective Action,' p. 263.

existential possibilities available to an engaged subjectivity. In Chapter Six, I read Heidegger as arguing that the collective good around which the subjectivity of seventeenth-century experimental philosophers organised itself was the autonomous First Thing. In the dispute between Robert Boyle and Francis Line, the Galilean First Thing served as the measure according to which Boyle successfully excluded Line from the community of experimentalists. Yet, I said nothing about the emotional dynamics which — on the present account — must have played a fundamental role in this dispute and its resolution. As a consequence, the story remains half-told. More work still needs to be done in order to complete the historical picture.[56]

In his later writings, Heidegger argued, with von Weizsäcker's agreement, that the collective good around which the community of modern physics organises itself is enframing. We should expect, on this argument, that the social interactions of modern physicists will be such that they defer to those who promote enframing and disregard those who do not. This is a sociological conjecture which requires empirical testing in order to determine its truth value. Nevertheless, it suggests an explanation for the difference in attitude between Heidegger and von Weizsäcker vis-à-vis enframing. It suggests that von Weizsäcker's professed 'obligation' to carry the systematisation of natural philosophy through to its end was, tacitly, an affective response to the deference-emotion dynamic characteristic of the status group to which he belonged. By manifesting this feeling of obligation in his behaviour, he showed himself to be an honourable member of his community.

The later Heidegger, in contrast, by failing to promote enframing, signalled his distance from this community, his relative immunity to its internal emotional dynamic. Indeed, for him, this dynamic, and the collective good it sustains, were a topic, rather than a resource, for investigation. In his earlier work, on which this book has chiefly focussed, Heidegger addresses that topic from the standpoint of a scientific philosophy, employing the methods of phenomenology. Like

---

56   A natural departure point for this would be Steven Shapin's work on trust and honour in seventeenth-century experimental philosophy (Steven Shapin (1994), *A Social History of Truth: Civility and Science in Seventeenth-Century England* (Chicago: University of Chicago Press)). See also Steven Shapin (2012), 'The Sciences of Subjectivity,' *Social Studies of Science* 42(2), 170–84, and a discussion thereof in Kochan (2013), 'Subjectivity and Emotion in Scientific Research.'

SSK, he sought, in part, to develop a scientific account of science, an account based on an analysis of the subjectivity of the scientific subject. In this analysis, the deference-emotion dynamic of modern science is treated as the combined result of the subject's affectivity and its being-with-others — as co-affectivity. The collective good sustained by co-affectivity is, in turn, treated as the combined result of the subject's understanding, or projectivity, and its being-with-others — as co-understanding. Yet, the collective good can be experienced as 'good' only with the support of co-affectivity. Without co-affectivity, by which the collective good is constituted as good, co-understanding will project only an image, absent the normative force necessary for it to be experienced as a compelling image. Hence, on this account, the compelling nature, the goodness, of the mathematical projection of nature, of *mathēsis* — indeed, of enframing — entails co-affectivity. Without co-affectivity, co-understanding will lack normative weight.

Inversely, without co-understanding, co-affectivity will lack order and direction. We saw the extreme consequence of this in Chapter Three, with Heidegger's discussion of anxiety. Anxiety arises in the face of unintelligibility, a global breakdown in meaning, our failure to make sense of, to understand, the things around us. I argued that, in such rare situations, we experience the thing as a thing-in-itself: we know *that* it is, but not *what* it is.

Heidegger argues that the thing in the face of which we feel anxiety is 'completely indefinite'; it is a thing 'essentially incapable of having an end-directedness [*Bewandtnis*].'[57] In other words, without the benefit of projectivity, we lack a final cause, an image by which to let things be what they are, by which to get our bearings in the world. Co-understanding gives us a collective image, co-affectivity lends that image normative weight, turning it into a measure. Hence, co-affectivity grounds the normativity of a practice, and co-understanding grounds the intelligibility, or order, of that practice. This reprises a key claim from Chapter Five, that a practice stripped of the final cause which gives it order and direction will amount to no more than random behaviour. Thus, on its own, the deference-emotion dynamic produces only random behaviour, never practice. Only when that dynamic

---

57   Heidegger (1962a), *Being and Time*, p. 231 [186]; translation modified, my brackets. Cf. Martin Heidegger (1927), *Sein und Zeit* (Tübingen: Max Niemeyer Verlag), p. 186.

coalesces around and affirms an end, giving that end weight, will it produce collective practice and social order. Enframing, as a normative force, a collective good, is thus constituted, on Heidegger's account, in the interactive, emotion-infused group dynamic of the modern physics community. By carrying enframing along in this way, the community becomes what it is. Indeed, on this account, every community becomes what it is by so affirming and carrying along the collective good, or goods, which give it its characteristic shape. These are the conditions of possibility for social identity and order as such. The virtue of SSK and Heidegger's existential conception of science, indeed, the virtue of the two combined, is that they equip us with fundamental tools by which to topicalise and examine those conditions. In so doing, they promise to throw new explanatory light on the workings of society, in general, and the sciences, in particular.

The pursuit of this exciting research prospect has been, and will continue to be, met with resistance. Paradoxically, as we have seen, some of the strongest criticism has come from those who themselves promote a practice-based theory of science. For them, a turn to practice entails the rejection of the scientific subject as an explanatory resource. Because SSK practitioners and the early Heidegger do not reject the explanatory importance of the subject, these critics conclude that neither account is compatible with a practice-based theory. As I have argued, this conclusion is false, and the evidence educed in its favour collapses under scrutiny. This suggests that the critics' motivation may be not so much to champion the practice-based study of science, as to ensure that such studies proceed without critical attention to the contribution of the subject. The subject has not been eliminated in these studies, but suppressed. Hence, subjectivity continues to circulate, tacitly and obscurely, in these practice-based accounts of science, while inquisitive scholars are denied the critical tools by which to address, analyse, and understand its role. To pinch a phrase from Charles Baudelaire, under such circumstances, the scientist — including the social scientist — becomes 'a prince who everywhere rejoices in his incognito.'[58]

---

58 Charles Baudelaire (1964 [1863]), 'The Painter of Modern Life,' in *The Painter of Modern Life and Others Essays*, by Charles Baudelaire, trans. by Jonathan Mayne (London: Phaidon Press), pp. 1–40 (p. 9).

Yet, what good is this incognito? I suggested earlier that the suppression of inquiry into subjectivity may, in effect, grant free licence to an unrestrained constructivity, a subjectivity which knows neither social nor historical limits, even while it remains (perhaps unwittingly) conditioned by those limits. This amounts to a particular conception of human freedom, one challenged by SSK practitioners and the early Heidegger alike, both of whom, as we saw in Chapter Three, emphasise the finitude of human cognition, and so espouse a policy of epistemic humility. We might now say that, in challenging this conception of human freedom, SSK practitioners and the early Heidegger signal their distance from a scholarly community in which that conception has acquired the normative weight of a collective good. Their challenge displays their relative immunity to the deference-emotion dynamic characteristic of that community. But another, deeper observation can also be made. Through the suppression of subjectivity, critical attention is deflected from the emotion-infused interactions by which knowledge is generated and sustained. Indeed, as noted earlier, Barnes argues that 'cultural taboos' may prevent explicit recognition of the largely non-deliberative, but nonetheless pervasively present deference-emotion dynamic of a community. These taboos against thematising co-affectivity would seem stronger than similar taboos against thematising co-understanding. With the suppression of subjectivity, co-affectivity seems to have been buried more deeply than co-understanding, perhaps because we are taught that feelings are more private than thoughts. This may help to explain why SSK practitioners and the early Heidegger both give more attention to thinking than to feeling, even as their respective accounts directly challenge this very distinction. As a consequence, the work of both remains incomplete. This is good news for us, as it means there is still much work left to be done. Instead of maintaining the established architecture of a familiar intellectual territory, we have a chance to explore new ground, to build new structures within fresh horizons, to come together around these still largely unarticulated possibilities in our irrepressibly motley social existence.

Appendix

This appendix supplements footnote 51 (p. 375) in this chapter.

Barry Barnes's interactionist attention to the emotional dynamics of 'status groups' complements, but has never yet been connected with, an earlier strand of research on scientific emotion by feminist scholars:

> Alison M. Jagger (1989), 'Love and Knowledge: Emotion in Feminist Epistemology,' in *Women, Knowledge, and Reality: Explorations in Feminist Epistemology*, ed. by Ann Garry & Marilyn Pearsall (Boston: Unwin Hyman), pp. 129–55.
>
> Evelyn Fox Keller (1982), 'Feminism and Science,' *Signs* 7(3), 589–602.
>
> Evelyn Fox Keller (1983), *A Feeling for the Organism: The Life and Work of Barbara McClintock* (San Francisco: W. H. Freeman).
>
> Helen E. Longino (1993), 'Subject, Power, and Knowledge: Description and Prescription in Feminist Philosophies of Science,' in *Feminist Epistemologies*, ed. by Linda Alcoff and Elizabeth Potter (London: Routledge), pp. 101–20.
>
> Sharon Traweek (1992), *Beamtimes and Lifetimes: The World of High Energy Physicists* (Cambridge, MA: Harvard University Press).

In a recent article, feminist epistemologist Lorraine Code laments the 'entrenched image' of the knower from whom 'affectivity is excised,' describing this as 'a curiously implausible conception of subjectivity':

> Lorraine Code (2015), 'Care, Concern, and Advocacy: Is There a Place for Epistemic Responsibility?,' *Feminist Philosophy Quarterly* 1(1), 1–20 (p. 9).

Despite this entrenched neglect, however, some recent work on the topic has been done in the psychological and sociological philosophy of science:

> Carlo Celluci (2013), *Rethinking Logic: Logic in Relation to Mathematics, Evolution, and Method* (Dordecht: Springer), chpt. 14.
>
> Jeff Kochan (2013), 'Subjectivity and Emotion in Scientific Research,' *Studies in History and Philosophy of Science* 44(3), 354–62.

Jeff Kochan (2015e), 'Reason, Emotion, and the Context Distinction,' *Philosophia Scientiae* 19(1), 35–43.

James W. McAllister (2002), 'Recent Work on Aesthetics of Science,' *International Studies in the Philosophy of Science* 16(1), 7–11.

James W. McAllister (2005), 'Emotion, Rationality, and Decision Making in Science,' in *Logic, Methodology, and Philosophy of Science: Proceedings of the Twelfth International Congress*, ed. by Petr Jájek, Luis Valdés-Villanueva and Dag Westerståhl (London: King's College Publications), pp. 559–76.

James W. McAllister (2007), 'Dilemmas in Science: What, Why, and How,' in *Knowledge in Ferment: Dilemmas in Science, Scholarship and Society*, ed. by Adriaan in't Groen, Henk Jan de Jonge, Eduard Klasen Hilje Papma and Piet van Slooten (Leiden: Leiden University Press), pp. 13–24.

James W. McAllister (2014), 'Methodological Dilemmas and Emotion in Science,' *Synthese* 191, 3143–58.

Lisa M. Osbeck and Nancy J. Nersessian (2011), 'Affective Problem-Solving: Emotion in Research Practice,' *Mind & Society* 10(1), 57–78.

Lisa M. Osbeck and Nancy J. Nersessian (2013), 'Beyond Motivation and Metaphor: "Scientific Passions" and Anthropomorphism,' in *EPSA 11: Perspectives and Foundational Problems in Philosophy of Science*, ed. by V. Karakostas & D. Dieks (Dordecht: Springer), pp. 455–66.

Lisa M. Osbeck, Nancy J. Nersessian, Wendy C. Newstetter and Kareen R. Malone (2011), *Science as Psychology: Sense-Making and Identity in Scientific Practice* (Cambridge: Cambridge University Press).

Sabine Roeser (2012), 'Emotional Engineers: Toward Morally Responsible Design,' *Science and Engineering Ethics* 18(1), 103–15.

Paul Thagard (2002), 'The Passionate Scientist: Emotion in Scientific Cognition,' in *The Cognitive Basis of Science*, ed. by Peter Carruthers, Stephen Stich and Michael Siegal (Cambridge: Cambridge University Press), pp. 235–50.

Paul Thagard (2006a), 'How to Collaborate: Procedural Knowledge in the Cooperative Development of Science,' *The Southern Journal of Philosophy* 44, 177–96.

And, much less recently:

Ludwik Fleck (1979 [1935]), *Genesis and Development of a Scientific Fact*, trans. by Fred Bradley and Thaddeus J. Trenn, ed. by Thaddeus J. Trenn and Robert K. Merton (Chicago: University of Chicago Press)

Michael Polanyi (1958), *Personal Knowledge: Towards a Post-Critical Philosophy* (London: Routledge & Kegan Paul).

As well as, in the history of science:

Otniel E. Dror, Bettina Hitzer, Anjy Laukötter and Pilar León-Sanz, eds. (2016), *History of Science and the Emotions* (*Osiris* 31) (Chicago: University of Chicago Press), and Paul White, ed. (2009), *The Emotional Economy of Science*, focus section in *Isis* 100(4), 792–851.

For more general discussions of epistemic emotion, see:

Georg Brun and Dominique Kuenzle (2008), 'A New Role for Emotions in Epistemology?,' in *Epistemology and Emotions*, ed. by Georg Brun, Ulvi Doğuoğlu and Dominique Kuenzle (Aldershot: Ashgate), pp. 1–31.

Louis C. Charland (1998), 'Is Mr. Spock Mentally Competent? Competence to Consent and Emotion,' *Philosophy, Psychiatry, & Psychology* 5(1), 67–81.

Antonio R. Damasio (1994), *Descartes' Error: Emotion, Reason, and the Human Brain* (New York: G. P. Putnam's Sons).

Antonio R. Damasio (1999), *The Feeling of What Happens: Body and Emotion in the Making of Consciousness* (New York: Harcourt Brace & Company).

Ronald de Sousa (1987), *The Rationality of Emotion* (Cambridge, MA: The MIT Press).

Ronald de Sousa, (2008), 'Epistemic Feelings,' in *Epistemology and Emotions*, ed. by Brun, Doğuoğlu and Kuenzle, pp. 185–204.

Sabine A. Döring (2010), 'Why Be Emotional?,' in *The Oxford Handbook of the Philosophy of Emotion*, ed. by Peter Goldie (Oxford: Oxford University Press), pp. 283–301.

George Downing (2001), 'Emotion Theory Revisited,' in *Heidegger, Coping, and Cognitive Science: Essays in Honor of Hubert L. Dreyfus*, vol. 2, ed. by Mark A. Wrathall and Jeff Malpas (Cambridge, MA: The MIT Press), pp. 245–70.

Catherine Z. Elgin (2008), 'Emotion and Understanding,' in *Epistemology and Emotions*, ed. by Brun, Doğuoğlu and Kuenzle, pp. 33–49.

Nico H. Frijda, Antony S. R. Manstead and Sacha Bem, eds. (2000), *Emotions and Beliefs: How Feelings Influence Thoughts* (Cambridge: Cambridge University Press).

Nico H. Frijda and Louise Sundararajan (2007), 'Emotion Refinement: A Theory Inspired by Chinese Poetics,' *Perspectives on Psychological Science* 2(3), 227–41.

Peter Goldie (2004), 'Emotion, Reason, and Virtue,' in *Emotion, Evolution, and Rationality*, ed. by Dylan Evans & Pierre Cruse (Oxford: Oxford University Press), pp. 249–67.

Patricia Greenspan (1988), *Emotions and Reasons: An Inquiry into Emotional Justification* (London: Routledge).

Patricia Greenspan (2000), 'Emotional Strategies and Rationality,' *Ethics* 110(3), 469–87.

Christopher Hookway (1998), 'Doubt: Affective States and the Regulation of Inquiry,' in *Pragmatism*, ed. by C. J. Misak (*Canadian Journal of Philosophy, Supplementary Volume* 24), pp. 203–25.

Christopher Hookway (2002), 'Emotions in Epistemic Evaluations,' in *The Cognitive Basis of Science*, ed. by Peter Carruthers, Steven Stich and Michael Siegel (Cambridge: Cambridge University Press), pp. 251–61.

Christopher Hookway (2008), 'Epistemic Immediacy, Doubt and Anxiety: On a Role for Affective States in Epistemic Evaluation,' in *Epistemology and Emotions*, ed. by Brun, Doğuoğlu and Kuenzle, p. 51–65.

Adam Morton (2010), 'Epistemic Emotions,' in *The Oxford Handbook of the Philosophy of Emotion*, ed. by Goldie, pp. 385–99.

Robert C. Solomon (1976), *The Passions: Emotions and the Meaning of Life* (Garden City, NY: Doubleday Press).

Robert C. Solomon (1977), 'The Logic of Emotion,' *Noûs* 11, 41–49.

Robert C. Solomon (1992), 'Existentialism, Emotions, and the Cultural Limits of Rationality,' *Philosophy East & West* 42(4), 597–621.

Robert C. Solomon (2003), 'Emotions, Thoughts and Feelings: What is a "Cognitive Theory" of the Emotions and Does It Neglect Affectivity?,' in *Philosophy and the Emotions*, ed. by Anthony Hatzimoysis (Cambridge: Cambridge University Press), pp. 1–18.

Michael Stocker (2010), 'Intellectual and Other Nonstandard Emotions,' in *The Oxford Handbook of the Philosophy of Emotion*, ed. by Goldie, pp. 401–23.

Paul Thagard (2001), 'How to Make Decisions: Coherence, Emotion, and Practical Inference,' in *Varieties of Practical Reasoning*, ed. by Elijah Millgram (Cambridge, MA: The MIT Press), pp. 355–71.

Paul Thagard, in collaboration with Fred Kroon, Josef Nerb, Baljinder Sahdra, Cameron Shelley and Brandon Wager (2006b), *Hot Thought: Mechanisms and Applications of Emotional Cognition* (Cambridge, MA: The MIT Press).

Paul Thagard (2008), 'How Cognition Meets Emotion: Beliefs, Desires and Feelings as Neural Activity,' in *Epistemology and Emotions*, ed. by Brun, Doğuoğlu and Kuenzle, pp. 167–84.

Acknowledgements

As I neared completion of this book, I was reminded of the idea that an isolated monkey tapping randomly on a keyboard for eternity would produce all possible written works. How depressing it was to suddenly realise that this book might just as well have been written by a monkey. I nevertheless draw some comfort from the fact that this task has taken me somewhat less than an eternity, and that I have not had to work entirely in isolation. Small, sometimes almost imperceptible acts of friendship — both intellectual and otherwise — have had a way of accumulating over time to tremendous effect. I acknowledge this happy circumstance, and express my thanks to all those concerned, even as I neglect to name names. There are also other, larger debts. I thank Trish Glazebrook and Simon Schaffer for their interest and encouragement at an early stage of this project. The penultimate draft was read in its entirety, or near entirety, by David Bloor, Andrew Buskell, Martin Kusch, and Estheranna Stäuble. I am very grateful to them all for their valuable comments. Although the labour that went into this book did not benefit from any direct financial support, over the span of the book's production I did nevertheless receive financial and institutional support from several sources. Without this indirect benefit, I may not have had the grit and dosh to keep on tapping. In this regard, I thank Wolfgang Freitag, Michael Hampe, Hans Bernhard Schmid, Robert W. Smith, Marcel Weber, and, most especially, the Director, Staff, and Fellows of the Zukunftskolleg, University of Konstanz, where most of this book was written.

While doing the research for this book, I relied, essentially, on the publically owned libraries of the University of Konstanz, the Swiss Federal Institute of Technology (ETH-Zürich), and the Central Library of Zürich. I may never have written a book at all if I had not first, as a child and youth, spent countless hours in the public libraries of the City of Edmonton. Especially in the sub-zero winter months, these libraries offered warm refuge for all city inhabitants, including the poor and homeless, whom I often saw reading newspapers and books, listening to music, or testing one another at games of chess. This spirit of a barrier-free knowledge commons is present also in the not-for-profit, open-access publishing model of Open Book Publishers. I am delighted to be able to contribute to their admirable social enterprise.

It behoves me to thank Estheranna Stäuble a second time, for reasons she and I know best.

Bibliography

Alderman, Harold (1978), 'Heidegger's Critique of Science and Technology,' in *Heidegger and Modern Philosophy*, ed. by Michael Murray (New Haven: Yale University), pp. 35–50.

Amsterdamska, Olga (1990), 'Surely You Are Joking, Monsieur Latour!' *Science, Technology, and Human Values* 15(4), 495–504. https://doi.org/10.1177/016224399001500407

Aristotle (1941a), *De Interpretatione*, trans. by E. M. Edghill, in *The Basic Works of Aristotle*, ed. by Richard KcKeon (New York: Random House), pp. 38–61.

Aristotle (1941b), *Categoriae*, trans. by E. M. Edghill, in *The Basic Works of Aristotle*, ed. by Richard KcKeon (New York: Random House), pp. 3–37.

Aristotle (1941c), *Physica*, trans. by R. P. Hardie and R. K. Gaye, in *The Basic Works of Aristotle*, ed. by Richard KcKeon (New York: Random House), pp. 213–394.

Aristotle (1941d), *On the Heavens*, trans. by J. L. Stocks, in *The Basic Works of Aristotle*, ed. by Richard KcKeon (New York: Random House), pp. 398–466.

Aristotle (1941e), *On the Generation of Animals*, trans. by Arthur Platt, in *The Basic Works of Aristotle*, ed. by Richard KcKeon (New York: Random House), pp. 665–80.

Aristotle (1941f), *Metaphysics*, trans. by W. D. Ross, in *The Basic Works of Aristotle*, ed. by Richard KcKeon (New York: Random House), pp. 689–926.

Aubrey, John (2014), *Brief Lives, with an Apparatus for the Lives of Our English Mathematical Writers*, vol. 1, ed. by Kate Bennett (Oxford: Oxford University Press).

Averill, James R. (1980a), 'Emotion and Anxiety: Sociocultural, Biological, and Psychological Determinates,' in *Explaining Emotions*, ed. by Amélie O. Rorty (Berkeley: University of California Press), pp. 37–72.

Averill, James R. (1980b), 'A Constructivist View of Emotion,' in *Theories of Emotion*, ed. by Robert Plutchik and Henry Kellerman (New York: Academic Press), pp. 305–39. https://doi.org/10.1016/B978-0-12-558701-3.50018-1

Babich, Babette E. (1995), 'Heidegger's Philosophy of Science: Calculation, Thought, and Gelassenheit,' in *From Phenomenology to Thought, Errancy, and Desire: Essays in Honor of William J. Richardson, SJ*, ed. by Babette E. Babich (Dordecht: Kluwer Academic Publishers), pp. 589–99. https://doi.org/10.1007/978-94-017-1624-6_36

Bagni, Giorgio T. (2010), 'Mathematics and Positive Sciences: A Reflection Following Heidegger,' *Education Studies in Mathematics* 73(1), 75–85. https://doi.org/10.1007/s10649-009-9214-0

Ballard, Edward G. (1971), 'Heidegger's View and Evaluation of Nature and Natural Science,' in *Heidegger and the Path of Thinking*, ed. by John Sallis (Pittsburgh: Duquesne University Press), pp. 37–64.

Bambach, Charles (2013), *Thinking the Poetic Measure of Justice: Hölderlin-Heidegger-Celan* (Albany: SUNY Press).

Barnes, Barry (1974), *Scientific Knowledge and Sociological Theory* (London: Routledge & Kegan Paul). https://doi.org/10.4324/9780203706541

Barnes, Barry (1977), *Interests and the Growth of Knowledge* (London: Routledge & Kegan Paul). https://doi.org/10.4324/9781315763576

Barnes, Barry (1981), 'On the "Hows" and "Whys" of Cultural Change (Response to Woolgar),' *Social Studies of Science* 11(4), 481–98. https://doi.org/10.1177/030631278101100404

Barnes, Barry (1982), *T. S. Kuhn and Social Science* (London: Macmillan). https://doi.org/10.1007/978-1-349-16721-0

Barnes, Barry (1983), 'Social Life as Bootstrapped Induction,' *Sociology* 17(4), 524–45. https://doi.org/10.1177/0038038583017004004

Barnes, Barry (1992a), 'Relativism, Realism and Finitism,' in *Cognitive Realism and Social Science*, ed. by Diederick Raven, Lietke van Vucht and Jan de Wolf (New Brunswick, NJ: Transaction), pp. 131–47.

Barnes, Barry (1992b), 'Status Groups and Collective Action,' *Sociology* 26(2), 259–70. https://doi.org/10.1177/0038038592026002008

Barnes, Barry (1995), *The Elements of Social Theory* (London: UCL Press). https://doi.org/10.1515/9781400864355

Barnes, Barry (2001), 'Practice as Collective Action,' in *The Practice Turn in Contemporary Theory*, ed. by Theodore R. Schatzki, Karin Knorr Cetina and Eike von Savigny (London: Routledge).

Barnes, Barry (2004), 'On Social Constructivist Accounts of the Natural Sciences,' in *Knowledge and the World: Challenges beyond the Science Wars*, ed. by Martin Carrier, Johannes Roggenhofer, Günter Küppers and Philippe Blanchard (Berlin: Springer), pp. 105–36. https://doi.org/10.1007/978-3-662-08129-7_5

Barnes, Barry (2011), 'Relativism as a Completion of the Scientific Project,' in *The Problem of Relativism in the Sociology of (Scientific) Knowledge*, ed. by Richard Schantz and Markus Seidel (Frankfurt: Ontos Verlag), pp. 23–39. https://doi.org/10.1515/9783110325904.23

Barnes, Barry and David Bloor (1982), 'Relativism, Rationalism and the Sociology of Knowledge,' in *Rationality and Relativism*, ed. by Martin Hollis and Steve Lukes (Oxford: Blackwell), pp. 21–47.

Barnes, Barry, David Bloor and John Henry (1996), *Scientific Knowledge: A Sociological Analysis* (London: Athlone).

Bartels, Christoph (2010), 'The Production of Silver, Copper, and Lead in the Harz Mountains from Late Medieval Times to the Onset of Industrialization,' in *Materials and Expertise in Early Modern Europe: Between Market and Laboratory*, ed. by Ursula Klein and Emma C. Spary (Chicago: University of Chicago Press), pp. 71–100. https://doi.org/10.7208/chicago/9780226439709.003.0004

Bast, Rainer A. (1986), *Der Wissenschaftsbegriff Martin Heideggers im Zusammenhang seiner Philosophie* (Stuttgart-Bad Cannstatt: Frommann-Holzboog Verlag).

Baudelaire, Charles (1964 [1863]), 'The Painter of Modern Life,' in *The Painter of Modern Life and Other Essays*, by Charles Baudelaire, trans. by Jonathan Mayne (London: Phaidon Press), pp. 1–40.

Bijker, Wiebe E. and Trevor Pinch (2012), 'Preface to the Anniversary Edition,' in *The Social Construction of Technological Systems: New Directions in Sociology and History of Technology (25th Anniversary Edition)*, ed. by Wiebe E. Bijker, Thomas P. Hughes and Trevor Pinch (Cambridge MA: MIT Press), pp. xi–xxxiv. https://books.google.de/books?id=B_Tas3u48f8C

Blackburn, Simon (2004), 'Lights! Camera! Being!' *New Republic* (February 23), 23.

Blattner, William D. (1994), 'Is Heidegger a Kantian Idealist?,' *Inquiry* 37(2), 185–201. https://doi.org/10.1080/00201749408602348

Blattner, William D. (1995), 'Decontextualization, Standardization, and Deweyan Science,' *Man and World* 28, 321–39. https://doi.org/10.1007/bf01273736

Blattner, William D. (2004), 'Heidegger's Kantian Idealism Revisited,' *Inquiry* 47(4), 321–37. https://doi.org/10.1080/00201740410004160

Bloor, David (1973), 'Wittgenstein and Mannheim on the Sociology of Mathematics,' *Studies in History and Philosophy of Science* 4(2), 173–91. https://doi.org/10.1016/0039-3681(73)90003-4

Bloor, David (1983), *Wittgenstein: A Social Theory of Knowledge* (London: Macmillan). https://doi.org/10.1007/978-1-349-17273-3

Bloor, David (1984), 'The Sociology of Reasons: Or Why "Epistemic Factors" Are Really "Social Factors,"' in *Scientific Rationality: The Sociological Turn*, ed. by James Robert Brown (Dordrecht: Reidel), pp. 295–324. https://doi.org/10.1007/978-94-015-7688-8_12

Bloor, David (1991 [1976]), *Knowledge and Social Imagery*, second edn (Chicago: University of Chicago Press).

Bloor, David (1996), 'Idealism and the Sociology of Knowledge,' *Social Studies of Science* 26(4), 839–56. https://doi.org/10.1177/030631296026004005

Bloor, David (1997a), 'Remember the Strong Program?,' *Science, Technology, & Human Values* 22(3), 373–85. https://doi.org/10.1177/016224399702200307

Bloor, David (1997b), *Wittgenstein, Rules and Institutions* (London: Routledge).

Bloor, David (1998), 'A Civil Scepticism,' *Social Studies of Science* 28(4), 655–65. https://doi.org/10.1177/030631298028004009

Bloor, David (1999a), 'Anti-Latour,' *Studies in History and Philosophy of Science* 30(1), 81–112. https://doi.org/10.1016/s0039-3681(98)00038-7

Bloor, David (1999b), 'Reply to Bruno Latour,' *Studies in History and Philosophy of Science* 30(1), 131–36. https://doi.org/10.1016/S0039-3681(98)00040-5

Bloor, David (2000), 'Whatever Happened to "Social Constructiveness"?,' in *Bartlett, Culture & Cognition*, ed. by Akiko Saito (London: Psychology Press), pp. 194–215. https://doi.org/10.4324/9780203457504

Bloor, David (2001), 'What Is a Social Construct?,' *Facta Philosophica* 3, 141–56.

Bloor, David (2003), 'Skepticism and the Social Construction of Science and Technology: The Case of the Boundary Layer,' in *The Skeptics: Contemporary Essays*, ed. by Steven Luper (Aldershot: Ashgate), pp. 249–65.

Bloor, David (2004a), 'Sociology of Scientific Knowledge,' in *Handbook of Epistemology*, ed. by I. Niiniluoto, M. Sintonen and J. Woleński (Dordecht: Kluwer), pp. 919–62. https://doi.org/10.1007/978-1-4020-1986-9_25

Bloor, David (2004b), 'Institutions and Rule-Scepticism: A Reply to Martin Kusch,' *Social Studies of Science* 34(4), 593–601. https://doi.org/10.1177/0306312704044910

Bloor, David (2007), 'Epistemic Grace: Antirelativism as Theology in Disguise,' *Common Knowledge* 13(2–3), 250–80. https://doi.org/10.1215/0961754x-2007-007

Bloor, David (2011), *The Enigma of the Aerofoil: Rival Theories in Aerodynamics, 1909–1930* (Chicago: Chicago University Press). https://doi.org/10.7208/chicago/9780226060934.001.0001

Bloor, David and David Edge (2000), 'Knowing Reality through Society,' *Social Studies of Science* 30(1), 158–60. http://journals.sagepub.com/doi/pdf/10.1177/030631200030001009

Boghossian, Paul A. (2006), *Fear of Knowledge: Against Relativism and Constructivism* (Oxford: Clarendon Press).

Borgmann, Albert (1978), 'Heidegger and Symbolic Logic,' in *Heidegger and Modern Philosophy*, ed. by Michael Murray (New Haven: Yale University Press), pp. 3–22.

Bourgeois, Patrick L. and Sandra B. Rosenthal (1988), 'Heidegger and Peirce: Beyond "Realism or Idealism,"' *Southwest Philosophy Review* 4(1), 103–10. https://doi.org/10.5840/swphilreview19884111

Boyle, Robert (1660 [1966]), *New Experiments Physico-Mechanical, Touching on the Spring of the Air, and its Effects; Made, for the Most Part, in an New Pneumatical Engine*, in *The Works of Robert Boyle*, vol. 1, by Robert Boyle, ed. by Thomas Birch (Hildesheim: Georg Olms), pp. 1–117.

Boyle, Robert (1662 [1966]), *A Defence of the Doctrine Touching the Spring and Weight of the Air, Proposed by Mr. R. Boyle, in his New Physico-Mechanical Experiments; Against the Objections of Franciscus Linus. Wherewith the Objector's Funicular Hypothesis is also Examined*, in *The Works of Robert Boyle*, vol. 1, by Robert Boyle, ed. by Thomas Birch (Hildesheim: Georg Olms), pp. 118–85.

Boyle, Robert (1685 [1966]), *An Essay of the Great Effects of Even Languid and Unheeded Motion*, in *The Works of Robert Boyle*, vol. 5, by Robert Boyle, ed. by Thomas Birch (Hildesheim: Georg Olms), pp. 1–37.

Bracken, Patrick J. (2002), *Trauma: Culture, Meaning and Philosophy* (London: Whurr).

Brandom, Robert (1983), 'Heidegger's Categories in *Being and Time*,' *Monist* 6(3), 387–409. https://doi.org/10.5840/monist198366327

Brown, James Robert (2001), *Who Rules in Science? An Opinionated Guide to the Science Wars* (Cambridge, MA: Harvard University Press).

Brun, Georg and Dominique Kuenzle (2008), 'A New Role for Emotions in Epistemology?,' in *Epistemology and Emotions*, ed. by Georg Brun, Ulvi Doğuoğlu and Dominique Kuenzle (Aldershot: Ashgate), pp. 1–31. https://doi.org/10.4324/9781315580128

Burnyeat, Miles F. (1982), 'Gods and Heaps,' in *Language and Logos*, ed. by Malcolm Schofeld and Martha Craven Nussbaum (Cambridge: Cambridge University Press), pp. 315–38. https://doi.org/10.1017/cbo9780511550874.017

Burtt, Edwin Arthur (1925), *The Metaphysical Foundations of Modern Physical Science: A Historical and Critical Essay* (London: Kegan Paul, Trench, Trubner & Co.). https://doi.org/10.4324/9781315822754

Caputo, John D. (1995), 'Heidegger's Philosophy of Science: The Two Essences of Science,' in *From Phenomenology to Thought, Errancy, and Desire: Essays in Honor of William J. Richardson, SJ*, ed. by Babette E. Babich (Dordecht: Kluwer Academic Publishers), pp. 43–60. https://doi.org/10.1007/978-94-009-4362-9_3

Caputo, John D. (2000), 'Hermeneutics and the Natural Sciences: Heidegger, Science, and Essentialism,' in *More Radical Hermeneutics*, by John D. Caputo (Bloomington: Indiana University Press), pp. 151–71, 281–84.

Carson, Cathryn (2010), 'Science as Instrumental Reason: Heidegger, Habermas, Heisenberg,' *Continental Philosophy Review* 42(4), 483–509. https://doi.org/10.1007/s11007-009-9124-y

Cartwright, Nancy (1999), *The Dappled World: A Study of the Boundaries of Science* (Cambridge: Cambridge University Press). https://doi.org/10.1017/cbo9781139167093

Celluci, Carlo (2013), *Rethinking Logic: Logic in Relation to Mathematics, Evolution, and Method* (Dordecht: Springer). https://doi.org/10.1007/978-94-007-6091-2

Cerbone, David R. (1995), 'World, World-Entry, and Realism in Early Heidegger,' *Inquiry* 38(4), 401–21. https://doi.org/10.1080/00201749508602397

Cerbone, David R. (2005), 'Realism and Truth,' in *A Companion to Heidegger*, ed. by Hubert L. Dreyfus and Mark Wrathall (Oxford: Blackwell), pp. 248–64. https://doi.org/10.1002/9780470996492.ch15

Cerbone, David R. (2012), 'Lost Belongings: Heidegger, Naturalism, and Natural Science,' in *Heidegger on Science*, ed. by Trish Glazebrook (Albany: SUNY Press), pp. 131–55.

Charland, Louis C. (1998), 'Is Mr. Spock Mentally Competent? Competence to Consent and Emotion,' *Philosophy, Psychiatry, & Psychology* 5(1), 67–81. http://muse.jhu.edu/article/28212

Charleton, Walter (1654 [1966]), *Physiologia Epicuro-Gassendo-Charltoniana: Or a Fabrick of Science Natural, Upon the Hypothesis of Atoms* (New York: Johnson Reproductions).

Chevalley, Catherine (1992), 'Heidegger and the Physical Sciences,' in *Martin Heidegger: Critical Assessments*, vol. 4: *Reverberations*, ed. by Christopher Macaan (London: Routledge), pp. 342–64.

Code, Lorraine (2015), 'Care, Concern, and Advocacy: Is There a Place for Epistemic Responsibility?,' *Feminist Philosophy Quarterly* 1(1), 1–20. http://ir.lib.uwo.ca/fpq/vol1/iss1/1/

Collins, Harry M. (1982), 'Special Relativism — The Natural Attitude,' *Social Studies of Science* 12(1), 139–43. https://doi.org/10.1177/030631282012001009

Collins, Harry M. (1992 [1985]), *Changing Order: Replication and Induction in Scientific Practice*, with a new foreword (Chicago: University of Chicago Press).

Collins, Harry M. (1999), 'The Science Police,' *Social Studies of Science* 29(2), 287–94. https://doi.org/10.1177/030631299029002012

Collins, Harry M. (2002), 'The Experimenter's Regress as Philosophical Sociology,' *Studies in History and Philosophy of Science* 33(1), 153–60. https://doi.org/10.1016/s0039-3681(01)00033-4

Collins, Harry M. and Trevor Pinch (2007), 'Who Is to Blame for the Challenger Explosion?,' *Studies in History and Philosophy of Science* 38(1), 254–55. https://doi.org/10.1016/j.shpsa.2006.12.006

Collins, Harry M. and Steven Yearley (1992), 'Journey into Space,' in *Science as Practice and Culture*, ed. by Andrew Pickering (Chicago: University of Chicago Press), pp. 369–89. https://doi.org/10.7208/chicago/9780226668208.001.0001

Crease, Robert P. (1992), 'The Problem of Experimentation,' in *Phenomenology of Natural Science*, ed. by Lee Hardy and Lester Embree (Dordecht: Kluwer Academic Publishers), pp. 215–35. https://doi.org/10.1007/978-94-011-2622-9_11

Crease, Robert P. (1993), *The Play of Nature: Experimentation as Performance* (Bloomington: Indiana University Press).

Crease, Robert P. (2009), "Covariant Realism,' *Human Affairs* 19(2), 223–32. https://doi.org/10.2478/v10023-009-0035-7

Crombie, Alistair C. (1962), *Robert Grosseteste and the Origins of Experimental Science (1100–1700)* (Oxford: Clarendon Press).

Crouthamel, Jason (2002), 'War Neurosis versus Saving Psychosis: Working-Class Politics and Psychological Trauma in Weimar Germany,' *Journal of Contemporary History* 37(2), 163–82. https://doi.org/10.1177/00220094020370020101

Crowell, Steven Galt (1992), 'Lask, Heidegger, and the Homelessness of Logic,' *Journal of the British Society for Phenomenology* 23(3), 222–39. https://doi.org/10.1080/00071773.1992.11006996

Crowell, Steven Galt (1994), 'Making Logic Philosophical Again (1912–1916),' in *Reading Heidegger from the Start: Essays in his Early Thought*, ed. by Theodore Kisiel and John van Buren (Albany: SUNY Press), pp. 55–72.

Dahlstrom, Daniel O. (2013), *The Heidegger Dictionary* (London: Bloomsbury).

Damasio, Antonio R. (1994), *Descartes' Error: Emotion, Reason, and the Human Brain* (New York: G. P. Putnam's Sons).

Damasio, Antonio R. (1999), *The Feeling of What Happens: Body and Emotion in the Making of Consciousness* (New York: Harcourt Brace & Company).

Daston, Lorraine and Peter Galison (2010), *Objectivity* (New York: Zone Books).

Davidson, Peter (2009), 'Francis Line S. J., *Explicatio horologii* (1673),' in *Jesuit Books in the Low Countries (1540–1773): A Selection from the Maurits Sabbe Library*, ed. by Paul Begheyn, S. J., Bernard Deprez, Rob Faesen, S. J., and Leo Kenis (Leuven: Uitgeverij Peeters), pp. 187–90.

Dea, Shannon (2009), 'Heidegger and Galileo's Slippery Slope,' *Dialogue: Canadian Philosophical Review* 48(1), 59–76. https://doi.org/10.1017/s0012217309090040

Dear, Peter (1990), 'Miracles, Experiments, and the Ordinary Course of Nature,' *Isis* 81(4), 663–83. https://doi.org/10.1086/355544

Dear, Peter (1995), *Discipline & Experience: The Mathematical Way in the Scientific Revolution* (Chicago: University of Chicago Press). https://doi.org/10.7208/chicago/9780226139524.001.0001

Dear, Peter (1998), 'Method and the Study of Nature,' in *The Cambridge Companion of Seventeenth-Century Philosophy*, vol. 1, ed. by Daniel Garber and Michael Ayers (Cambridge: Cambridge University Press), pp. 147–77. https://doi.org/10.1017/CHOL9780521307635.009

Dear, Peter (2001), *Revolutionizing the Sciences: European Knowledge and Its Ambitions, 1500–1700* (Basingstoke: Palgrave).

Dear, Peter (2006a), 'The Meanings of Experience,' in *The Cambridge History of Science*, vol. 3: *Early Modern Science*, ed. by Katharine Park and Lorraine Daston (Cambridge: Cambridge University Press), pp. 106–31. https://doi.org/10.1017/chol9780521572446.005

Dear, Peter (2006b), *The Intelligibility of Nature: How Science Makes Sense of the World* (Chicago: University of Chicago Press). https://doi.org/10.7208/chicago/9780226139500.001.0001

de Bruyckere, Anna and Maarten Van Dyck (2013), 'Being in or Getting at the Real: Kochan on Rouse, Heidegger and Minimal Realism,' *Perspectives of Science* 21(4), 453–62. https://doi.org/10.1162/posc_a_00112

Descartes, René (1969a [1644]), *The Principles of Philosophy*, trans. by Elizabeth S. Haldane and G. R. T. Ross, in *The Philosophical Works of Descartes*, vol. 1, by René Descartes (Cambridge: Cambridge University Press), pp. 201–302. https://doi.org/10.1017/cbo9780511805042.007

Descartes, René (1969b [1644]), *Rules for the Direction of the Mind*, trans. by Elizabeth S. Haldane and G. R. T. Ross, in *The Philosophical Works of Descartes*, vol. 1, by René Descartes (Cambridge: Cambridge University Press), pp. 1–77. https://doi.org/10.1017/cbo9780511805042.007

de Sousa, Ronald (1987), *The Rationality of Emotion* (Cambridge, MA: The MIT Press).

de Sousa, Ronald (2008), 'Epistemic Feelings,' in *Epistemology and Emotions*, ed. by Georg Brun, Ulvi Doğuoğlu and Dominique Kuenzle (Aldershot: Ashgate), pp. 185–204. https://doi.org/10.4324/9781315580128

Desroches, Dennis (2003), 'Phenomenology, Science Studies, and the Question of Being,' *Configurations* 11(3), 383–416. https://doi.org/10.1353/con.2004.0025

Dijksterhuis, Eduard Jan (1961 [1950]), *The Mechanization of the World Picture*, trans. by C. Dikshoorn (London: Oxford University Press).

Döring, Sabine A. (2010), 'Why Be Emotional?,' in *The Oxford Handbook of the Philosophy of Emotion*, ed. by Peter Goldie (Oxford: Oxford University Press), pp. 283–301. https://doi.org/10.1093/oxfordhb/9780199235018.001.0001

Douglas, Mary (1980), *Evans-Pritchard* (Sussex: Harvester Press).

Downing, George (2001), 'Emotion Theory Revisited,' in *Heidegger, Coping, and Cognitive Science: Essays in Honor of Hubert L. Dreyfus*, vol. 2, ed. by Mark A. Wrathall and Jeff Malpas (Cambridge, MA: The MIT Press), pp. 245–70.

Dreyfus, Hubert L. (1991a), 'Heidegger's Hermeneutic Realism,' in *The Interpretive Turn: Philosophy, Science, Culture*, ed. by David R. Hiley, James F. Bohman and Richard Shusterman (Ithaca: Cornell University Press), pp. 25–41. https://doi.org/10.1093/oso/9780198796220.003.0005

Dreyfus, Hubert L. (1991b), *Being-in-the-World: A Commentary on Heidegger's Being and Time, Division I* (Cambridge, MA: The MIT Press).

Dreyfus, Hubert L. (2001), 'How Heidegger Defends the Possibility of a Correspondence Theory of Truth with Respect to the Entities of Natural Science,' in *The Practice Turn in Contemporary Theory*, ed. by Theodore R. Schatzki, Karin Knorr D. Cetina and Eike von Savigny (London: Routledge), pp. 151–62. https://doi.org/10.1093/oso/9780198796220.003.0006

Dreyfus, Hubert L. and Charles Spinosa (1999), 'Coping with Things-in-Themselves: A Practice-Based Phenomenological Argument for Realism,' *Inquiry* 42(1), 49–78. https://doi.org/10.1080/002017499321624

Dror, Otniel E., Bettina Hitzer, Anjy Laukötter and Pilar León-Sanz, eds. (2016), *History of Science and the Emotions* (*Osiris* 31) (Chicago: University of Chicago Press). https://doi.org/10.1086/687590

DSM-IV-TR (2000), *Diagnostic and Statistical Manual of Mental Disorders*, fourth edn, text revision (Arlington: American Psychiatric Association).

Elgin, Catherine Z. (2008), 'Emotion and Understanding,' in *Epistemology and Emotions*, ed. by Georg Brun, Ulvi Doğuoğlu and Dominique Kuenzle (Aldershot: Ashgate), pp. 33–49. https://doi.org/10.4324/9781315580128

Evans-Pritchard, Edward E. (1937), *Witchcraft, Oracles and Magic among the Azande* (Oxford: Clarendon). https://doi.org/10.4324/9781912128297

Evans-Pritchard, Edward E. (1965), *Theories of Primitive Religion* (Oxford: Clarendon).

Evans-Pritchard, Edward E. (1970 [1934]), 'Lévy-Bruhl's Theory of Primitive Mentality,' *Journal of the Anthropological Society of Oxford* 1(2), 39–60. [First appearance: *Bulletin of the Faculty of Arts*, vol. ii (Cairo: Egyptian University, 1934).]

Falcon, Andrea (2015), 'Aristotle on Causality,' in *The Stanford Encyclopedia of Philosophy*, ed. by Edward N. Zalta (Spring 2015 Edition). https://plato.stanford.edu/archives/spr2015/entries/aristotle-causality/

Fay, Thomas A. (1977), *Heidegger: The Critique of Logic* (The Hague: Martinus Nijhoff). https://doi.org/10.1007/978-94-010-1057-3

Feenberg, Andrew (1992), 'Subversive Rationalization: Technology, Power, and Democracy,' *Inquiry* 35(3/4), 301–22. https://doi.org/10.1080/00201749208602296

Feenberg, Andrew (2005), *Heidegger and Marcuse: The Catastrophe and Redemption of History* (London: Routledge). https://doi.org/10.4324/9780203489000

Feenberg, Andrew (2006), 'Symmetry, Asymmetry, and the Real Possibility of Radical Change: A Reply to Kochan,' *Studies in the History and Philosophy of Science* 37(4), 721–27. https://doi.org/10.1016/j.shpsa.2006.09.003

Fell, Joseph P. (1989), 'The Familiar and the Strange: On the Limits of Praxis in the Early Heidegger,' in *Heidegger and Praxis*, ed. by Thomas J. Nenon (*The Southern Journal of Philosophy* 28, Spindel Conference Supplement), pp. 23–41. https://doi.org/10.1111/j.2041-6962.1990.tb00563.x

Fleck, Ludwik (1979 [1935]), *Genesis and Development of a Scientific Fact*, trans. by Fred Bradley and Thaddeus J. Trenn, ed. by Thaddeus J. Trenn and Robert K. Merton (Chicago: University of Chicago Press).

Forman, Paul (1971), 'Weimar Culture, Causality, and Quantum Theory, 1918–1927: Adaptation by German Physicists and Mathematicians to a Hostile Intellectual Environment,' *Historical Studies in the Physical Sciences* 3, 1–115. https://doi.org/10.2307/27757315

Foucault, Michel (1970), *The Order of Things: An Archaeology of the Human Sciences* (New York: Pantheon Books).

Freud, Sigmund (1985 [1919]), 'The "Uncanny,"' in *Art and Literature*, vol. 14 of *The Penguin Freud Library*, ed. by Albert Dickson (London: Penguin Books), pp. 339–76.

Freydberg, Bernard (2002), 'What Becomes of Science in "The Future of Phenomenology?"' *Research in Phenomenology* 32(1), 219–29. https://doi.org/10.1163/15691640260490656

Friedman, Michael (2000), *A Parting of the Ways: Carnap, Cassirer, and Heidegger* (Chicago: Open Court).

Frijda, Nico H., Antony S. R. Manstead and Sacha Bem, eds. (2000), *Emotions and Beliefs: How Feelings Influence Thoughts* (Cambridge: Cambridge University Press). https://doi.org/10.1017/cbo9780511659904

Frijda, Nico H. and Louise Sundararajan (2007), 'Emotion refinement: A theory inspired by Chinese poetics,' *Perspectives on Psychological Science* 2(3), 227–41. https://doi.org/10.1111/j.1745-6916.2007.00042.x

Galen of Pergamon (1944), *On Medical Experience*, first edn of the Arabic version, with English trans. by Richard Walzer (Oxford: Oxford University Press).

Ginev, Dimitri (2005), 'Against the Politics of Postmodern Philosophy of Science,' *International Studies in the Philosophy of Science* 19(2), 191–208. https://doi.org/10.1080/02698590500249522

Ginev, Dimitri (2011), *The Tenets of Cognitive Existentialism* (Athens, OH: Ohio University Press).

Ginev, Dimitri (2012), 'Two Accounts of the Hermeneutic Fore-structure of Scientific Research,' *International Studies in the Philosophy of Science* 26(4), 423–45. https://doi.org/10.1080/02698595.2012.748498

Ginev, Dimitri (2015), 'The Battle for Mathematical Existentialism and the War of the Heideggerian Succession: Rejoinder to Kochan,' in *Debating Cognitive Existentialism*, ed. by Dimitri Ginev (Leiden: Brill Rodopi), pp. 168–93. https://doi.org/10.1163/9789004299191_008

Glazebrook, Trish (1998), 'Heidegger on the Experiment,' *Philosophy Today* 42(3), 250–61. https://doi.org/10.5840/philtoday199842326

Glazebrook, Trish (2000a), *Heidegger's Philosophy of Science* (New York: Fordham University Press).

Glazebrook, Trish (2000b), 'From φύσις to Nature, τέχνη to Technology: Heidegger on Aristotle, Galileo, and Newton,' *The Southern Journal of Philosophy* 38, 95–118. https://doi.org/10.1111/j.2041-6962.2000.tb00892.x

Glazebrook, Trish (2001a), 'Heidegger and Scientific Realism,' *Continental Philosophy Review* 34, 361–401. https://doi.org/10.1023/a:1013148922905

Glazebrook, Trish (2001b), 'The Role of the *Beiträge* in Heidegger's Critique of Science,' *Philosophy Today* 45(1), 24–32. https://doi.org/10.5840/philtoday200145139

Glazebrook, Trish, ed. (2012a), *Heidegger and Science* (Albany: SUNY Press).

Glazebrook, Trish (2012b), 'Why Read Heidegger on Science?,' in *Heidegger on Science*, ed. by Trish Glazebrook (Albany: SUNY Press), pp. 13–26.

Godin, Benoît and Yves Gingras (2002), 'The Experimenter's Regress: From Skepticism to Argumentation,' *Studies in History and Philosophy of Science* 33(1), 133–48. https://doi.org/10.1016/s0039-3681(01)00032-2

Goffman, Erving (1967), *Interaction Ritual* (New York: Anchor).

Goldie, Peter (2004), 'Emotion, Reason, and Virtue,' in *Emotion, Evolution, and Rationality*, ed. by Dylan Evans & Pierre Cruse (Oxford: Oxford University Press), pp. 249–67. https://doi.org/10.1093/acprof:o so/9780198528975.003.0013

Goldie, Peter (2010), *Continental Divide: Heidegger, Cassirer, Davos* (Cambridge, MA: Harvard University Press).

Greenspan, Patricia (1988), *Emotions and Reasons: An Inquiry into Emotional Justification* (London: Routledge).

Greenspan, Patricia (2000), 'Emotional Strategies and Rationality,' *Ethics* 110(3), 469–87. https://doi.org/10.4324/9781315831848

Gross, Alan G. (2006), 'The Verbal and the Visual in Science: A Heideggerian Perspective,' *Science in Context* 19(4), 443–74. https://doi.org/10.1017/s0269889706001037

Gründer, Karlfried (1963), 'Heidegger's Critique of Science in Its Historical Background,' *Philosophy Today* 7(1), 15–32. https://doi.org/10.5840/philtoday1963719

Hacking, Ian (1992), '"Style" for Historians and Philosophers,' *Studies in History and Philosophy of Science* 23, 1–20. https://doi.org/10.1016/0039-3681(92)90024-z

Hacking, Ian (1999), *The Social Construction of What?* (Cambridge MA: Harvard University Press).

Hacking, Ian (2012), '"Language, Truth and Reason" 30 Years Later,' *Studies in History and Philosophy of Science* 43(4), 599–609. https://doi.org/10.1016/j.shpsa.2012.07.002

Haugland, John (2013), *Dasein Disclosed: John Haugeland's Heidegger*, ed. by Joseph Rouse (Cambridge, MA: Harvard University Press). https://doi.org/10.4159/harvard.9780674074590

Heelan, Patrick (1987), 'Husserl's Later Philosophy of Natural Science,' *Philosophy of Science* 54 (3), 368–90. https://doi.org/10.1086/289389

Heelan, Patrick (1995), 'Heidegger's Longest Day: Twenty-Five Years Later,' in *From Phenomenology to Thought, Errancy, and Desire: Essays in Honor of William J. Richardson, SJ*, ed. by Babette E. Babich (Dordecht: Kluwer Academic Publishers), pp. 579–87. https://doi.org/10.1007/978-94-017-1624-6_35

Heidegger, Martin (1927), *Sein und Zeit* (Tübingen: Max Niemeyer Verlag).

Heidegger, Martin (1949), *Über den Humanismus* (Frankfurt: Vittorio Klostermann).

Heidegger, Martin (1954a), 'Die Frage nach der Technik,' in *Vorträge und Aufsätze*, by Martin Heidegger (Pfullingen: Verlag Günther Neske), pp. 9–40.

Heidegger, Martin (1954b), 'Das Ding,' in *Vorträge und Aufsätze*, by Martin Heidegger (Pfullingen: Verlag Günther Neske), pp. 157–79.

Heidegger, Martin (1962a [1927]), *Being and Time*, trans. by John Macquarrie and Edward Robinson (Oxford: Blackwell).

Heidegger, Martin (1962b [1949]), 'Letter on Humanism,' trans. by Edgar Lohner, in *Philosophy in the Twentieth Century*, vol. 3, ed. by William Barrett and Henry D. Aiken (New York: Random House), pp. 271–302.

Heidegger, Martin (1967 [1962]), *What Is a Thing?*, trans. by William B. Barton, Jr., and Vera Deutsch (Chicago: Henry Regnery).

Heidegger, Martin (1969), 'Das Ende der Philosophie und die Aufgabe des Denkens,' in *Zur Sache des Denkens*, by Martin Heidegger (Tübingen: Max Niemeyer Verlag), pp. 61–80.

Heidegger, Martin (1971a), '"…Poetically Man Dwells…,"' in *Poetry, Language, Thought*, by Martin Heidegger, trans. by Albert Hofstadter (New York: Harper & Row), pp. 213–29.

Heidegger, Martin (1971b [1954]), 'The Thing,' in *Poetry, Language, Thought*, by Martin Heidegger, trans. by Albert Hofstadter (New York: Harper & Row), pp. 165–86.

Heidegger, Martin (1976 [1967]), 'On the Being and Conception of φύσις in Aristotle's Physics B, 1,' *Man and World* 9(3), 219–70. https://doi.org/10.1007/bf01249371

Heidegger, Martin (1977a [1952]), 'The Age of the World Picture,' in *The Question Concerning Technology*, by Martin Heidegger, trans. by William Lovitt (New York: Harper & Row), pp. 115–54. https://doi.org/10.1007/978-1-349-25249-7_3

Heidegger, Martin (1977b [1954]), 'The Question Concerning Technology,' in *The Question Concerning Technology and Other Essays*, by Martin Heidegger, trans. by William Lovitt (New York: Harper & Row), pp. 3–35.

Heidegger, Martin (1982a [1975]), *Basic Problems of Phenomenology*, trans. by Albert Hofstadter (Bloomington: Indiana University Press).

Heidegger, Martin (1982b), *Parmenides* (*Gesamtausgabe*, vol. 54) (Frankfurt: Vittorio Klostermann).

Heidegger, Martin (1984a [1978]), *The Metaphysical Foundations of Logic*, trans. by Michael Heim (Bloomington: Indiana University Press).

Heidegger, Martin (1984b [1962]), *Die Frage nach dem Ding: zu Kants Lehre von den transzendentalen Grundsätzen* (*Gesamtausgabe*, vol. 41) (Frankfurt: Vittorio Klostermann).

Heidegger, Martin (1985), *History of the Concept of Time*, trans. by Theodore Kisiel (Bloomington: Indiana University Press).

Heidegger, Martin (1987/88), 'Brief an Jean Beaufret / Letter to Jean Beaufret,' trans. by Steven Davis, *Heidegger Studies* 3/4, 3–6. https://www.pdcnet.org/pdc/bvdb.nsf/purchase?openform&fp=heideggerstud&id=heideggerstud_1987_0003_0000_0005_0006

Heidegger, Martin (1992 [1982]), *Parmenides*, trans. by André Schuwer and Richard Rojcewicz (Bloomington: Indiana University Press).

Heidegger, Martin (1993a [1978]), 'What Is Metaphysics?,' trans. by David Farrell Krell, in *Basic Writings*, revised and expanded edn, by Martin Heidegger, ed. by David Farrell Krell (New York: HarperCollins), pp. 93–110. https://doi.org/10.1017/CBO9780511812637.006

Heidegger, Martin (1993b [1947]), 'Letter on Humanism,' trans. by Frank A. Capuzzi and J. Glenn Gray, in *Basic Writings*, revised and expanded edn, ed. by David Farrell Krell (New York: HarperCollins), pp. 217–65.

Heidegger, Martin (1993c [1969]). 'The End of Philosophy and the Task of Thinking,' trans. by Joan Stambaugh, in *Basic Writings*, revised and expanded edn, by Martin Heidegger, ed. by David Farrell Krell (New York: HarperCollins), pp. 431–49.

Heidegger, Martin (1995a [1983]), *The Fundamental Concepts of Metaphysics: World, Finitude, Solitude*, trans. by William McNeill and Nicholas Walker (Bloomington: Indiana University Press).

Heidegger, Martin (1995b [1985]). *Aristotle's* Metaphysics Θ 1–3: *On the Essence and Actuality of Force*, trans. Walter Brogan & Peter Warnek (Bloomington: Indiana University Press).

Heidegger, Martin (1996 [1984]), *Hölderlin's Hymn 'The Ister,'* trans. by William McNeill and Julia Davis (Bloomington: Indiana University Press).

Heidegger, Martin (1997 [1929]), *Kant and the Problem of Metaphysics*, fifth edn, enlarged, trans. by Richard Taft (Bloomington: Indiana University Press).

Heidegger, Martin (1998), 'Traditional Language and Technological Language,' trans. by Wanda Torres Gregory, *Journal of Philosophical Research* 23, 129–45. https://doi.org/10.5840/jpr_1998_16

Heidegger, Martin (1999 [1989]), *Contributions to Philosophy (From Enowning)*, trans. by Pravis Emad and Kenneth Maly (Bloomington: Indiana University Press).

Heidegger, Martin (2000 [1953]), *Introduction to Metaphysics*, trans. by Gregory Fried and Richard Polt (New Haven: Yale University Press).

Heidegger, Martin (2009 [1998]), *Logic as the Question Concerning the Essence of Language*, trans. by Wanda Torres Gregory and Yvonne Unna (Albany: SUNY Press).

Heidegger, Martin (2010a [1976]), *Logic: The Question of Truth*, trans. by Thomas Sheehan (Bloomington: Indiana University Press).

Heidegger, Martin (2010b), *Being and Time*, trans. by Joan Stambaugh, revised edn (Albany: SUNY Press).

Heisenberg, Werner (1959), 'Grundlegende Voraussetzungen in der Physik der Elementarteilchen,' in *Martin Heidegger zum Siebzigsten Geburtstag. Festschrift*, ed. by Günther Neske (Pfullingen: Verlag Günther Neske), pp. 291–97.

Heisenberg, Werner (1977), 'Brief an Martin Heidegger zum 80. Geburtstag,' in *Dem Andenken Martin Heideggers. Zum 26. Mai 1976* (Frankfurt: Vittorio Klostermann), pp. 44–45.

Hempel, Hans-Peter (1990), *Natur und Geschichte. Der Jahrhundertdialog zwischen Heidegger und Heisenberg* (Frankfurt: Hain).

Hoffman, Piotr (2000), 'Heidegger and the Problem of Idealism,' *Inquiry* 43(4), 403–12. https://doi.org/10.1080/002017400750051215

Hookway, Christopher (1998), 'Doubt: Affective States and the Regulation of Inquiry,' in *Pragmatism*, ed. by C. J. Misak (*Canadian Journal of Philosophy, Supplementary Volume* 24), pp. 203–25. http://dx.doi.org/10.1080/00455091.1998.10717500

Hookway, Christopher (2002), 'Emotions in Epistemic Evaluations,' in *The Cognitive Basis of Science*, ed. by Peter Carruthers, Steven Stich and Michael Siegel (Cambridge: Cambridge University Press), pp. 251–61. https://doi.org/10.1017/CBO9780511613517.014

Hookway, Christopher (2008), 'Epistemic Immediacy, Doubt and Anxiety: On a Role for Affective States in Epistemic Evaluation,' in *Epistemology and Emotions*, ed. by Georg Brun, Ulvi Doğuoğlu and Dominique Kuenzle (Aldershot: Ashgate), p. 51–65. https://doi.org/10.4324/9781315580128

Howson, Colin (2009), 'Sorites Is No Threat to *Modus Ponens*: A Reply to Kochan,' *International Studies in the Philosophy of Science* 23(2), 209–12. https://doi.org/10.1080/02698590903007188

Husserl, Edmund (1965 [1910]), 'Philosophy as Rigorous Science,' trans. by Quentin Lauer, in *Phenomenology and the Crisis of Philosophy*, by Edmund Husserl (New York: HarperCollins), pp. 71–147.

Husserl, Edmund (1970 [1954]), *The Crisis of European Sciences and Transcendental Phenomenology*, trans. by David Carr (Evanston: Northwestern University Press).

Hyder, David and Hans-Jörg Rheinberger, eds. (2010), *Science and the Life-World: Essays on Husserl's* Crisis of European Sciences (Stanford: University of Stanford Press). https://doi.org/10.11126/stanford/9780804756044.001.0001

Jagger, Alison M. (1989), 'Love and Knowledge: Emotion in Feminist Epistemology,' in *Women, Knowledge, and Reality: Explorations in Feminist Epistemology*, ed. by Ann Garry & Marilyn Pearsall (Boston: Unwin Hyman), pp. 129–55. https://doi.org/10.1080/00201748908602185

Jardine, Nicholas (1976), 'Galileo's Road to Truth and the Demonstrative Regress,' *Studies in History and Philosophy of Science* 7(4), 277–318. https://doi.org/10.1016/0039-3681(76)90009-1

Jünger, Ernst (1929), *The Storm of Steel*, trans. by Basil Creighton (London: Chatto & Windus).

Kahn, Charles H. (1966), 'The Greek Verb "To Be" and the Concept of Being,' *Foundations of Language* 2, 245–65.

Kant, Immanuel (1956), *Critique of Practical Reason*, trans. by Lewis White Beck (New York: Macmillan). https://doi.org/10.1017/CBO9780511809576

Kant, Immanuel (1998 [1781/1787]), *Critique of Pure Reason*, ed. and trans. by Paul Guyer and Allen W. Wood (Cambridge: Cambridge University Press). https://doi.org/10.1017/CBO9780511804649

Käufer, Stephan (2001), 'On Heidegger on Logic,' *Continental Philosophy Review* 34(4), 455–76. https://doi.org/10.1023/a:1013110920041

Käufer, Stephan (2005), 'Logic,' in *A Companion to Heidegger*, ed. by Hubert L Dreyfus and Mark A. Wrathall (Oxford: Blackwell), pp. 141–55. https://doi.org/10.1002/9780470996492.ch9

Kaufman, Doris (1999), 'Science as Cultural Practice: Psychiatry in the First World War and Weimar Germany,' *Journal of Contemporary History* 34(1), 125–44. https://doi.org/10.1177/002200949903400107

Keller, Evelyn Fox (1982), 'Feminism and Science,' *Signs* 7(3), 589–602. https://doi.org/10.1086/493901

Keller, Evelyn Fox (1983), *A Feeling for the Organism: The Life and Work of Barbara McClintock* (San Francisco: W. H. Freeman).

Kessler, Eckhard (1988), 'The Intellective Soul,' in *The Cambridge History of Renaissance Philosophy*, ed. by Charles B. Schmitt, Quentin Skinner, Eckhard Kessler and Jill Kraye (Cambridge: Cambridge University Press), pp. 485–534. https://doi.org/10.1017/CHOL9780521251044.017

Kisiel, Theodore J. (1970), 'Science, Phenomenology, and the Thinking of Being,' in *Phenomenology and the Natural Sciences, Essays and Translations*, ed. by Joseph J. Kockelmans and Theodore J. Kisiel (Evanston: Northwestern University Press), pp. 167–83.

Kisiel, Theodore J. (1992), 'Heidegger and the New Images of Science,' in *Martin Heidegger: Critical Assessments*, vol. 4: *Reverberations*, ed. by Christopher Macaan (London: Routledge), pp. 325–41. https://doi.org/10.1163/156916477x00112

Kitcher, Philip (1998), 'A Plea for Science Studies,' in *A House Built on Sand: Exposing Postmodern Myths about Science*, ed. by Noretta Koertge (Oxford: Oxford University Press), pp. 32–56. https://doi.org/10.1093/0195117255.003.0004

Kiverstein, Julian, ed. (2012), *Heidegger and Cognitive Science* (Basingstoke: Palgrave Macmillan).

Klein, Ursula and Emma C. Spary (2010), 'Introduction: Why Materials?,' in *Materials and Expertise in Early Modern Europe: Between Market and Laboratory*, ed. by Ursula Klein and Emma C. Spary (Chicago: University of Chicago Press), pp. 1–23. https://doi.org/10.7208/chicago/9780226439709.003.0001

Knorr-Cetina, Karin D. (1981), *The Manufacture of Knowledge: An Essay on the Constructivist and Contextual Nature of Science* (Oxford: Pergamon Press). https://doi.org/10.1016/B978-0-08-025777-8.50014-1

Kochan, Jeff (2006), 'Feenberg and STS: Counter-Reflections on Bridging the Gap,' *Studies in History and Philosophy of Science* 37(4): 702–20. https://doi.org/10.1016/j.shpsa.2006.06.001

Kochan, Jeff (2008), 'Realism, Reliabilism, and the "Strong Programme" in the Sociology of Scientific Knowledge,' *International Studies in the Philosophy of Science* 22(1), 21–38. https://doi.org/10.1080/02698590802280878

Kochan, Jeff (2009a), 'The Exception Makes the Rule: Reply to Howson,' *International Studies in the Philosophy of Science* 23(2), 213–16. https://doi.org/10.1080/02698590903007196

Kochan, Jeff (2009b), 'Popper's Communitarianism,' in *Rethinking Popper*, ed. by Zuzana Parusniková and Robert S. Cohen (*Boston Studies in the Philosophy of Science* 272) (Berlin: Springer), pp. 287–303. https://doi.org/10.1007/978-1-4020-9338-8_22

Kochan, Jeff (2010a), 'Contrastive Explanation and the "Strong Programme" in the Sociology of Scientific Knowledge,' *Social Studies of Science* 40(1), 127–44. https://doi.org/10.1177/0306312709104780

Kochan, Jeff (2010b), 'Latour's Heidegger,' *Social Studies of Science* 40(4), 579–98. https://doi.org/10.1177/0306312709360263

Kochan, Jeff (2011a), 'Getting Real with Rouse and Heidegger,' *Perspectives on Science* 19(1), 81–115. https://doi.org/10.1162/posc_a_00026

Kochan, Jeff (2011b), 'Husserl and the Phenomenology of Science,' *Studies in History and Philosophy of Science* 42(3), 467–71. https://doi.org/10.1016/j.shpsa.2011.02.003

Kochan, Jeff (2013), 'Subjectivity and Emotion in Scientific Research,' *Studies in History and Philosophy of Science* 44(3), 354–62. https://doi.org/10.1016/j.shpsa.2013.05.003

Kochan, Jeff (2015a), 'Scientific Practice and Modes of Epistemic Existence,' in *Debating Cognitive Existentialism*, ed. by Dimitri Ginev (Leiden: Brill Rodopi), pp. 95–106. https://doi.org/10.1163/9789004299191_007

Kochan, Jeff (2015b), 'Putting a Spin on Circulating Reference, or How to Rediscover the Scientific Subject,' *Studies in History and Philosophy of Science* 49, 103–07. https://doi.org/10.1016/j.shpsa.2014.10.004

Kochan, Jeff (2015c), 'Circles of Scientific Practice: *Regressus, Mathēsis, Denkstil,*' in *Critical Science Studies after Ludwik Fleck*, ed. by Dimitri Ginev (Sophia: St. Kliment Ohridski University Press), pp. 83–99.

Kochan, Jeff (2015d), 'Objective Styles in Northern Field Science,' *Studies in History and Philosophy of Science* 52, 1–12. https://doi.org/10.1016/j.shpsa.2015.04.001

Kochan, Jeff (2015e), 'Reason, Emotion, and the Context Distinction,' *Philosophia Scientiae* 19(1), 35–43. https://doi.org/10.4000/philosophiascientiae.1036

Kochan, Jeff (2016), 'Circles of Scientific Practice: *Regressus, Mathēsis, Denkstil,*' in *Fleck and the Hermeneutics of Science* (*Collegium Helveticum Heft* 14), ed. by Erich Otto Graf, Martin Schmid and Johannes Fehr (Zürich, 2016), pp. 85–93. (Reprint of Kochan (2015c).)

Kochan, Jeff and Hans Bernhard Schmid (2011), 'Philosophy of Science,' in *The Routledge Companion to Phenomenology*, ed. by Sebastian Luft and Søren Overgaard (London: Routledge), pp. 461–72. https://doi.org/10.4324/9780203816936.ch42

Kockelmans, Joseph J. (1970), 'Heidegger on the Essential Difference and Necessary Relationship between Philosophy and Science,' in *Phenomenology and the Natural Sciences, Essays and Translations*, ed. by Joseph J. Kockelmans and Theodore J. Kisiel (Evanston: Northwestern University Press), pp. 147–66.

Kockelmans, Joseph J. (1985), *Heidegger and Science* (Lanham: The Center for Advanced Research in Phenomenology, and The University Press of America).

Kohler, Robert E. (2002), *Landscapes and Labscapes: Exploring the Lab-Field Border in Biology* (Chicago: University of Chicago Press). https://doi.org/10.7208/chicago/9780226450117.001.0001

Kolb, David A. (1983), 'Heidegger on the Limits of Science,' *Journal of the British Society for Phenomenology* 14(1), 50–64. https://doi.org/10.1080/00071773.1983.11007608

Kovacs, George (1990), 'Philosophy as Primordial Science (*Urwissenschaft*) in the Early Heidegger,' *Journal of the British Society for Phenomenology* 21(2), 121–35. https://doi.org/10.1080/00071773.1990.11006890

Kovacs, George (1994), 'Philosophy as Primordial Science in Heidegger's Courses of 1919,' in *Reading Heidegger from the Start: Essays in His Earliest Thought*, ed. by Theodore Kisiel & John van Buren (Albany: SUNY Press), pp. 91–107, 429–31.

Koyré, Alexandre (1931), 'L'introduction du "*Qu'est-ce que la métaphysique?*"' *Bifur* 8, 5–8.

Koyré, Alexandre (1943a), 'Galileo and Plato,' *Journal of the History of Ideas* 4(4), 400–28. https://doi.org/10.2307/2707166

Koyré, Alexandre (1943b), 'Galileo and the Scientific Revolution of the Seventeenth Century,' *Philosophical Review* 52(4), 333–48. https://doi.org/10.2307/2180668

Kuhn, Thomas (1977), 'Mathematical versus Experimental Traditions in the Development of Physical Science,' in *The Essential Tension: Selected Studies in Scientific Tradition and Change*, by Thomas Kuhn (Chicago: University of Chicago Press), pp. 31–65. https://doi.org/10.2307/202372

Kusch, Martin (2004a), 'Reply to My Critics,' *Social Studies of Science* 34(4), 615–20. https://doi.org/10.1177/0306312704046600

Kusch, Martin (2004b), 'Rule-Scepticism and the Sociology of Scientific Knowledge: The Bloor-Lynch Debate Revisited,' *Social Studies of Science* 34(4), 571–91. https://doi.org/10.1177/0306312704044168

Kusch, Martin (2018), 'Scientific Realism and Social Epistemology,' in *Routledge Handbook of Scientific Realism*, ed. by Juha Saatsi (London: Routledge), pp 261–75.

Langton, Rae (1998), *Kantian Humility: Our Ignorance of Things in Themselves* (Oxford: Clarendon Press). https://doi.org/10.1093/0199243174.001.0001

Latour, Bruno (1983), 'Give Me a Laboratory and I Will Raise the World,' in *Science Observed: Perspectives on the Social Study of Science*, ed. by Karin D. Knorr-Cetina and Michael Mulkay (London: SAGE), pp. 141–70.

Latour, Bruno (1987), *Science in Action: How to Follow Scientists and Engineers through Society* (Cambridge, MA: Harvard University Press).

Latour, Bruno (1992), 'One More Turn after the Social Turn…,' in *The Social Dimensions of Science*, ed. by Ernan McMullin (Notre Dame: University of Notre Dame Press), pp. 272–94.

Latour, Bruno (1993), *We Have Never Been Modern* (Harvard: Harvard University Press).

Latour, Bruno (1999a), 'For David Bloor … and Beyond: A Reply to David Bloor's "Anti-Latour,"' *Studies in History and Philosophy of Science* 30(1), 113–29. https://doi.org/10.1016/S0039-3681(98)00039-9

Latour, Bruno (1999b), *Pandora's Hope: Essays on the Reality of Science Studies* (Cambridge, MA: Harvard University Press).

Latour, Bruno (2002), 'The Science Wars: A Dialogue,' *Common Knowledge* 8(1), 71–79. https://doi.org/10.1215/0961754x-8-1-71

Leung, Ka-wing (2006), 'Heidegger on the Problem of Reality,' *The New Yearbook for Phenomenology and Phenomenological Philosophy* 6, 169–84. Available at https://www.pdcnet.org/pdc/bvdb.nsf/purchase?openform&fp=nyppp&id=nyppp_2006_0006_0169_0184

Lewens, Tim (2005), 'Realism and the Strong Program,' *British Journal for the Philosophy of Science* 56(3), 559–77. https://doi.org/10.1093/bjps/axi125

Linus, Franciscus (1661), *Tractatus de Corporum Inseparabilitate in quo Experimenta de Vacuo, tam Torriculliana, quam Magdeburgica, et Boyliana examinantur, veraque eorum causa detecta, ostenditur, vacuum naturaliter dari non posse: unde et Aristotelica de Rarefactione sententia tam contra Assertores Vacuitatum, quam Corpusculorum demonstratur* (London: Typis Thomas Roycroft, impensis Joan. Martin, Ja. Allestry & Tho. Ditas). Available at https://books.google.co.uk/books?id=KQFRnQEACAAJ

Longino, Helen E. (1993), 'Subject, Power, and Knowledge: Description and Prescription in Feminist Philosophies of Science,' in *Feminist Epistemologies*, ed. by Linda Alcoff and Elizabeth Potter (London: Routledge), pp. 101–20.

Lukács, György (1971 [1923]), *History and Class Consciousness: Studies in Marxist Dialectics*, trans. by Rodney Livingstone (London: Merlin Press).

Ma, Lin and Jaap van Brakel (2014), 'Heidegger's Thinking on the "Same" of Science and Technology,' *Continental Philosophy Review* 47(1), 19–43. https://doi.org/10.1007/s11007-014-9287-z

Magomed-Eminov, Madrudin S. (1997), 'Post-Traumatic Stress Disorders as a Loss of the Meaning of Life,' in *States of Mind: American and Post-Soviet Perspectives on Contemporary Issues in Psychology*, ed. by Diane F. Halpern and Alexander E. Voiskounsky (Oxford: Oxford University Press), pp. 238–50.

Mannheim, Karl (1984 [1925]), *Konservatismus. Ein Beitrag zur Soziologie des Wissens*, ed. by David Kettler, Volker Meja and Nico Stehr (Frankfurt: Suhrkamp).

Mannheim, Karl (1953 [1925]), 'Conservative Thought,' in *Essays on Sociology and Social Psychology*, by Karl Mannheim, ed. by Paul Kecskemeti (London: Routledge), pp. 74–164.

May, Reinhard (1996), *Heidegger's Hidden Sources: East Asian Influence on His Work*, trans. by Graham Parkes (London: Routledge). https://doi.org/10.4324/9780203979792

Mayr, Otto (1986), *Authority, Liberty & Automatic Machinery in Early Modern Europe* (Baltimore: John Hopkins University Press).

McAllister, James W. (2002), 'Recent Work on Aesthetics of Science,' *International Studies in the Philosophy of Science* 16(1), 7–11. https://doi.org/10.1080/02698590120118783

McAllister, James W. (2005), 'Emotion, Rationality, and Decision Making in Science,' in *Logic, Methodology, and Philosophy of Science: Proceedings of the Twelfth International Congress*, ed. by Petr Jájek, Luis Valdés-Villanueva and Dag Westerståhl (London: King's College Publications), pp. 559–76.

McAllister, James W. (2007), 'Dilemmas in Science: What, Why, and How,' in *Knowledge in Ferment: Dilemmas in Science, Scholarship and Society*, ed. by Adriaan in't Groen, Henk Jan de Jonge, Eduard Klasen Hilje Papma and Piet van Slooten (Leiden: Leiden University Press), pp. 13–24.

McAllister, James W. (2014). 'Methodological Dilemmas and Emotion in Science,' *Synthese* 191, 3143–58. https://doi.org/10.1007/s11229-014-0477-3

McGuire, James E. and Barbara Tuchanska (2000), *Science Unfettered: A Philosophical Study in Sociohistorical Ontology* (Athens, OH: Ohio University Press).

McManus, Denis (2007), 'Heidegger, Measurement and the "Intelligibility" of Science,' *European Journal of Philosophy* 15(1), 82–105. https://doi.org/10.1111/j.1468-0378.2007.00243.x

McManus, Denis (2012), *Heidegger and the Measure of Truth: Themes from his Early Philosophy* (Oxford: Oxford University Press). https://doi.org/10.1093/acprof:oso/9780199694877.001.0001

McNeill, William (1999), *The Glance of the Eye: Heidegger, Aristotle, and the Ends of Theory* (Albany: SUNY Press).

Merchant, Carolyn (1980), *The Death of Nature: Women, Ecology and the Scientific Revolution* (San Francisco: Harper & Row).

Mohanty, Jitendranath (1988), 'Heidegger on Logic,' *Journal of the History of Philosophy* 26(1), 107–35. https://doi.org/10.1353/hph.1988.0014

Morton, Adam (2010), 'Epistemic Emotions,' in *The Oxford Handbook of the Philosophy of Emotion*, ed. by Peter Goldie (Oxford: Oxford University Press), pp. 385–99. https://doi.org/10.1093/oxfordhb/9780199235018.003.0018

Mosse, George L. (2000), 'Shell-Shock as a Social Disease,' *Journal of Contemporary History* 35(1), 101–08. https://doi.org/10.1177/002200940003500109

Mumford, Lewis (1964), 'Authoritarian and Democratic Technics,' *Technology and Culture* 5(1), 1–8. https://doi.org/10.2307/3101118

Nagel, Thomas (1987), *What Does It All Mean? A Very Short Introduction to Philosophy* (New York: Oxford University Press).

Nicholson, Graeme (2008), 'Heidegger, Descartes and the Mathematical,' in *Descartes and the Modern*, ed. by Neil Robertson, Gordon McOuat and Tom Vinci (Newcastle: Cambridge Scholars Publishing), pp. 216–34.

Nieli, Russell (1987), *Wittgenstein: From Mysticism to Ordinary Language* (Albany: SUNY Press).

Norris, Christopher (1997), *Against Relativism: Philosophy of Science, Deconstruction and Critical Theory* (Oxford: Blackwell).

Norton, John D. (2000), 'How We Know about Electrons,' in *After Popper, Kuhn and Feyerabend*, ed. by Robert Nola and Howard Sankey (Dordecht: Kluwer), pp. 67–97. https://doi.org/10.1007/978-94-011-3935-9_2

O'Brien, Tim (1990), *The Things They Carried* (New York: Broadway Books).

Osbeck, Lisa M. and Nancy J. Nersessian (2011), 'Affective Problem-Solving: Emotion in Research Practice,' *Mind & Society* 10(1), 57–78. https://doi.org/10.1007/s11299-010-0074-1

Osbeck, Lisa M. and Nancy J. Nersessian (2013), 'Beyond Motivation and Metaphor: "Scientific Passions" and Anthropomorphism,' in *EPSA 11: Perspectives and Foundational Problems in Philosophy of Science*, ed. by V. Karakostas & D. Dieks (Dordecht: Springer), pp. 455–66. https://doi.org/10.1007/978-3-319-01306-0_37

Osbeck, Lisa M., Nancy J. Nersessian, Wendy C. Newstetter and Kareen R. Malone (2011), *Science as Psychology: Sense-Making and Identity in Scientific Practice* (Cambridge: Cambridge University Press). https://doi.org/10.1017/cbo9780511933936

Pickering, Andrew (2009), 'The Politics of Theory: Producing Another World, With Some Thoughts on Latour,' *Journal of Cultural Economy* 2(1–2), 197–212. https://doi.org/10.1080/17530350903064204

Pickering, Andrew (2010), *The Cybernetic Brain: Sketches of Another Future* (Chicago: University of Chicago Press). https://doi.org/10.7208/chicago/9780226667928.001.0001

Polanyi, Michael (1958), *Personal Knowledge: Towards a Post-Critical Philosophy* (London: Routledge & Kegan Paul). https://doi.org/10.7208/chicago/9780226232768.001.0001

Popkin, Richard H. (1979), *The History of Scepticism from Erasmus to Spinoza* (Berkley: University of California Press).

Potter, Elizabeth (2001), *Gender and Boyle's Law of Gases* (Bloomington: Indiana University Press).

Psillos, Stathis (1999), *Scientific Realism: How Science Tracks the Truth* (London: Routledge). https://doi.org/10.4324/9780203979648

Randall, John Herman, Jr (1940), 'The Development of the Scientific Method in the School of Padua,' *Journal of the History of Ideas* 1(2), 177–206. https://doi.org/10.2307/2707332

Randall, John Herman Jr (1961), *The School of Padua and the Emergence of Modern Science* (Padua: Editrice Antenore).

Ratcliffe, Matthew (2002), 'Heidegger's Attunement and the Neuropsychology of Emotion,' *Phenomenology and the Cognitive Sciences* 1(3), 287–312. https://doi.org/10.1023/a:1021312100964

Reilly, Conor (1969), *Francis Line S. J.: An Exiled English Scientist 1595–1675* (Rome: Institutum Historicum).

Rheinberger, Hans-Jörg (1997), *Towards a History of Epistemic Things: Synthesizing Proteins in the Test Tube* (Stanford: University of Stanford Press).

Rheinberger, Hans-Jörg (2010a), *An Epistemology of the Concrete: Twentieth-Century Histories of Life* (Durham: Duke University Press). https://doi.org/10.1215/9780822391333

Rheinberger, Hans-Jörg (2010b), *On Historicizing Epistemology: An Essay* (Stanford: Stanford University Press).

Richardson, William J. (1968), 'Heidegger's Critique of Science,' *New Scholasticism* 42(4), 511–36. https://doi.org/10.5840/newscholas196842444

Roeser, Sabine (2012), 'Emotional Engineers: Toward Morally Responsible Design,' *Science and Engineering Ethics* 18(1), 103–15. https://doi.org/10.1007/s11948-010-9236-0

Rorty, Richard (1991), 'Is Natural Science a Natural Kind?,' in *Objectivity, Relativism, and Truth* (*Philosophical Papers*, vol. 1), by Richard Rorty (Cambridge: University of Cambridge Press), pp. 46–62. https://doi.org/10.1017/CBO9781139173643.004

Roubach, Michael (1997), 'Heidegger, Science, and the Mathematical Age,' *Science in Context* 10(1), 199–206. https://doi.org/10.1017/s0269889700000326

Rouse, Joseph (1985), 'Science and the Theoretical "Discovery" of the Present-at-Hand,' in *Descriptions*, ed. by Don Ihde and Hugh J. Silverman (Albany: SUNY Press), pp. 200–10.

Rouse, Joseph (1987a), *Knowledge and Power: Toward a Political Philosophy of Science* (Ithaca: Cornell University Press).

Rouse, Joseph (1987b), 'Husserlian Phenomenology and Scientific Realism,' *Philosophy of Science* 54(2), 222–32. https://doi.org/10.1086/289371

Rouse, Joseph (1993), 'What Are Cultural Studies of Scientific Knowledge?,' *Configurations* 1(1), 57–94. https://doi.org/10.1353/con.1993.0006

Rouse, Joseph (1996a), *Engaging Science: How to Understand Its Practices Philosophically* (Ithaca: Cornell University Press).

Rouse, Joseph (1996b), 'Feminism and the Social Construction of Scientific Knowledge,' in *Feminism, Science, and the Philosophy of Science*, ed. by L. H. Nelson and J. Nelson (Dordecht: Kluwer), pp. 195–215. https://doi.org/10.1007/978-94-009-1742-2_10

Rouse, Joseph (1998), 'Heideggerian Philosophy of Science,' in *Routledge Encyclopedia of Philosophy*, vol. 4, ed. by Edward Craig (London: Routledge), pp. 323–27.

Rouse, Joseph (1999), 'Understanding Scientific Practices: Cultural Studies of Science as a Philosophical Program,' in *The Science Studies Reader*, ed. by Mario Biagioli (London: Routledge), pp. 442–56.

Rouse, Joseph (2002a), 'Vampires: Social Constructivism, Realism, and Other Philosophical Undead,' *History and Theory* 41, 60–78. https://doi.org/10.1111/1468-2303.00191

Rouse, Joseph (2002b), *How Scientific Practices Matter: Reclaiming Philosophical Naturalism* (Chicago: University of Chicago Press).

Rouse, Joseph (2005a), 'Heidegger and Scientific Naturalism,' in *Continental Philosophy of Science*, ed. by Gary Gutting (Oxford: Blackwell), pp. 123–41. https://doi.org/10.1002/9780470755501.ch10

Rouse, Joseph (2005b), 'Heidegger's Philosophy of Science,' in *A Companion to Heidegger*, ed. by Hubert L. Dreyfus and Harrison Hall (Oxford: Blackwell), pp. 173–89. https://doi.org/10.1002/9780470996492.ch11

Rouse, Joseph (2015), *Articulating the World: Conceptual Understanding and the Scientific Image* (Chicago: University of Chicago Press). https://doi.org/10.7208/chicago/9780226293707.001.0001

Sallis, John (1971), 'Toward the Movement of Reversal: Science, Technology, and the Language of Homecoming,' in *Heidegger and the Path of Thinking*, ed. by John Sallis (Pittsburgh: Duquesne University Press), pp. 138–68.

Schatzki, Theodore R. (1992), 'Early Heidegger on Being, the Clearing, and Realism,' in *Heidegger: A Critical Reader*, ed. by Hubert L. Dreyfus and Harrison Hall (Oxford: Blackwell), pp. 81–98.

Scheff, Thomas J. (1988), 'Shame and Conformity: The Deference-Emotion System,' *American Sociological Review* 53(3), 395–406. https://doi.org/10.2307/2095647

Schmitt, Charles B. (1969), 'Experience and Experiment: A Comparison of Zabarella's View with Galileo's in *De Motu*,' *Studies in the Renaissance* 16, 80–138. https://doi.org/10.2307/2857174

Schwendtner, Tibor (2005), *Heideggers Wissenschaftsauffasung: Im Spiegel der Schriften 1919–29* (Frankfurt: Peter Lang).

Schyfter, Pablo (2012), 'Standing Reserves of Function: A Heideggerian Reading of Synthetic Biology,' *Philosophy & Technology* 25(2), 199–219. https://doi.org/10.1007/s13347-011-0053-4

Seigfried, Hans (1980), 'Scientific Realism and Phenomenology,' *Zeitschrift für philosophische Forschung* 34, 395–404.

Seigfried, Hans (1990), 'Autonomy and Quantum Physics: Nietzsche, Heidegger, and Heisenberg,' *Philosophy of Science* 57(4), 619–30.

Serjeantson, Richard W. (2004), 'Casaubon, Méric,' in *Oxford Dictionary of National Biography*, vol. 10, ed. by H. C. G. Matthews and Brian Harrison (Oxford: Oxford University Press), pp. 464–66.

Shapin, Steven (1988), 'Robert Boyle and Mathematics: Reality, Representation, and Experimental Practice,' *Science in Context* 2(1), 23–58. https://doi.org/10.1017/s026988970000048x

Shapin, Steven (1992), 'Discipline and Bounding: The History and Sociology of Science as Seen through the Externalism-Internalism Debate,' *History of Science* 30(4), 333–69. https://doi.org/10.1177/007327539203000401

Shapin, Steven (1994), *A Social History of Truth: Civility and Science in Seventeenth-Century England* (Chicago: University of Chicago Press).

Shapin, Steven (1995), 'Here and Everywhere: Sociology of Scientific Knowledge,' *Annual Review of Sociology* 21, 289–321. https://doi.org/10.1146/annurev.soc.21.1.289

Shapin, Steven (1996), *The Scientific Revolution* (Chicago: University of Chicago Press). https://doi.org/10.7208/chicago/9780226750224.001.0001

Shapin, Steven (2012), 'The Sciences of Subjectivity,' *Social Studies of Science* 42(2), 170–84. https://doi.org/10.1177/0306312711435375

Shapin, Steven and Simon Schaffer (1985), *Leviathan and the Air-Pump: Hobbes, Boyle, and the Experimental Life* (Princeton: Princeton University Press).

Shaw, Robert (2013), 'The Implications for Science Education of Heidegger's Philosophy of Science,' *Education Philosophy and Theory* 45(5), 546–70. https://doi.org/10.1111/j.1469-5812.2011.00836.x

Sherover, Charles (1967), 'Heidegger's Ontology and the Copernican Revolution,' *Monist* 51(4), 559–73. https://doi.org/10.5840/monist196751432

Shirley, Greg (2010), *Heidegger and Logic: The Place of Lógos in Being and Time* (London: Continuum). https://doi.org/10.5040/9781472546661

Skinner, Quentin (2002), 'Hobbes and the Politics of the Early Royal Society,' in *Visions of Politics*, vol. 3: *Hobbes and Civil Science*, by Quentin Skinner (Cambridge: Cambridge University Press), pp. 324–45. https://doi.org/10.1017/CBO9780511613784.015

Skulsky, Harold (1968), 'Paduan Epistemology and the Doctrine of the One Mind,' *Journal of the History of Philosophy* 6(4), 341–61. https://doi.org/10.1353/hph.2008.1596

Smith, Pamela H. (2004), *The Body of the Artisan: Art and Experience in the Scientific Revolution* (Chicago: University of Chicago Press). https://doi.org/10.7208/chicago/9780226764269.001.0001

Sokal, Alan and Jean Bricmont (1998), *Fashionable Nonsense: Postmodern Intellectuals' Abuse of Science* (New York: Picador).

Sokolowski, Robert (1979), 'Exact Science and the World in which We Live,' in *Lebenswelt und Wissenschaft in der Philosophie Edmund Husserls*, ed. by Elisabeth Ströker (Frankfurt: Vittorio Klostermann), pp. 92–106.

Solomon, Robert C. (1976), *The Passions: Emotions and the Meaning of Life* (Garden City, NY: Doubleday Press).

Solomon, Robert C. (1977), 'The Logic of Emotion,' *Noûs* 11, 41–49. https://doi.org/10.2307/2214329

Solomon, Robert C. (1992), 'Existentialism, Emotions, and the Cultural Limits of Rationality,' *Philosophy East & West* 42(4), 597–621. https://doi.org/10.2307/1399671

Solomon, Robert C. (2003), 'Emotions, Thoughts and Feelings: What is a "Cognitive Theory" of the Emotions and Does It Neglect Affectivity?,' in *Philosophy and the Emotions*, ed. by Anthony Hatzimoysis (Cambridge: Cambridge University Press), pp. 1–18. https://doi.org/10.1017/CBO9780511550270.002

Spiller, Michael R. G. (1980), *'Concerning Natural Experimental Philosophie': Méric Casaubon and the Royal Society* (The Hague: Martinus Nijhoff). https://doi.org/10.1007/978-94-009-8913-9

Spinosa, Charles (2000), 'Heidegger on Living Gods,' in *Heidegger, Coping, and Cognitive Science: Essays in Honor of Hubert L. Dreyfus*, ed. by Mark A. Wrathall and Jeff Malpas (Cambridge: Cambridge University Press), pp. 209–28.

Steiner, Carol J. (1999), 'Constructive Science and Technology Studies: On the Path to Being?,' *Social Studies of Science* 29(4), 583–616. https://doi.org/10.1177/030631299029004005

Steiner, Carol J. (2008), 'Ontological Dance: A Dialogue between Heidegger and Pickering,' in *The Mangle in Practice: Science, Society, and Becoming*, ed. by Andrew Pickering and Keith Guzik (Durham: Duke University Press), pp. 243–65. https://doi.org/10.1215/9780822390107

Stepanich, Lambert V. (1991), 'Heidegger: Between Idealism and Realism,' *The Harvard Review of Philosophy* 1(1), 20–28. https://doi.org/10.5840/harvardreview1991113

Sterelny, Kim (2012), 'Language, Gesture, Skill: the Co-Evolutionary Foundations of Language,' in *Philosophical Transactions of the Royal Society B* 367, 2141–51. https://doi.org/10.1098/rstb.2012.0116

Stocker, Michael (2010), 'Intellectual and Other Nonstandard Emotions,' in *The Oxford Handbook of the Philosophy of Emotion*, ed. by Peter Goldie (Oxford: Oxford University Press), pp. 401–23. https://doi.org/10.1093/oxfordhb/9780199235018.003.0019

Ströker, Elisabeth (1997), *The Husserlian Foundations of Science* (Dordecht: Kluwer Academic Publishers). https://doi.org/10.1007/978-94-015-8824-9

Tanzer, Mark Basil (1995), 'Heidegger's Critique of Realism,' *Southwest Philosophy Review* 11(2), 145–59. https://doi.org/10.5840/swphilreview199511228

Tanzer, Mark Basil (1998), 'Heidegger on Realism and Idealism,' *Journal of Philosophical Research* 23, 95–111. https://doi.org/10.5840/jpr_1998_14

Thagard, Paul (2001), 'How to Make Decisions: Coherence, Emotion, and Practical Inference,' in *Varieties of Practical Reasoning*, ed. by Elijah Millgram (Cambridge, MA: The MIT Press), pp. 355–71.

Thagard, Paul (2002), 'The Passionate Scientist: Emotion in Scientific Cognition,' in *The Cognitive Basis of Science*, ed. by Peter Carruthers, Stephen Stich and Michael Siegal (Cambridge: Cambridge University Press), pp. 235–50. https://doi.org/10.1017/CBO9780511613517.013

Thagard, Paul (2006a), 'How to Collaborate: Procedural Knowledge in the Cooperative Development of Science,' *The Southern Journal of Philosophy* 44, 177–96. https://doi.org/10.1111/j.2041-6962.2006.tb00038.x

Thagard, Paul, in collaboration with Fred Kroon, Josef Nerb, Baljinder Sahdra, Cameron Shelley and Brandon Wager (2006b), *Hot Thought: Mechanisms and Applications of Emotional Cognition* (Cambridge, MA: The MIT Press).

Thagard, Paul (2008), 'How Cognition Meets Emotion: Beliefs, Desires and Feelings as Neural Activity,' in *Epistemology and Emotions*, ed. by Georg Brun, Ulvi Doğuoğlu and Dominique Kuenzle (Aldershot: Ashgate), pp. 167–84. https://doi.org/10.4324/9781315580128

Tiez, John (1993), 'Heidegger on Realism and the Correspondence Theory of Truth,' *Dialogue* 32(1), 59–76. https://doi.org/10.1017/s0012217300014980

Tiez, John (2005), 'Heidegger on Science, Realism, and the Transcendence of the World: *Being and Time*, Section 69,' *Idealistic Studies* 35(1), 1–20. https://doi.org/10.5840/idstudies20053518

Tomasello, Michael (2014), *A Natural History of Human Thinking* (Cambridge, MA: Harvard University Press). https://doi.org/10.4159/9780674726369

Tosh, Nick (2007), 'Science, Truth and History, Part II. Metaphysical Bolt-holes for the Sociology of Scientific Knowledge?,' *Studies in History and Philosophy of Science* 38(1), 185–209. https://doi.org/10.1016/j.shpsa.2006.12.002

Traweek, Sharon (1992), *Beamtimes and Lifetimes: The World of High Energy Physicists* (Cambridge, MA: Harvard University Press).

Tugendhat, Ernst (1967), *Der Wahrheitsbegriff bei Husserl und Heidegger* (Berlin: de Gruyter). https://doi.org/10.1515/9783110826890

van Fraassen, Bas C. (1980), 'A Re-Examination of Aristotle's Philosophy of Science,' *Dialogue* 19(1): 20–45. https://doi.org/10.1017/s0012217300024719

von Weizsäcker, Carl Friedrich (1977a), 'Begegnungen in vier Jahrzehnten,' in *Erinnerung an Martin Heidegger*, ed. by Günther Neske (Pfullingen: Verlag Neske), pp. 239–47.

von Weizsäcker, Carl Friedrich (1977b), 'Heidegger und die Naturwissenschaft,' in *Der Garten des Menschlichen: Beiträge zur geschichtlichen Anthropologie*, by Carl Friedrich von Weizsäcker (Munich: Carl Hanser Verlag), pp. 413–31.

von Weizsäcker, Carl Friedrich (1990 [1949]), 'Beziehungen der theoretischen Physik zum Denken Heideggers,' in *Zum Weltbild der Physik*, thirteenth edn, by Carl Friedrich von Weizsäcker (Stuttgart: S. Hirzel Verlag), pp. 243–45.

Weber, Max (1968), *Economy and Society*, trans. by Günther Roth and Claus Wittich (Berkeley: University of California Press).

White, Paul, ed. (2009), *The Emotional Economy of Science*, focus section in *Isis* 100(4), 792–851. https://doi.org/10.1086/652019

Winner, Langdon (1980), 'Do Artifacts Have Politics?,' *Daedalus* 109(1), 121–36.

Wittgenstein, Ludwig (1958), *Philosophical Investigations*, third edn, trans. by G. E. M. Anscombe (Oxford: Basil Blackwell).

Young, Julian (2006), 'The Fourfold,' in *The Cambridge Companion to Heidegger*, second edn, ed. by Charles B. Guignon (Cambridge: Cambridge University Press), pp. 373–92. https://doi.org/10.1017/CCOL0521821363.015

Zabarella, Jacopo (1995), *Über die Methoden (De Methodis); Über den Rückgang (De Regressu)*, trans. by Rudolf Schicker (Munich: Wilhelm Fink Verlag).

Zabarella, Jacopo (1985), *De Methodis; De Regressus*, ed. by Cesare Vasoli (Bologna: Cooperativa Libraria Universitaria Editrice).

Zilsel, Edgar (1942), 'The Sociological Roots of Science,' *American Journal of Sociology* 47(4), 544–62. https://doi.org/10.1086/218962

Zimmerman, Michael E. (1990), *Heidegger's Confrontation with Modernity: Technology, Politics, Art* (Bloomington: University of Indiana Press).

Zoller, David J. (2012), 'Realism and Belief Attribution in Heidegger's Phenomenology of Religion,' *Continental Philosophy Review* 45(1), 101–20. https://doi.org/10.1007/s11007-011-9208-3

Index

This book need not end here…

You may also be interested in:

The Scientific Revolution Revisited

By *Mikuláš Teich*

https://www.openbookpublishers.com/product/334

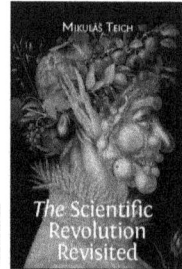

Measuring the Master Race
Physical Anthropology in Norway, 1890-1945

By *Jon Røyne Kyllingstad*

https://www.openbookpublishers.com/product/123

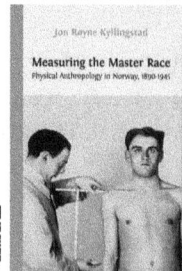

Behaviour, Development and Evolution

By *Patrick Bateson*

https://www.openbookpublishers.com/product/490

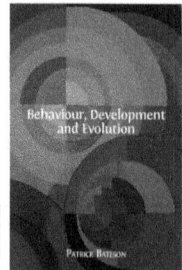

Democracy and Power
The Delhi Lectures

By *J. Noam Chomsky*

https://www.openbookpublishers.com/product/300

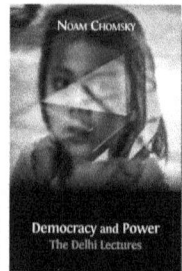